# Evolution of Innovation Management

*Also by Alexander Brem:*

PERSPECTIVES ON SUPPLIER INNOVATION

Theories, Concepts and Empirical Insights on Open Innovation and the Integration of Suppliers (*with J. Tidd*)

THE BOUNDARIES OF INNOVATION AND ENTREPRENEURSHIP

Conceptual Background and Selected Theoretical and Empirical Aspects

*Also by Éric Viardot:*

THE TIMELESS PRINCIPLES OF SUCCESSFUL BUSINESS STRATEGY

SUCCESSFUL MARKETING STRATEGY FOR HIGH TECH FIRMS (3rd edition)

# Evolution of Innovation Management

## Trends in an International Context

Edited by

**Alexander Brem**
*University of Erlangen-Nuremberg, Germany*

and

**Éric Viardot**
*EADA Business School, Barcelona, Spain*

First published 2013 by
PALGRAVE MACMILLAN

Palgrave Macmillan in the UK is an imprint of Macmillan Publishers Limited, registered in England, company number 785998, of Houndmills, Basingstoke, Hampshire RG21 6XS.

Palgrave Macmillan in the US is a division of St Martin's Press LLC, 175 Fifth Avenue, New York, NY 10010.

Palgrave Macmillan is the global academic imprint of the above companies and has companies and representatives throughout the world.

Palgrave® and Macmillan® are registered trademarks in the United States, the United Kingdom, Europe and other countries

ISBN: 978–0–230–36896–5

This book is printed on paper suitable for recycling and made from fully managed and sustained forest sources. Logging, pulping and manufacturing processes are expected to conform to the environmental regulations of the country of origin.

A catalogue record for this book is available from the British Library.

A catalog record for this book is available from the Library of Congress.

Transferred to Digital Printing in 2013

# Contents

# List of Figures

# List of Tables

# Preface

Innovation provides a strong competitive advantage and is one of the best ways for a firm to speed up the rate of change and adaptation to the global environment. According to a recent survey by Capgemini, 76 percent of executives indicated that innovation is among the top three priorities of their organization and is the main lever for growth. Concurrently, the innovation topic is also gaining more and more visibility and interest among the academic community worldwide.

Although some of the challenges of innovation are remarkably consistent, the recent period has seen the emergence of new ways of stimulating and managing the innovation process, especially from an international perspective. Even if these processes are taking place in very different industries, there are many parallels in ways to successfully manage them.

The purpose of this book is to inventory those new routes, to explore them in depth, and to assess their value both for markets and for companies. More specifically, the book is organized around three themes:

- Innovation strategies
- Innovation management tools
- International perspectives

These three core themes are enriched with the contributions of people in professional service firms who have a strong expertise in innovation. Each of the contributor comes with a perspective about the evolution of innovation management that is slightly different from that of the academics. We welcome their contributions to this book, as they illustrate the necessary collaboration between academics and practitioners when it comes to working out new ways of innovating in an international context.

We seek to provide an overview of current trends in an international innovation management context. All the chapters received peer reviews to ensure the highest quality.

## Summary

The book starts with a chapter from Robert J. Thomas and Yoram (Jerry) Wind about symbiotic innovation. They consider a model of symbiotic innovation from a biological perspective and explore five types of collaboration that can be used to define a network to drive innovation. They stress the importance of network orchestration across the different types of collaboration and the critical role of a network orchestrator to manage such symbiotic innovation.

Involving consumers from all over the world to jointly develop new products in so-called co-creation initiatives has become a widespread instrument in corporate ideation processes. Based on the principle of crowdsourcing, idea contests have been widely experimented within practice to utilize the creativity and knowledge of consumers. Still, the value of idea contests and their diverse benefit dimensions have not been investigated in detail. Companies neglect manifold powerful indirect effects of idea contests on branding and relationship building with consumers. In Chapter 2, Volker Bilgram aims to address this gap by adapting Brown and Svenson's (1988) performance measurement framework designed for "closed" research and development (R&D) innovation projects to the open innovation environment of idea contests.

Change and the implementation of new strategies always correspond with increased risk, which leads to the immanent need for measuring success. Traditional management measures and methods are, however, obsolete when applying the open innovation strategy. Managing the open innovation process demands new measures that allow managers to monitor and control the innovation process and the associated risks and opportunities. In Chapter 3, Erik Brau, Ronny Reinhardt, and Sebastian Gurtner develop qualitative and quantitative measures and adapt existing methods to manage the inflows and outflows of knowledge and, more generally, the implementation of the open innovation strategy.

Innovation management is a multidisciplinary field of study. However, the management of innovation is often considered equivalent to technology management or the management of research and development. The same confusion arises when general-purpose management tools are also confronted with innovation management tools. In Chapter 4, Jon Mikel Zabala-Iturriagagoitia sets out to determine the characteristics of innovation management tools. In particular, he focuses on one of these tools: creativity. Based on experience from the application of the creative process in several machine tool companies, he defines a set of generic stages and introduces the most relevant roles for their implementation.

Scenarios are means to sensitize people to alternative future directions and paradigm shifts. However, complex network structures in open innovation projects, dynamic collaborative arrangements, and the fragmentation of communication systems have led to the necessity of developing a new framework for applying scenario-based learning as a daily practice. In Chapter 5, Nicole Pfeffermann and Henning Breuer present a framework – exemplified by a case in telecommunications – with managerial implications and limitations for innovation management. They provide new perspectives on scenario planning and innovation communications. In sum, efficiency in management of innovation communication is a prerequisite for communication management in scenario planning.

Why are some organizational structures and interactions more likely to generate novel and useful ideas, convert them into products or services, and move them to market? To date, there have only been a few studies that specifically illuminate the challenges and experiences companies face when establishing or joining networks and integrating these activities into their own and specific innovation processes. The aim of Chapter 6 is not to broadly discuss the role of social networking in innovation, but rather to focus on a series of tools that can be used by an innovation manager to understand the depth and breadth of the interactions that the manager's team members have. Using network analysis of internal and external networks, Gerhard Drexler and Bernard Janse provide data that highlight the evolution, functionality, and benefit of social network structures to idea generation and to building effective collaborative ties.

Mobile social networks are increasing in popularity as people utilize mobile communication technology to communicate with their social network. This form of community interactivity represents a technological innovation that has a strong business potential. Mobile social networks are a form of computer-mediated communication that will impact the way technology is marketed to people. In Chapter 7, Vanessa Ratten aims to review the existing research on mobile social networks by providing a set of research propositions that can be used to understand the adoption of mobile social networks.

The purpose of Chapter 8, by Federico Frattini, Gabriele Colombo, and Claudio Dell'Era, is twofold. It aims to study the role that early adopters have in the commercialization of high-technology innovations and, specifically, the influence they exert on the innovation's commercial success. The chapter also means to disclose the commercialization approaches that are most effective for stimulating a positive acceptance of the innovation among early adopters. Chapter 8 reports on eleven case studies concerning technological innovations commercialized in industrial high-tech markets. The findings show that the ability to raise a positive appraisal in the early market is fundamental for the commercial success of market-related radical innovations, i.e., those entailing a deep change in the clients' behaviors and consumption patterns. Moreover, the chapter shows that targeting, communication, distribution, and whole-product configuration are the most important dimensions of a commercialization strategy when it comes to stimulating a positive appraisal of the innovation among the early adopters.

Drawing on Zyc Chemical's experience, Stefano Borzillo and Renata Kaminska (Chapter 9) offer insights into how managers can use communities of practice (CoPs) to help their organizations innovate. The main conclusion drawn from their study is that supporting the development of innovation capability involves four driving forces that interact dynamically with one another within CoPs: management sponsorship, leadership, enhanced cooperation, and boundary spanning.

Among the many problems that Japanese and Korean universities face, they have one common shortcoming: the inability to convince domestic and foreign firms to switch from North American or European Union universities to Japanese or Korean universities for R&D partnerships. This is one reason that world-class multinational enterprises (MNEs) have avoided working with prestigious East Asian universities for R&D partnerships and new technology commercialization. Using the concept of barriers to interorganizational learning in innovation, in Chapter 10, Ingyu Oh discusses how Japanese and Korean universities can attract their home country firms and foreign firms to joint innovation efforts, using both feed-forward and feedback learning. Based on comparable archival data, he shows how separating feed-forward from feedback learning can motivate Japanese and Korean MNEs to rely on their domestic universities for knowledge development.

By reconciling theory and practice, Fabian Ariel Salum, Rosana Silveira Reis, and Hugo Ferreira Braga Tadeu aim in Chapter 11 to provide the reader with a view of the innovation process from the perspective of an emerging nation. The case studies presented illustrate both the theoretical references and the empirical analysis running throughout the sections. The chapter presents, for instance, a brief context about innovation in Brazil and the case study of Azul Brazilian Airlines, named one of the world's 50 most innovative companies by *Fast Company* magazine (2011). Another example is the Embraer case study, in which the authors address the importance of customers and of the supply chain.

Innovation champions are increasingly being recognized as potential catalysts for identifying and supporting new business ideas and ensuring their implementation. However, much less is understood about what role champions play in China. Acknowledging what is known about these specialist individuals and then probing aspects and variants in China is timely for those seeking answers to managing innovation more effectively. As this conceptual article highlights, champions in China confront important philosophical, geographical, cultural, and organizational factors. Adding within-cultural or emic insights in nontraditional domains is important, as thus far this aspect has been largely overlooked. Important issues in China relating to observance of hierarchy, personal connections (or *guanxi*), and giving "face" are keys. In Chapter 12, Anton Kriz, Courtney Molloy, and Bonnie Denness highlight the fact that innovation champions are just as likely as in firms found in government and in universities in China. There is also a rationale for developing champion "teams" in China, rather than relying on individuals.

Frugal innovation, targeted to developing countries, has become a recent phenomenon of interest for academics, practitioners, and policy makers. In

Chapter 13, using an analysis of English-language newspaper articles, Preeta M. Banerjee finds that achieving frugality does have social and environmental aspects – namely, the entrepreneurial concept of *bricolage* becomes essential (utilizing people and their knowledge, skills, and capabilities in novel ways). In regard to the environment, frugality can cause concerns because of the lack of robustness of some things created in this way. Frugal innovation does not need to be systematized, but should be supported by the incorporation of sustainable design. Thus, Banerjee concluded that in an ideal practice of frugal innovation should integrate *jugaad* (which means that solutions are typically based on imaginative problem-solving rather than on technological inventions), human capital bricolage, and sustainable design.

Chapters 14 and 15 are written from the view of professional service firms, to enrich the academically focused papers of the preceding chapters.

Against the background of Microsoft, Ramon Costa-i-Pujol (Chapter 14) discusses the role of mobile workers in our workplace: their different needs not only change the use of tools and work (where ICT, information communications and technology, become key), but also create new needs for the organization and management of "virtual teams" and changes in work areas and, ultimately, lead to a new concept – flexible working. Although the launch of this new way of working, with a high degree of mobility, implies the changes in three areas (space, personnel change management, and technology), this chapter emphasizes the third aspect (ICT), makes a superficial analysis of new workspaces, and points out the part of change management.

Koen Klokgieters and Robin Chu from Capgemini focus on managerial implications in Chapter 15, and elaborate on the formal and informal levers that provide an environment for successful innovation. Also, they explicate the challenges for implementing an innovation environment and reveal that the basic principles of strategic management and change are still valid in our ever-faster evolving business world. Finally, they describe the implications for innovation success of business model innovation and discuss the growing importance of collaboration.

The last chapter is our conclusion, in which we sum up the lessons learned from this book, supplemented with our view on the evolution of innovation management.

## References

Fast Company (2011) The 50 Most Innovative Companies. http://www.fastcompany.com/most-innovative-companies/2011/.

Duppen, F. & Innis, D. (2010) Global-innovation Survey 2010, Capgemini Consulting. http://www.capgemini.com/insights-and-resources/by-publication/global-innovation-survey-2010/.

# Acknowledgments

We would like to thank the reviewers for their constructive comments that helped to improve the quality of the contributions. We thank the editorial team of Palgrave Macmillan, Keri Dickens and Virginia Thorp, for their support and encouragement. We look forward to receiving any feedback. Please feel free to contact either Alexander Brem (brem@idee-innovation.de) or Éric Viardot (eviardot@eada.edu).

# Notes on Contributors

**Preeta M. Banerjee** is an assistant professor of strategy at Brandeis University. She obtained her PhD from the Wharton School, University of Pennsylvania. Her research interests are strategy, technology and innovation management, and entrepreneurship. She has written numerous research articles, which can be found in journals such as *R&D Management, Technovation,* and *IEEE-TEM.* She is a recipient of the Fulbright-Nehru scholarship to research technology entrepreneurship in West Bengal, India, the Aspen Institute Rising Star Finalist 2011, an IBM Innovation Award for 2010–2011, and she was one of the recipients of the National Institute of Standards and Technology's advanced technology program awards in 2008–2009. She is a contributing member of the Strategic Management Society, the Academy of Management, and the Indus Entrepreneurs (TiE).

**Volker Bilgram** is a research associate at the TIM Group, Rheinisch-Westfaelische Technische Hochschule (RWTH), Aachen University. He is also Project Manager at HYVE AG, a co-creation and open innovation enabler based in Munich. He graduated in international business law from the University of Erlangen-Nuremberg. The main focus of his research is on new product development, open innovation, user innovation, and the use of co-creation as an empowerment and branding tool. He blogs about current open innovation and social media topics.

**Stefano Borzillo** is a professor at Skema Business School, where he currently holds the Chair of Microsoft Research. His main research fields are the strategic balancing of autonomy and control mechanisms within communities of practice, the processes of innovation, and knowledge creation. He is responsible for steering the research activities within the Microsoft–Skema partnership. He has a PhD in organizational behavior and organizational theory. His research has been funded by the Fonds National Suisse de la Recherche scientifique and by Microsoft for the management of the research chair.

**Erik Brau** received his diploma degree in industrial engineering and management from the University of Technology, Dresden, Germany. His research interests are focused on open innovation. For his diploma thesis, he worked on the issue "Open Innovation and Controlling." He is now working for Volkswagen AG in the procurement department.

**Alexander Brem** is Professor of Idea and Innovation Management at the University of Erlangen-Nuremberg, Germany. He is founder and partner of VEND Consulting GmbH, Nuremberg. His research interests include technology and innovation management as well as entrepreneurship. He is a reviewer and editorial board member of various international journals such as *Technovation* and *International Journal of Innovation Management*, and editor of the *International Journal of Technology Marketing*.

**Henning Breuer** is Managing Director of the consultancy company Bovacon – Designing Business Interaction. He has a PhD in media and organizational psychology and degrees in law and philosophy. He focuses on innovation management, futures research, and business modeling. He teaches at the University for Applied Sciences in Potsdam. His research interests are human-centered innovations for informal learning environments. Henning currently manages a project field for user-driven innovation at Telekom Innovation Laboratories.

**Robin Chu** is a management consultant in strategy and business innovation at Capgemini Consulting. He has a master's in supply chain and strategic management from Tilburg University. His research interests include business models, operating models in innovation management, strategic management, supply chain management, business analysis, consultancy, and entrepreneurship.

**Gabriele Colombo** is a research fellow in management, economics, and industrial engineering at Politecnico di Milano. His research interests include innovation strategies and strategic management of knowledge in an open environment. His research focuses on how firms improve and enlarge their existing knowledge bases through collaborating with different external subjects.

**Ramon Costa-i-Pujol** is an engineer in computer science. He holds a master's in pedagogy from the Universitat Politècnica de Catalunya (UPC) and a diploma in general management from the Escuela de Alta Dirección y Administración (EADA). His research interests include information workers, project management, and Web 2.0. He is member of the Project Management Institute and the Junior Chamber International. He is currently business productivity advisor at the Microsoft Innovation Center and is also an associate professor of project management and information systems at the UPC and Escuela Universitaria Gimbernat (Universitat Autónoma de Barcelona, UAB) and EADA.

**Claudio Dell'Era** is an assistant professor in the Department of Management, Economics and Industrial Engineering at the Politecnico di Milano. He is co-director of MaDe In Lab, the Laboratory of Management of Design and Innovation of MIP Politecnico di Milano. His research interests include design management, innovation management, and project management. He has published articles in journals such as *Journal of*

*Product Innovation Management, R&D Management,* and *International Journal of Innovation Management.*

**Bonnie Denness** is a research assistant at the University of Newcastle, Australia. Her research interests include the detailed investigation of the role of innovation champions in organizations and implications for regional development. She is currently working on a regional development project and undertaking a detailed investigation into the behavior of champions of innovation and innovation champions.

**Gerhard Drexler** is Head of R&D at Mondi UFP. He graduated in environmental management and process management at the Donau-University in Krems. He has an MBA in general management from the Joseph Schumpeter Institute. His research interests include open innovation, idea management, analysis of patent networks and industry–university collaboration. He won the Gepard Process Award in 2006 and the Best Innovator Award in 2005.

**Federico Frattini** is an assistant professor in the Department of Management, Economics and Industrial Engineering at the Politecnico di Milano, Italy. He is Vice-Director of the executive MBA program. His research interests include performance measurement of R&D, technology and open and collaborative innovation, and diffusion of innovation. He has authored two books and published more than 90 papers in journals such as the *Journal of Product Innovation Management* and *California Management Review.*

**Sebastian Gurtner** is a research assistant and chair for entrepreneurship and innovation at the University of Dresden. His research interests are innovation and product management and technology and financing. He has published eight articles and is a member of the Academy of Management, the Health Technology Assessment International, and the International Society for Pharmaeconomics and Outcomes Research. He is a recipient of the Academy of Management Scholarship for 2010, 2011, and 2012.

**Bernard Janse** is Director of Innovation at Buckman International, USA. His responsibilities include new product development and innovations in the fields of biotechnology and chemistry. He has 15 years of experience in developing systems to support innovation, mostly in an international context. His research interests include creating a culture of innovation in multinational corporations. He considers the link between culture and innovation an important aspect. He has a PhD in Microbiology.

**Renata Kaminska** is a professor of strategy and organization at Skema Business School and also a visiting professor at the Krakow University of Economics and Kozminski Business School in Warsaw. Her research interests include strategy process, innovation, and organizational dynamics with specific focus on economic transition in Eastern European countries. She is former director of one of CERAM Business School's continuing education programs. She is a member of the Academy of Management, the European Group for Organization Studies, and GREDEG Research Laboratory.

**Koen Klokgieters** is Vice President of Global Leader R&D and Business Innovation at Capgemini Consulting. He has more than 20 years of experience as a management consultant in different roles.

**Anton Kriz** is a senior lecturer at the University of Newcastle, Australia. He holds a PhD in Interpersonal Trust in Chinese business. His research interests focus on China and international business, with a strong emphasis on cross-culture relationships, services marketing, and trust. He has received a Churchill Fellowship and a Chinese Fellowship from Zhejiang University's National Institute of Innovation Management. He is a member of the Australian Institute of Management.

**Courtney Molloy** is a PhD candidate at the University of Newcastle, Australia. She is studying key motivating behaviors of innovation champions. She has several years of experience working for Westfield and IBM in management roles.

**Ingyu Oh** is a researcher at IOM MRTC (the International Organization for Migration Research and Training Centre) and a former associate professor at Solbridge International School of Business. His research interests include innovation in the entertainment industry in Korea and university–industry R&D collaborations. He has researched the safety of nuclear power plants in Korea. He is an editorial board member of the *Asian Journal of Innovation and Policy*.

**Nicole Pfeffermann** is a management consultant and researcher at ISEIC Pfeffermann Consulting. She has a diploma in business economics and a PhD from the University of Bremen. Her field of interest focuses on innovation strategy, entrepreneurship, and innovation communication.

**Vanessa Ratten** is a senior lecturer in strategic management and entrepreneurship at Deakin University, Australia. Her research interests include technological innovation (including m-commerce, social cognitive theory, and strategic alliances) and entrepreneurship (including sport entrepreneurship, international entrepreneurship, and social entrepreneurship). She is a member of the Academy of Management, the Academy of International Business, and the Australian and New Zealand Academy of Management.

**Ronny Reinhardt** is a research assistant at the Chair for Entrepreneurship and Innovation, University of Dresden. He is the co-author of an article about disruptive innovations through the use of the customer analysis method. He received the Erich-Glowatzky-Award to honor his outstanding achievement in his research field.

**Rosana Silveira Reis** is a professor and researcher at ISG Paris, France, and visiting professor and researcher at Fundação Dom Cabral, Brazil. She has an MBA from the Federal University of Santa Catarina, Brazil, and a PhD in Management from the University of Bologna, Italy. Her research interests include creativity and innovation, new product development, globally distributed teams, international management and culture, and human resources management. She is a member of the Academy of Management, the European Academy of Management, and the Brazilian Association of Human Resources, which she directed between 1994 and 1997.

**Fabian Ariel Salum** is a professor and researcher in engineering and management in Brazil. He has an MBA from Fundação Pedro Leopoldo of Belo Horizonte, Brazil. His research interests include innovation, growth strategy, and corporate strategy. He has worked in large companies such as Unilever, Toshiba, and Fiat and for several business schools. He is currently a permanent professor and researcher at the Fundação Dom Cabral, Brazil.

**Hugo Ferreira Braga Tadeu** is a professor and researcher at Centro Universitário UNA, Brazil, and visiting professor and researcher at Fundação Dom Cabral, Brazil. He has a master's in electrical engineering and a PhD in mechanical engineering from the Pontifícia Universidade Católica de Minas Gerais, Brazil. He has a postdoctoral qualification in transportation from the Sauder School of Business, Canada. His research interests include supply chain management, operations, strategy, and innovation. He has several years' experience in management and teaching management in graduate, MBA, and master's courses.

**Robert J. Thomas** is a professor of marketing at the McDonough School of Business at Georgetown University in Washington, DC. His research interests include product development and forecasting, market segmentation, organizational buying behavior, innovation, and strategic marketing management. In his areas of research he has over 50 publications. He is on the editorial board of the *Journal of Product Innovation Management* and is a distinguished research fellow at the Institute for the Study of Business Markets.

**Éric Viardot** is a professor of marketing and strategy at the Escuela de Alta Dirección y Administración (EADA) in Barcelona. He graduated from the HEC Business School, Paris, and the Institute of Political Sciences, Paris, and has a doctorate in management. His research interests include strategic management and marketing, with a strong focus on technology and innovation management. He has published various books and articles and teaches in executive programs in Europe. He is editor of the *International Journal of Technology Marketing* and is on several editorial boards. He is an active consultant and trainer working with a variety of major multinational companies.

**Yoram (Jerry) Wind** is Lauder Professor at the Wharton School, University of Pennsylvania. His research interests include global marketing strategy and growth strategies, marketing and marketing-driven corporate strategy, and new product and business development. Having written 22 books and more than 250 articles, he is one of the most cited authors in marketing. He is a founding editor of Wharton School Publishing. He has won numerous academic awards, including the Charles Coolidge Parlin Award, the AMA/ Irwin Distinguished Educator Award, and the Paul D. Converse Award.

**Jon Mikel Zabala-Iturriagagoitia** is an assistant professor at Lund University, Sweden and a professor at Deusto Business School in San Sebastian, University of Deusto. He graduated in Industrial Engineering at the University of Mondragon, Spain. He has lecturing experience in R&D and innovation indicators, project management, and technology and innovation management. His research interests include analysis of regional/ national innovation systems, science and technology policies and evaluation, innovation support policies, and innovation management tools. He has a PhD in Engineering and Innovation Projects from the Polytechnic University of Valencia.

# 1
# Symbiotic Innovation: Getting the Most Out of Collaboration

*Robert J. Thomas and Yoram (Jerry) Wind*

## 1 Introduction

During 2010 and 2011, IBM conducted research globally on the challenges facing over 3000 chief executives and chief marketing officers. The main finding of these studies was that these leaders believed themselves to be in a world that was substantially more volatile, uncertain, and complex than anything they had seen before (IBM, 2010, 2011). This finding was somewhat expected; what was surprising was that these executives revealed they were not at all prepared to cope with this rapidly changing environment. The study concluded that to make any progress whatsoever, a new kind of creative leadership was needed to develop breakthrough thinking that encouraged experimentation and innovation.

The studies also revealed that the survival path to innovation in an increasingly complex world involved intimately engaging with and empowering consumers as never before. It was clear that organizations could no longer succeed alone when faced with the complexity of an unstable world system of customers, markets, governments, and institutions; *collaboration* with consumers and with a variety of strategic partners in some form of symbiotic relationship would be required. For example, in 1999 Nike developed an online service called NikeiD (www. nike.com), which offered individuals the opportunity to personally design and purchase their own clothing and shoes. Using the increasing availability and sophistication of online communications, consumers interacted with Nike by selecting from a variety of shapes, colors, and materials, much as they had with traditional bespoke cobblers and tailors – however, with a much larger selection of options. Nike's online success (Brohan, 2010) eventually led to the development of in-store design studios, which created additional intimacy with consumers.

Early applications, such as Nike's experiment, demonstrated the importance of collaboration, especially with consumers, and revealed that it can be strategically sustainable over time. The literature in the fields of marketing, strategy, and innovation reflect this growing importance and has begun recommending some form of collaboration or interaction between and among a firm, its consumers, and other stakeholders. Consider the emerging lexicon of terms and concepts such as *co-creation* (Prahaladad and Ramaswamy, 2000), *open innovation* (Chesbrough, 2003), *open source* (von Hippel and Krogh, 2003), *service-dominant logic* (Vargo and Lusch, 2004), and *networked innovation* (Nambisan and Sawhney, 2007). As might be expected, these concepts build on earlier ones, including *lead-users* (von Hippel, 1986), *one-to-one marketing* (Peppers and Rogers, 1993), and *customer-centric marketing* (Sheth et al., 2000), to name a few.

To help cope with this multiplicity of innovation-based concepts and put . the chapter into focus, we consider the following central question: if organizations are facing highly complex and unstable environments, how can they more effectively use collaboration to develop and manage innovation? We explore collaboration through the concept of symbiosis,[1] which recognizes that different organisms can form persistent associations, or collaborations, that benefit those involved to varying degrees. We use a model of symbiosis to help understand how the dynamic balance among the participants may contribute to different kinds of innovation. We then recognize that this model can exist at the core of many different types of collaboration that may be important for the development of innovation.

In the context of business, we consider five types of collaboration, which we propose must be orchestrated properly to contribute to innovation. The first type, *internal collaboration*, is at the core of an organization's ability to function with multiple units in order to facilitate innovation. Different individuals and/or functions within the firm, such as research and development, marketing, and operations must interact and collaborate to innovate, whether for new products or for other business activities.[2] To further facilitate the process of innovation, marketing and other functions within the organization may also seek targeted interaction with external partners outside the firm.

The most important type of external collaboration is with consumers. Most successful innovations eventually involve some form of *consumer collaboration*, if not to generate new ideas, at least to refine them into new offerings. A third type of collaboration involves the *value network* which requires a firm to engage trade partners, suppliers, and other value chain partners to increase the chance of innovation success with consumers. Subsequently, with the help of Internet access many firms and market stakeholders have realized that innovation might be sparked from a more *open collaboration* involving anyone, even crowds, to generate ideas and other sources of value. Ultimately, the realization that the potential for innovation can occur beyond specific individuals or organizations recognizes the importance of *ecological collaboration*, which includes structured communities and their environmental surroundings.

Given the five types of collaboration noted above and the recognition that some form of symbiosis may exist among collaborating participants, several important questions emerge around how best to develop and manage innovation, especially when breakthroughs are needed more and more in business and society. To address these questions, we propose a network-based model of symbiotic innovation that begins with collaboration from the internal perspective of the firm and then expands to include other forms of collaboration and ultimately a more encompassing ecological collaboration. This sets the stage for a discussion of the needed network orchestration to implement and manage this kind of symbiotic innovation. The summary model of symbiotic innovation we propose has five key components: (1) stay focused on the consumer, (2) employ all five types of collaboration, recognizing various symbiotic relationships within each, (3) design the network orchestration and select the orchestrator, (4) design and implement the organizational and network architecture to support the orchestration, and (5) use adaptive experimentation to define the best collaboration approaches.

## 2   Symbiosis

To better understand how participants might collaborate for innovation, we borrow the concept of *symbiosis* from biology. According to Douglas (2010), biologist Heinrich Anton de Bary is credited with coining the term "symbiosis" in 1879 to describe any association between *different* species. He suggested that not all of these associations were positive, and Douglas refines this view by arguing that it is a matter of degree. That is, the focus of a symbiotic relationship should be on the benefits derived by each participant, and that these benefits can be subject to variation. Variation can occur from the nature of the participants, their history, their relationship, and/ or to environmental factors. Such relationships can sustain imbalances or conflict in the benefits received by each party, but if too severe, imbalances and conflicts can lead to dissolution of the relationship. Douglas (2010, p. 1) therefore defines symbiosis in its broadest context as relationships in which "individuals of different species form persistent associations from which they all benefit."

Figure 1.1 provides an illustration of how biologists view symbiosis in terms of interaction patterns between participants, which we show with possible relationships between a firm and a consumer. As Douglas (2010) notes, the core motivating principle of biological symbiosis, and the collaboration within it, is reciprocity. The costs to provide a service and the benefits gained to each participant can be illustrated in terms of arbitrary units. For example if each party gives 10 units of service and receives 30 units in return, each has mutually benefited in the same magnitude, or a net gain of 20 units by each.[3] When both participants realize a positive and near equal benefit from a persistent relationship, it is considered to be symbiotic *mutualism* in biological terminology.

Mutualism is the most frequently interpreted meaning of a symbiotic relationship. For example, when consumers go online to customize a wristwatch, pet food, shoes, cereal or some other offering, they are co-creating or innovating with a firm. If a transaction occurs, each potentially derives a benefit. However, if one participant gains more than the other, then in biological terms a condition of *commensalism* may exist. Commensalism is a relationship among participants with an imbalance in the benefits realized in the relationship. More specifically, as shown in Figure 1.1, the benefits can become negative for one party or the other. Either participant may gain at the expense of the other, but depending on the knowledge and severity of the difference or imbalance in benefits, the relationship may continue because the entire relationship still benefits from the collaboration.

In a simple example, when a consumer purchases a needed product or service and believes he or she is paying more for it than believed to be its worth (i.e., the transaction is in the negative benefit zone), and the firm believes it is benefitting more than consumers, they are in a state of imbalance that is firm-beneficial. Alternatively, if the consumer believes he or she is getting more value for the money paid, and the firm believes it is getting less, it is consumer-beneficial. Notably, as transparency between the participants increases, especially in terms of consumers empowered with search engines and comparative shopping tools (especially on their smart phones), they have more information about the products and services they are considering, including prices. The greater this transparency for a consumer, the higher the likelihood of an imbalanced relationship with a firm. One has only to consider the airline industry, in which consumers have access

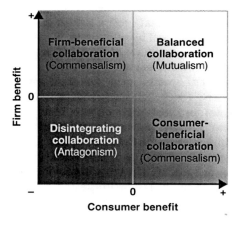

*Figure 1.1*  Symbiotic interaction patterns and collaboration

to a variety of information tools to obtain the best price for a specific flight. For a firm to prevent this imbalance or reinstate a balanced situation, it may need to offer a truly win-win offering to consumers.

Depending on the extent and variability of the benefits gained or lost, any symbiotic relationship can degenerate into one of *antagonism,* in which the benefits from the relationship for each participant begins to deteriorate. Unless there is a compelling reason for the participants to remain together, the relationship may be on a path to disintegration. The outcome of antagonistic relations can be a struggle to regain a mutually beneficial relationship with the same partner or, alternatively, it may lead to completely new partnerships and benefits being formed – in either case it can be a source of innovation.

As one example, consider the business-to-business customer relationship between Apple and Samsung, in which Apple purchased phone and computer components from Samsung. In April 2010, Apple launched its iPad tablet computer. In September 2010 Samsung launched its Galaxy Tab tablet computer, a competitor with similar characteristics to Apple's iPad. In April 2011, Apple sued Samsung for violation of certain patents related to the Apple's innovative iPad tablet (Apple Inc. v. Samsung Electronics Co., Ltd. 2012). Clearly, this is a pattern of antagonistic actions by each party. However, as Barrett (2012) reports, during 2011 Apple purchased some $7 billion for components made by Samsung, which represents 7.6 percent of Samsung's 2011 revenue. Despite their antagonisms, they are deeply involved in a symbiotic relationship in which each depends on the other.

At issue is how the various relationships in Figure 1.1 might lead to innovation. As to the first issue, biologists generally agree that symbiosis is related to innovation. According to Douglas (2010, p. 24), "the symbiotic habit is a significant source of evolutionary innovation and is ecologically important." Similarly, Sapp (2009, p. 115) considers symbiosis to be "a means of evolutionary innovation." While there are many ways in which innovation might occur in this context, one way to visualize it is to begin with a relationship that is mutually beneficial to both participants. Assuming a relatively steady state, one can imagine that a mutually beneficial relationship would continue, but with perhaps minor reason to innovate. In the context of Robertson's (1967) classification, these might be considered as rather "continuous" or incremental innovations, with little disruption of established patterns of behavior.

However, biologists believe environmental turbulence, random or otherwise (as well as the behavior of a specific organism), can put sufficient stress on a relationship of mutualism to push it out of a balanced state into commensalism or even into antagonism. Depending on the benefit balance in the commensal relationship, either participant may work to regain balance, thereby requiring some form of innovation to change the direction of the imbalance. In Robertson's terminology, innovation arising in these conditions may be termed dynamically continuous, or may be the outcome of somewhat disruptive behavior, revealed by or a result of the imbalance in the benefits.

A collaborative relationship that is in, or moving into, a state of antagonism reveals more disruptive behaviors which can potentially lead to discontinuous innovation, in Robertson's classification. As noted in the Apple example discussed above, the collaboration is evident from their buyer-seller relationship, but the intensity of the disruptive behavior may lead to either or both parties seeking an innovative resolution. For example, Apple may decide to innovate its next generation iPad with non-Samsung components, thereby leaving the relationship – or Apple may depend on its attorneys to develop a legal innovation that returns both parties to a collaborative relationship that is beneficial to both. Of course, there are other possibilities from Samsung's perspective, as well as other options for each. The point is that the state of collaborative antagonism may lead to completely discontinuous innovation that returns the relationship to a balanced state or one that involves leaving the relationship entirely (e.g., seeking a new partner).

To summarize, when one thinks of "symbiosis" one imagines a collaboration in which both partners benefit in relatively balanced amounts; i.e., mutualism in biological terms. Ironically, as discussed above, it is possible that partnerships in balance may not produce significant innovation, but only minor or incremental improvements, because they are largely benefiting from the harmony of their existing relationship. However when collaboration becomes imbalanced in terms of benefits to each participant (commensalism), one or both parties may engage in a variety of innovative actions to regain balance or may possibly leave the collaboration for other options. Finally, when the collaboration finds both parties with negative benefits (antagonism), the drive for innovation to regain a more beneficial and balanced state may become more intense, or the parties may completely abandon the collaboration for other options.

As becomes apparent, there are a variety of symbioses that may work in different ways to produce different types of innovation outcomes. As biologists note, several factors may influence the symbiosis and the dynamic relationship within it. For example, they note factors such as the historical relationship of the collaborators, their pre-existing capabilities, the internal or external drivers of variation in their makeup, how they address conflicts, how they select future collaborators, how they network, and their persistence in the face of environmental risk – factors that can occur in any business or market relationship. Consequently the types and structure of collaborators and collaborations involved, the factors that drive them, as well as the ability of someone or some entity to orchestrate them are central to better understanding how symbiotic innovation may be managed in a productive way – topics we turn to in the next sections.

## 3   Structuring collaboration for innovation

There are many paths to innovation, all of which eventually require some form of collaboration, especially with consumers. Even the hero image of the lone genius working in solitude to create an invention ultimately requires collaboration to bring the invention to market (Cain, 2012). Consider Steve Jobs and his legendary status as the genius behind the many products driving Apple's success (Isaacson, 2011): suppliers, retailers, application providers, and a host of partners were necessary for the innovations to occur. In contrast, the early and frequently cited example of Dell Computer's 1997 innovative business model of engaging customers in the design of their own personal computers (within a set of practical parameters) demonstrated that individuals and organizations could collaborate more directly to meet their needs (Magretta, 1998). Dell referred to this as a type of "virtual integration" that brought together several collaborators, including suppliers, assemblers, customers, delivery firms, and service organizations to create value for all involved.

Although the logic of collaborating with consumers and other partners to create value and otherwise innovate is not particularly new (Alderson, 1957), the impact of information technology on innovation and collaboration and the way it empowers consumers is new. It has enabled all participants in an exchange to more effectively communicate virtually anytime, anywhere globally over a much broader set of options than has ever been available. It is part of the sea change in marketing that Wind et al. (2001) have described as "convergence marketing," based on empowered "hybrid" consumers who exhibit five key needs, briefly described here:

- "Customerization," or the consumer's need for uniqueness, personalization, and "it's made for me" products, services, and messages
- Community, or the desire for social interaction in a variety of real and virtual groups;
- Convenient access, or the need to seamlessly interact with firms through multiple channels to obtain the goods, services, and experiences consumers want
- Competitive value, or the desire to meet ones needs at the best price
- Choice, or the desire to have the tools to make better-informed decisions

What is evident from these needs is that convergence is not only applicable to consumers, but also to other potential collaborators. Whether they are suppliers, assemblers, channel partners, crowds, communities, or other stakeholders, the impact of converging technologies on the potential needs and behaviors of individuals and organizations creates opportunities for firms to engage collaborators in pursuing new business models, products and services, and marketing approaches. The message for innovation is straightforward: firms can no longer view consumers and other partners as passive sources of needs and information – traditional marketing research is not enough. Firms must learn how to better collaborate and interact with consumers to drive innovation.

We propose a model of symbiotic innovation with five types of collaboration as a basis for orchestrating a network of collaborators in rapidly changing and complex global environments. The diagram in Figure 1.2 provides a structure with which to envision the five types of collaboration. More specifically, we identify interactive sets of potential collaborators that must function together in some symbiotic way for innovation to occur – in much the same way that multiple cells function together symbiotically in a biological context to support the proper functioning of an organism.

In subsequent sections, each type of collaboration is briefly reviewed, with examples to reveal their interconnectedness for innovation. The key point of Figure 1.2, which is necessarily a simplified depiction of reality, is that all potential collaborations must be orchestrated in a network to create the greatest opportunity to develop the type of innovation desired. Although we focus on the firm as the core unit of analysis, our discussion easily

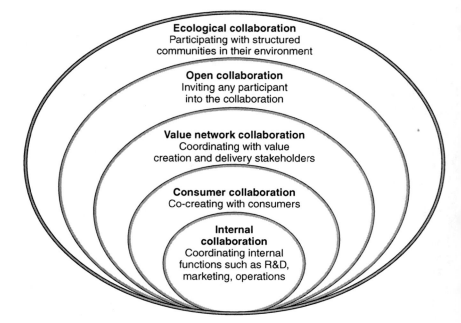

*Figure 1.2*   Network orchestration to achieve collaboration for innovation

extends to business and revenue models and other forms of organizations and institutions in society, including nonprofits, social enterprises, and governments.

## Internal collaboration

The origins of innovation are ideas that can emerge anytime from anywhere – by accident (Austin et al., 2012) or purposively (for example, through research and development competence, Prahalad and Hamel, 1990). More typically however, ideas that arise from within a firm and lead to innovation are a combination of purposive search, sometimes instigated by creativity or by chance, and are nurtured by the *internal collaboration* of various functions within the organization (Wind, 1982). For example, the relationship between marketing and R&D, though traditionally an area of conflict (Gupta et al., 1986), has increasingly benefited from the collaborative practices of cross-functional teams, as recommended in the literature (Slotegraaf and Atuahene-Gima, 2011).

Consider Visteon Corporation, which produces electronics, lighting, and other solutions for car interiors purchased by the major automotive manufacturers. According to Tim Yerdon, Visteon's global director of innovation and design, developing a culture of collaboration was most responsible for transforming Visteon's innovation capabilities (Jaruzelski et al., 2011). He attributes the collaboration that takes place across functions, geographies, and joint-venture partners as a key driver of the firm's innovation performance. In his view, the ways in which their customers manufacture cars as integrated systems have made Visteon's own internal collaboration a necessary capability to succeed.

As Wind (1981) has described in detail, the interaction of marketing with other business functions is, and should be, extensive across a number of areas for innovation, especially in new product development. Consider the development of DuPont's aramid fiber, branded as Kevlar, which began as a kind of accident that was part of an experimental process. Because the fiber exhibited properties making it five times stronger than steel when woven into fabric or mixed into a composite, it was believed that there might be a variety of uses for it. Consequently, several teams were assigned to identify applications. Customer visits and a variety of marketing research methods were employed to generate a prioritized set of opportunities. Several organizational functions, including marketing, R&D, finance, manufacturing, accounting, sales, customer service, and human resources, were required to collaborate to implement the prioritized development projects. The eventual design and development of manufacturing facilities, product formulations, the creation of sales and marketing communication messages, the establishment of pricing guidelines, implementing customer service, and developing aggressive launch plans required a significant amount of internal collaboration. The outcomes were several successful launches, including products such as tires, bulletproof vests and helmets, cable and rope, and sports equipment.

The Kevlar example is product-focused, but internal collaboration can also lead to innovation in marketing, operations, human resources, finance and other business functions where new approaches can lead to competitive advantage. Part of the success of internal collaboration depends not only on cross-functional teams, who may operate globally, but also on how well various tools are used to acquire relevant market knowledge for internal use. This may include traditional marketing research, conjoint analysis to measure consumer preferences (Green and Wind, 1975), the "voice of the customer" (Griffin and Hauser, 1993), lead user studies (von Hippel, 1986), and newer approaches such as text mining and related data analytic approaches. It can also include information about new technologies, new financing arrangements, and market trends.

In many cases this kind of information can facilitate internal collaboration; however, such collaboration is not always easy to achieve within organizations. For example, many firms often have multiple market units and divisions addressing the same customer base. The patient with a disease or health concern is often required to visit several specialists for different indications; what he really wants is an integrated solution or treatment program. Similarly, a pharmaceutical company often calls on the same doctor with different sales people, who represent different drugs from the same firm, when the doctor would really like one sales call about the various drugs from the same firm, or even one call about similar drugs from different firms, to save time and gain more relevant information.

Unfortunately, when managers vie for their brands or product categories within the same company, these silos prevent integration at the consumer level. For example a major personal care company has a manager for its face care range of products, a different one for its body care range, yet another for its deodorant range, and one for its hair care, even though all these products may potentially be applied on the same consumer. Without some form of internal collaboration, the result can be a lower share of the consumer's wallet for the firm.

One cannot ignore that a primary reason for organizations to form in the first place is to bring structure to the complexity of their various functions and operations. However, without careful management, this process of specialization can give way to the silo mentality, which often has to be bridged or broken for the organization to work more effectively. The traditional product development process, which normally begins with ideas turned into concepts, and eventually gets to launched products, illustrates how organizations try to manage innovation internally to bring the different functions together. Bridging these silos with some form of collaboration (which might vary from mutualism to commensalism in biological terms) is imperative for innovation to emerge from an organization, and it may be a necessary condition to deeply engage more effectively with other market participants in developing opportunities for innovation.

## Consumer collaboration

The second inner circle in the Figure 1.2 represents *consumer collaboration*, an organization's ability to deeply engage with and empower consumers in its markets for purposes of innovation. These consumers with access to new communication technologies from almost anywhere in the world, have become primary drivers of market collaboration. As Wind (2008, p. 22) summarized from his earlier work on convergence marketing (Wind et al., 2001), this type of collaboration is characterized by "empowered consumers" who have "become co-inventors, co-producers and even co-marketers." Consider the Build-A-Bear Workshop retail and online enterprise (*www.buildabear.com*), which allows consumers the opportunity to create their own teddy bear in an interactive, playful themed environment. The process begins with choosing a basic bear, then moving from station to station in the store, adding various options, such as prerecorded internal sounds, stuffing (even a heart), stitching (with a unique barcode identifier), clothing, naming, and certification (birth certificate). There are online opportunities for consumers to continue their interaction through the purchase of additional accessories and to engage in games and other activities.

The most frequently cited term to define this process has been *co-creation* between a firm and its consumers. As described by Prahaladad and Ramaswamy (2004), it involves four factors: (1) a deep dialogue with customers that involves interactivity, engagement, and a propensity to act on both sides, (2) access to tools and information needed to create, (3) risk assessment that leads to informed choice, especially by consumers, and (4) transparency of information on both sides. The authors argue that combining these four factors in creative ways can lead to new forms of value.

The example of NikeiD presented earlier illustrates this kind of market collaboration between a firm and its consumers. However there are other examples permeating the landscape of consumer-based innovation. Consider mymuesli.com (*www.mymuesli.com*), founded in Germany in 2007 as a website where consumers can mix and name their own organic muesli online and have it shipped directly to their homes. Given a choice of over 75 ingredients, it is possible for consumers to create some 566 quadrillion individual muesli mixes! In 2009 Coca-Cola began offering consumers its Freestyle vending machine, which is a soda fountain accessible by touch screen that lets consumer create their own soft drinks from over 125 custom flavors. Or consider redmoonpetfood.com (*www.redmoonpetfood.com*), which offers pet owners the opportunity to select ingredients to create their own customized pet food using fresh produce, meat, and various supplements.

While several examples of successful co-creative consumer collaborations are available, Echeverri and Skalén (2011) proposed that value co-creation has not been well documented by research, and there may even be a potential dark side they term value *co-destruction*. More specifically, through a study of transportation services, they find that value co-creation does occur between providers and customers, but so too does value destruction, or more precisely the diminishment of value that takes place when the customer better understands and even uses the product or service. A distinction should be made between real value loss from usage and psychological value loss due to deeper learning about the product or service under development (e.g., "It's not what I thought it was supposed to be!"). The latter type of loss can actually come into conflict with the value co-creation process and therefore must be carefully managed in the context of innovation. The authors offer five interaction value practices that can be used to better manage the co-creation process: informing, greeting, delivering, charging, and helping. The underlying message is that co-creation is a relatively novel process itself, and therefore requires additional research, understanding, and guidelines for practice.

Füller (2010) also sheds light on the complexity of the co-creation process in his findings that show that consumers' motivations can determine their expectations towards a virtual co-creation design; i.e., when you engage the consumer, you engage an entire range of possible behaviors. There is also the manufacturer's side of the co-creation process. The implication that co-creation is a move toward the one-to-one future espoused by Peppers and Rogers (1993) may lead to a greater need for some form of mass customization (Pine, 1993). The mass customization of products and services can be costly, and therefore not profitable; further, in some cases it may not even be possible with current technology, nor does it seriously consider the consumer. Wind and Rangaswamy (2001) propose "customerization" as an alternative and an enhanced approach that is more about customizing marketing to meet consumer needs than customizing manufacturing. As they point out, manufacturing can even be outsourced if it better supports the creation of value with consumers. The key point is that consumer collaboration, most closely associated with value co-creation, is a process like internal collaboration that is required for innovation, but must be thoughtfully developed and carefully managed to achieve it.

### Value network collaboration

The third inner circle in the Figure 1.2 represents *value network collaboration*, or an organization's ability to develop platforms that deeply engage targeted participants in its markets for purposes of innovation. These participants can be distributors, suppliers, or other relevant market stakeholders located anywhere in the world, all of whom may define a network that are part of the process required to generate and deliver value to consumers. A good example of this is the Starbucks Coffee Company (*www.starbucks.com*). To deliver the coffee experience they enjoy to consumers, the company must bring together a network of coffee and tea growers located in many countries, coffee roasters, dairy farmers, various suppliers of food, and over 15,000 retail stores throughout the world, each with a trained staff.

The value network also includes participants who recognize that their own business-to-business collaborations can improve each other's innovation performance. Consider the case of Moleskine (*www.moleskine.com*), an Italian manufacturer of classic notebooks for writers and artists. Traditionally their notebooks had solid black covers and are sold through retailers. To grow their business, Moleskine began collaborating more interactively with retail stores, such as Bloomingdales, inviting them to design new platforms on which to offer their products to shoppers. These platforms not only included customized designs with and for retailers, but creative displays and promotions that both parties found innovative and of value (Schmidt, 2011). Moleskine has also partnered with Lego to create books that included a Lego brick inserted in the cover as a foundation for Lego constructions, as well as Lego bookmarks and stickers.

The distinguishing factor about participants in the value network is that they are primarily involved in business-to-business relationships. Such relationships are complex and involve understanding organizational buying behavior and buying centers within each organization that may have different needs and require their own internal collaboration, as well as collaboration with others in their network constellation. Wind and Thomas (2010) explore this complexity as well as the consequences of such an interdependent network of organizations in terms of five key drivers: accelerating globalization, flattening networks of organizations, disrupting value chains, intensifying government involvement, and continuously fragmenting customer needs. These drivers create both opportunities and problems for firms that desire to collaborate for innovation, but also open possibilities for ideas and solutions beyond the value network into more open spaces.

### Open collaboration

The fourth circle in Figure 1.2 is *open collaboration*, or going beyond the firm, its consumers, and its value network to invite virtually anyone into the process of innovation. This type of collaboration can range from parties who know each other face-to-face to parties who may interact in a virtual collaboration mode, perhaps even indirectly. One of the earlier forms of open collaboration for purposes of value creation was the type associated with open-source software development (e.g., Linux), Internet-based communities of software developers who voluntarily collaborate to develop software that they or their organizations need. As von Hippel and Krogh (2003) note, Richard Stallman drove the development of open source software code in response to the Massachusetts Institute of Technology's action to license some of the code created by its "hacker employees" to a commercial firm. The source code was no longer accessible to these "hackers," who were not technically MIT employees. Stallman set about creating a legal process that would enable software coders who chose to allow anyone to use their code to do so, thereby creating a new concept of openness.

In effect, the objective of the open-source movement was not necessarily to develop innovation that enabled the capture of value, but rather to more fundamentally create value without necessarily capturing it. With increasing online access, this model was readily extended to a variety of potential virtual collaborators, who could contribute to innovation in a variety of contexts, including products, services, advertising, and general problem solving. Another classic example of this is Wikipedia (*www.wikipedia.com*), a collaborative Internet encyclopedia, which reports that it contains over 20 million articles written collaboratively by 100,000 contributors from around the world.

Several versions of open collaboration have developed in recent years. Chesbrough (2003) describes it through the traditional product development process from ideas to launch. In his view, the boundaries of the process are permeable and ideas can seep in and out at various stages of development. He contrasts this with earlier models of innovation, which are more or less "closed." He cites Procter & Gamble's (P&G's) 2001 "Connect and Develop" program as an example of the effectiveness of opening opportunities for innovation outside the firm. The original goal of delivering 50 percent of the company's innovation through external collaboration was met in 2005, two years ahead of schedule (Drummond, 2011). The program has grown in size and value and according to Drummond (2011), P&G says its Connect and Develop is "about collaborating for mutual value creation with anyone, anywhere to accelerate P&G's innovation and deliver growth goals."

Whereas Chesbrough's view of open innovation originally implied a fairly structured approach to selecting the sources of ideas in collaboration with organizational partners, others have invited just about anyone to the table to provide ideas, not just for innovation, but for any productive activity that can be outsourced to anyone. Howe (2006) originally defined this as "crowdsourcing," or "the act of taking a job traditionally performed by a designated agent (usually an employee) and outsourcing it to an undefined, generally large group of people in the form of an open call," or more briefly, applying open-source principles to fields outside of software, a view of open collaboration that can take many forms.

Based on the idea of the "wisdom of crowds," Surowiecki (2004) proposed that an aggregation or averaging of individuals in a group can produce a better decision than any single person in the group, even a group containing experts. One form of aggregated crowdsourcing based on wisdom is the prediction market, in which participants trade in contracts with payoffs that depend on unknown future events, and the market price is the best predictor of the event (Wolfers and Zitzewitz, 2004). Applications include predicting election outcomes, movies, new product forecasts, and even the likelihood that regulators would approve new pharmaceuticals. Another form of aggregated crowdsourcing is the tournament as described by Terwiesch and Ulrich's (2009), which involves a competition among a large set of opportunities that are voted down to the few winning ones selected by the "crowd." For example, Walkers of the United Kingdom (*www.walkers. com*), a snack food manufacturer owned by Frito-Lay, regularly conducts tournaments to name a flavor or find the best new flavor for one or more of its chips or snack foods.

Yet another approach of crowdsourcing is broadcasting a problem to the largest audience possible, usually online, and finding a solution. For example, InnoCentive (*www.innocentive.com*) provides an online platform that presents problems to a diverse crowd of more than 250,000 problem solvers from some 200 countries. Winners are selected from what are essentially prize-based competitions. The value of this approach to innovation is more than just the number of people who work on the challenges they accept, but their diversity (Page, 2007). According to Lakhani et al. (2007), who analyzed 166 problems processed through InnoCentive: "We found a positive and significant correlation between the self-assessed distance between the problem field and the solver's expertise and the probability of being a winning solver. The further the focal problem was from the solvers' field of expertise, the more likely they were to solve it."

A typical example experienced by Roche on a pharmaceutical challenge using this type of open collaboration with InnoCentive is reported by Birkinshaw and Crainer (2009). Roche struggled for over 15 years to improve the quality and volume measurement of a clinical specimen as it is passed through chemistry analyzers. They defined this problem as a challenge for InnoCentive and in two months they had some 1,000 unique solvers committing to the project, and eventually got 113 proposals from them. Roche not only solved their problem from these proposals, but in reviewing the submitted solutions, they realized that all the solutions they had tried over 15 years were also submitted!

Notably, open collaboration activities for innovation extend beyond product development and can include piecework. In this approach, small "intelligence" tasks are farmed out to anyone anywhere for a fee at Amazon's Mechanical Turk (*www.mturk.com*). Financial investing opportunities can be found at kickstarter.com (*www.kickstarter.com*), and problem solving at IBM Jams (*www.collaborationjam.com*). As an example of open collaboration in the area of marketing communication, consider the Victors & Spoils ad agency (*www.victorsandspoils.com*), or V&S. When Harley Davidson split from its ad agency, the CEO of newly formed victorsandspoils.com agency, Jon Winsor, used his online team of over 6,000 creative people to provide unsolicited ideas to Harley for a campaign. Compensation to the creative people depended on ideas selected by the agency and the client. The V&S CEO sent a selection of ideas to Harley CEO via Twitter. Ultimately, Harley selected V&S as their agency of choice and aired their "Cages" campaign in 2011. In every other respect the agency is traditional; however, by using its "crowd" of creative people it was able to provide a broader pool of innovative ideas to meet the communication challenges and win the business.

Finally we consider how open collaboration supports customized innovation in services. Consider the consumer who intends to purchase a tattoo. He or she may seek out a local tattooist who may offer a variety of options, or, alternatively, he may consider the online offering called Createmytattoo. com (*www.createmytattoo.com*), which provides the opportunity to be connected online with over 3500 tattoo designers. The consumer provides an initial idea to these designers and receives in return at least 10 designs from different tattooists. The consumer can then interact further with one or more of the designers to obtain the final design prior to implementation at a local tattooist's shop, who may be part of the Createmytattoo.com community.

In their book, *The Global Brain*, Nambisan and Sawhney (2007) provide structure to open collaboration to address the challenge of managing innovation. They propose two core dimensions to the problem: the structure of the players involved (centralized management vs. diffused communities) and the nature of the innovation problem (defined or structured vs. emergent or really new). This supports four different models of collaboration for innovation: the *orchestra* (centralized/defined), the *creative bazaar* (centralized/emergent), the *mod station* (diffused/defined), and *jam central* (diffused/ emergent). The latter two models are most consistent with open collaboration in that they are relatively open, less visible, and informal compared to the more centralized models. Victors & Spoils and InnoCentive are examples of either mod station or jam central, depending on the problem definition. Clearly, open collaboration can be productive for innovation, but it requires some thought to managing it for success.

The increasing popularity of open collaboration has been evident for some time, as described in a lead *Business Week* magazine article entitled "The Power of Us" in 2005. It can come in many forms, from real to virtual, from single participant to many, and from direct to indirect, where parties may never know each other. Nevertheless, there can be difficulties in establishing this capability. Should open collaboration involve anyone and everyone, or should targeted communities be the source of ideas? Is there a risk of "group think" – will like-minded people online eventually lead to mediocre innovations? In addition, how does one establish screens to filter hundreds and possibly thousands of ideas? Does failure of idea acceptance diminish the chance that a participant will submit future ideas? Who will own the intellectual property rights? How important is security of the outcomes and how will it be managed? While there are such concerns, preliminary case study research on open collaboration via crowdsourcing has shown that it can favorably complement traditional internal idea generation methods (Poetz and Schreier, 2012). This suggests that it should be required as part of the innovation process, but also carefully managed to create opportunities for success.

### Ecological collaboration

The encompassing fifth circle in Figure 1.2 is *ecological collaboration*. In the context of collaborative innovation we define "ecology" as a community of participants who assemble to meet their needs, and who can influence and be influenced by their environment. Consider an industry association, an academic community focused on a particular area of knowledge, or product users who communicate to solve problems and otherwise collaborate to meet their needs. The interaction of firms, their markets of suppliers, distributors, customers, and regulators, and other open sources of collaborators can create an ecology that generates the knowledge needed for substantial value creation and innovation (Dougherty and Dunne, 2011).

In many cases these ecologies are equivalent to structured communities that are sometimes driven by their environment, but must be controlled to succeed. An early example was the so-called "Wintel" community, which involved Microsoft Windows, Intel, and scores of software and hardware firms that formed an ecological system with common standards to meet the computing needs of individual and organizational users. Consider the case of Tiger21 (*www.tiger21.com*), which defines itself as a peer-to-peer learning group for high-net-worth investors (Hawthorne, 2010). The groups are made up of 12 people. They meet for a day once a month for an annual fee of $30,000 to discuss their financial problems and opportunities. In some way it can be likened to a support group that exists because the surrounding financial environment has created sufficient turbulence to create the need for knowledge, which the collaboration provides. What is unusual about these small ecologies of investors is that the members admit that they discuss issues in the group that they cannot discuss with anyone else, not even their close friends.

The ecological collaboration spurred by Apple, Inc. from its products such as iPhone (Laugesen and Yuan, 2010) provides a well-defined example of the formation of such ecologies. The phone was developed from *internal collaboration*, announced in January 2007, and was launched in June 2007. By December 2009, some 42 million iPhones had been sold. To facilitate this success, Apple had to create *consumer collaborations* through its own stores and phone retailers (such as AT&T) to educate consumers on product use. In July 2008, Apple launched its App Store, based on *open collaboration* with thousands of independent third party software suppliers who could design and create a variety of software applications for iPhone users, albeit within certain parameters. Consumers could then purchase and download the applications of their choice from the App Store. Similarly, Apple developed its iTunes site, which is a collaboration of music artists, publishers, and other related content producers (movies, TV shows, podcasts, etc.). This collaboration supports the iPod, and other Apple devices as well, including the iPhone and iPad.

The symbiotic aspects of Apple's innovation process should be evident in this basic relationship: the application developers are highly dependent on Apple for their success, and in turn, Apple's continued success with the iPhone and related devices is dependent on the relationship it has with its application developers. With their own retail stores to both sell products and directly support consumers, Apple has created a complete *ecological collaboration* to support innovation from product idea, to design and development, to manufacture, to software applications, to marketing and sales, and to solution of consumer problems. Whereas competitors can essentially duplicate the basic iPhone, Apple's advantage derives from the strength of its ecological collaboration. For the iPod, iPad, or whatever Apple's next new product might be, there is an ecological collaboration encompassing its success (iTunes, the App Store, iCloud). The outcome of the process is the creation of considerable value, not only for Apple and its developers, but also for business and consumer users of the phones and related devices.

There are also other communities that reveal the universality of the ecological collaboration experience. Consider Facebook, Twitter, Google, Wikipedia, LinkedIn, Groupon, and Amazon: all are online-based communities that exist to serve the needs of their various partners, whether consumers, advertisers, manufacturers, retailers, or even regulatory groups. Some communities are more loosely organized than others, but all exist to support the needs of participants. In terms of innovation, the value created from the knowledge gained, shared, or otherwise exchanged from the collaboration drives the continued survival of these ecologies.

Google is a primary example of an ecological community that represents a collaboration based on a fundamental human need – the search for information. Google has its own internal collaboration of software developers who work to improve its search engine and develop other offerings. Its new offerings are made available to consumers in beta-test format, a type of consumer collaboration that engages and empowers Google's visitors/consumers. The sources of revenue for Google depend primarily on its value network, which includes advertisers, advertising agencies, and research firms who collaborate to create innovative ways to gain the attention of consumers ("the number of eyeballs") for their own products and services. Google frequently uses open collaboration for special projects, such as its "Project ·10 to the 100," which presented a problem to the virtual world via Google and screened the ideas to fund solutions. In other ways Google has to carefully manage its citizenship in the global community of governments, regulators, and special interest groups concerned about privacy and other aspects of Google's ubiquity. It is evident that Google is in every way its own ecological system of collaborators that can interact with other ecological systems.

Not all such communities thrive. MySpace is an example of an early leader in the social community world that fell on hard times. While there are many reasons for its decline, one involved how MySpace chose to develop new offerings and services compared to their main competitor, Facebook. Facebook embraced a more ecological collaboration perspective than MySpace. As noted in a *Business Week* cover story (Gillette, 2011): "While Facebook focused on creating a robust platform that allowed outside developers to build new applications, MySpace did everything itself." The risks of limiting innovation to only internal collaboration in highly complex and turbulent markets highlights the need to better orchestrate potential collaboration for innovation.

## 4    Network orchestration for symbiotic innovation

Taken together, the five types of collaboration presented in Figure 1.2 represent a challenge to develop breakthrough innovation. The challenge is that all five types of collaboration are required and each entails some form of symbiotic relationship, but it is difficult for most firms to control them. Fung et al. (2007) describe this requirement for control as *network orchestration*. It is not enough to collaborate internally, with your market, with open crowds, or in an ecological community; these collaborations must be orchestrated to compete more effectively through innovation. Organizations such as MySpace, cited above, that do not understand that competition is "network against network" run the risk of damaging their relationships with potential partner organizations and can miss opportunities for breakthrough innovation. These potential partners may join with other networks and abandon a working relationship with an organization that fails to integrate, thereby putting the organization at risk to capture value in its markets. Firms can therefore become locked in to or locked out of networks, with considerable impact on their performance.

To orchestrate a network for innovation is therefore critical, but not easy. It takes considerable management skill, because it is not the same as managing internal collaboration, the most common form managers have found. Instead, it requires a more fluid approach that empowers partners, employees, customers, distributors, and other stakeholders, while also maintaining control. It is akin to assembling the musical composition (possibly being the composer), the orchestra of musicians, the conductor, the concert hall, and the communications and practice required to deliver an experience to the audience.

An example of creating and orchestrating such a network is GE Healthcare's Healthymagination effort (*www.healthymagination.com*). In September 2011,

GE committed $100 million to fund innovation in combating breast cancer, orchestrating a variety of participants. In addition to its own organization, GE, in cooperation with venture partners, announced an open call to action for oncology researchers and healthcare innovators seeking ideas to accelerate innovation for solutions to breast cancer. In less than two months, its first challenge involved some 4,000 people, who submitted over 500 ideas. Numerous other challenges are planned to fund advances that will empower doctors and patients and enable more personalized treatment for women worldwide. This effort clearly involved a well-orchestrated network of players for each type of collaboration in Figure 1.2. They are using internal collaboration (their own management structure) and consumer collaboration with patients in mind; they value network collaboration with doctors, hospitals, and other healthcare institutions as well as open collaboration (inviting research ideas from a broad array of external sources in the medical and scientific fields); and they use an ecological collaboration with various medical and cancer-concerned communities.

All five types of collaboration are essential to create significant innovation outcomes. Further, a network must be designed and orchestrated to produce breakthrough ideas, whether for products, services, communications, business processes, or other areas in need of innovation. To make this happen, a *network orchestrator* must be designated who has the responsibility to understand the type of innovation sought (incremental vs. breakthrough), the various collaborators and collaborations required to achieve it, and the kind and balance of symbiotic relationships within the collaboration necessary to bring about the desired outcome.

The network orchestrator has three critical roles to succeed (Fung et al., 2007). First, the network has to be designed and managed in a different way than a firm is managed. The firm must be considered, but in the context of the various types of collaborators and collaborations described in preceding sections. Second, the orchestrator must recognize the need to empower the various collaborators in the network. This is a different kind of incentive structure and motivation than happens within the firm. Here it becomes critical to truly understand if the collaborators are in mutually beneficial relationships, or are going into them, or are in ones that are imbalanced and need rebalancing. Finally, value does not necessarily come from internal competencies; it comes from integration, bridging borders, and from leveraging the company's value and intellectual property across the network. Successful network orchestration also gives a firm the needed dexterity and flexibility to operate in a dynamic and uncertain world. For example, if there is a crisis in one country or part of the world, the firm can readily shift R&D, people, or production capability to other countries.

Returning to the GE Healthymagination example, all three roles of the network orchestrator were evident: (1) the network was designed and managed not as a GE business unit, but as a network of multiple participants.

orchestrated to develop innovations for breast cancer cure; (2) control through empowerment is clearly evident to encourage external researchers to submit ideas, but it is flexible enough to allow an evaluation and ranking of ideas generated; and (3) there is creation of value through the integration of all parties by capturing the best ideas before competition and beginning deep research on the winning ones. Bringing all the pieces of a network together is not an easy task, but one that is essential to realize symbiotic innovation.

## 5  Implementing a model of symbiotic innovation

We are proposing a significant undertaking for firms who want to achieve a portfolio of innovations, from incremental to breakthrough in scope, which can be product and/or process in substance, and which can stretch beyond product to include innovations in other aspects of the business, such as new models of business and revenue, creative marketing approaches, and comprehensive networks of collaborators that will make a difference in their business growth and profitability, as well as contributions to society. We cannot present all the detailed steps necessary to implement this approach, but we provide a model of symbiotic innovation in Figure 1.3 to help rethink current innovation practices. We briefly review each of the five components of the model in the following sections.

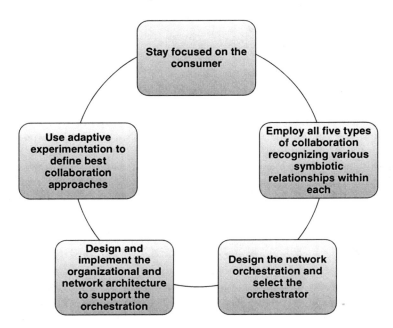

*Figure 1.3*  A model of symbiotic innovation

## Stay focused on the consumer

It may sound basic, but in structuring a network of collaborators, consumers must be at the center of activity. The next new wave that drives an economy is often a breakthrough innovation that benefits consumers and creates an infrastructure of enterprises to support that demand. Consumers may not be the inspiration for the next big innovative idea, but their ultimate purchase and usage of it are at the core of its success. This means incorporating the consumer not just at the stage of idea generation and concept development, but all through prototyping, product development, marketing, usage, and modification. Whatever methods are used – traditional marketing research, lead user, voice of the customer, co-creation, or crowdsourcing – collaboration with and among consumers should be evident throughout all types of collaboration.

The consumer must also be broadly defined to include his or her social structure. Couples, families, friends, acquaintances at work, participants in social networks, and other social units the consumer interacts with define the consumer outside the individual unit of analysis. How these significant others affect consumers' preferences and choices is central to innovation. For example, to drive down costs, hospitals recognize the need to move patients home as quickly as possible. However, the availability of products and services to assist families and others to care for the sick, or those recovering from surgery, is an important area for innovation that is yet to be developed. Understanding a variety of collaborations with the consumer at the center will be required.

## Employ all five types of collaboration

In the literature on innovation and new product development, the 1980s and 1990s witnessed a focus on internal collaboration through integrated new product development processes. Beginning at the turn of the twenty-first century, the role of market collaboration and co-creation became highly relevant and was made possible with the advent of more interactive communication technologies. By the end of the first decade of the new century, it has become clear that open collaboration and embracing communities of existing collaborators opens the door to more rapid and effective innovation.

In effect, it is not enough to innovate today with internal, consumer, and value network collaboration. Open and ecological collaboration must be engaged in and orchestrated to create significant innovation. Several examples of these new forms of collaborations exist; however, many firms may be reluctant to step out into these somewhat unchartered waters. For example, legal departments in organizations may battle against any form of collaboration for fear of losing ownership of patents or patented processes. Rather than back away from such threats, new forms of patents and intellectual property rights may need to be pursued. Similarly, marketing and manufacturing personnel in organizations may be unwilling to share data and information with collaborators; again, new ways of thinking that it can be mutually beneficial to share information to speed the development of innovation can have financial benefits for all.

### Design the network orchestration and select the orchestrator

Designing a network for symbiotic innovation in any detail goes beyond the limits of this chapter. However what is required may involve several considerations. First, who are the potential collaborators one would desire in a network for symbiotic innovation? What are their needs and potential benefits that might make them dependent on each other for successful innovation? If there is no basis for reciprocal collaboration, then it will be difficult to sustain the network. Second, what are the roles and expectations for the various collaborators? What is each expected to contribute? How will their roles be communicated and learned, if necessary? Third, what are the strengths and weaknesses, or unique competencies of each collaborator and how do they contribute to a competitive advantage for the network? These are questions that cannot be answered quickly, but require some study of potential collaborators, including their history, their likelihood of cooperating, and other factors.

Once potential collaborators are identified, the critical flows among them must be identified and mapped. Will it be primarily information flows, or will there be physical flows, financial flows, and/or workflows involved? Understanding these flows and how collaborators manage them will be central to defining the network. Finally, the collaboration outcome must be defined, if not in detail, at least in terms of the general objectives. For example, if the required outcome is a creative communication approach, then a communication brief describing the situation and the problem to be addressed must be written and communicated. Similarly, if a new product is the desired outcome, then an innovation brief or similar document should be prepared to describe the situation and general type of innovation required, perhaps from an incremental innovation to a major breakthrough or even a completely disruptive innovation.

Achieving the above tasks will require the work of a network orchestrator. In addition to defining the characteristics of such a person, recruiting, motivating, and empowering him or her to carry out the requirements of effective network structure will be a challenge. Recall the three key roles a collaborator must play, defined by Fung et al. (2007). First, the person must be network-centric, not firm or market-centric. Once the entire network is identified and understood, potential innovation collaborators within it can be identified and approached. Second, the person must have a sense of control through empowerment. That is, networks can involve independent players whose cooperation is needed, and it may be that the only way to gain their cooperation is to empower them to collaborate for purposes of innovation; which may mean that the network orchestrator must get the various collaborators to see the value of the innovation. Trust, communication, understanding, and other behavioral and social motivation may be more important than financial remuneration in this type of empowerment. Third, the network orchestrator must be able to facilitate and incentivize the creation of value through the collaboration. It is not enough to bring collaborators in the network together; the orchestrator must have a clear vision of how an innovation can come about through carefully managing the network and providing the necessary incentives. Finding and motivating such an orchestrator will not be a trivial task, but once found, this person should be strongly supported by the organization.

### Design and implement the organizational and network architecture to support the orchestration

If the leaders and managers in an organization decide that innovation is essential to grow and profit, if not survive, and that symbiotic innovation based on a network model of collaboration is required to do so, then a different mental model of how the organization and network function will be necessary to achieve the desired outcomes. Figure 1.4 provides a structure of how such an organizational and network architecture might be conceptualized. At the top are the vision, objectives, and strategies of the organization as articulated by its leaders. However this articulation will reflect the network of collaborators at its core (in the center of the chart). The network orchestrator must then rally the organization and it leaders to rethink how innovation can happen among the network participants. The right people, processes, facilities, and culture will be required to interact with, stimulate, and empower collaborators. In addition the organization's technology, the proper allocation of resources, clear communication of performance measures and incentives, and the organization's basic structure and ownership will not only influence internal people and processes, but collaborators as well.

Achieving the architecture implied in Figure 1.4 is not something that can be accomplished overnight for an organization seeking to create the organizational and network architecture among collaborators for purposes of innovation. It may take some years to achieve, depending on the current situation of an organization. Unfortunately, organizations that are quite successful in terms of technology often have difficulty capitalizing on it. One is reminded of Kodak, which apparently had a difficult time in appreciating the need to respond to external threats from a constantly changing environment.[4] For example, it was among the first to assemble the technology for digital photography (beginning in 1975), and even partnered with Kinko's, Microsoft, IBM, and Hewlett Packard to bring it to market, but not necessarily in a consumer-friendly form. Instead, Kodak chose to market

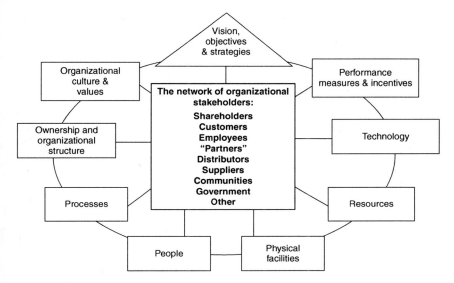

*Figure 1.4*   Architecture of networked organization

its technology in many cases to niche applications, licensing or selling their technology to other camera manufacturers.

After mobile phones came equipped with excellent electronic cameras, the market for cameras changed forever. This occurred some ten years before Kodak went into bankruptcy on January 19, 2012. The implication is that Kodak had time to do something about their situation, but apparently was unable to do so. While one may speculate on the reasons for this – they are inside-out driven, or they are still locked in the mindset of images on paper, etc. (Scheyder, 2012), the outcome is a threat to the existence of a 130-year-old firm with a classic brand name. Would a different organizational and network architecture have provided Kodak with the opportunity to survive with some form of symbiotic innovation, better capitalizing on its technology? While we won't know the answer to this question, it is clear that the way in which organizations respond to their environment influences their need for innovation, and it follows that the very architecture of their organization and network will influence their ability to innovate, and perhaps, to survive.

### Use adaptive experimentation to define best collaboration approaches

Serious managers will question the value of all the work required to achieve symbiotic innovation through multiple types of collaboration and network orchestration – or will ask for the return on investment of such an approach. Ideally one would like to compare the kind of innovation that might come out of one or a few types of collaboration compared with that derived from a more comprehensive network orchestration. However this kind of formal experimentation would be cost prohibitive, even if it were possible to design and implement. Instead, we recommend a more adaptive experimentation approach in which different types of collaboration are compared individually to each other or in combination, and the results tracked.

For example, imagine the problem of creating a new approach to care for postsurgical patients at home. After defining the problem as a challenge, one could imagine giving it to an internal team of collaborators to use brainstorming, morphological techniques, and other creative methods to develop as many ideas as possible. The same could be done in terms of generating ideas from a co-creation process with actual consumers and their families responsible for home care of patients. In addition, the problem could be sent to a crowdsourcing venue (e.g., InnoCentive) to obtain their ideas. Using predefined metrics and screening criteria, the number and quality of ideas can be compared to estimate the value of each type of collaboration. Essentially, adaptive experimentation is a characteristic of learning organizations, and applying it to the process of symbiotic innovation via different types of collaboration is a kind of learning that marks an organization as not only as a learning one, but also as one that is dynamic and is more likely to thrive in unstable environments than others are.

Given the complexity and uncertainty facing firms, it should be evident that there is no single optimal solution for successful innovation, and the only way to really succeed is to learn over time and acquire the knowledge necessary for innovation by using adaptive experimentation. This process can lead to breakthrough ideas, better decisions, faster learning, uncertainty and confusion among competitors about a firm's actions, and can significantly contribute to creating a culture of innovation.[5] It achieves the latter by recognizing that failure is an inevitable outcome of experimentation, which in turn leads to acceptance of failure and the subsequent focus on lessons learned from failures.

## 6   Conclusions

Innovation is not a local phenomenon. In 2012, PwC (Pricewaterhouse Coopers) published a study of over 1250 CEOs about the challenges they faced. While their overall findings about the new world of uncertainty and complexity were similar to those of IBM that were indicated at the outset of this chapter, they also indicated that in response to complex rapidly changing environments CEOs were not just exporting their current offerings to emerging markets; they were building entirely new businesses and reconfiguring operations to meet local market needs to pursue growth (PwC, 2012). The CEO's of reporting firms were (PwC, 2012, p. 8) "...simultaneously building local capabilities in important markets, extending operational footprints, building strategic alliances and creating new networks for new markets that include research and development (R&D), manufacturing and services support. They're adapting how they go to market, reconfiguring processes and at times entire operating models."

The PwC study describes the need for the kind of symbiotic innovation and collaboration we have discussed in this chapter. It is not an easy process that can be achieved in a stage-gate fashion, but rather a complex and cluttered one with many participants with different objectives and goals who must be orchestrated to achieve the harmony required for innovation to emerge. First it is essential to understand the various interaction patterns among collaborators from a symbiotic perspective (see Figure 1.1). With this conceptualization, it can be used to more deeply study collaboration patterns found in each of the five types of collaboration – from inside the innovating firm to the ecological communities (see Figure 1.2). With a concentrated focus on the consumer as a key beneficiary of innovation, all five types of collaboration can be evaluated to design and implement the network architecture that most efficiently and effectively defines the best symbiotic innovation process to achieve desired outcomes (see Figures 1.3 and 1.4).

Pursuing this kind of effort is substantial, and it leaves many unanswered questions that can be pursued in future research by both academics and practitioners. For example, which collaborators should be chosen, based on which criteria in the situation faced by the innovating organization (or entrepreneur)? What is the appropriate balance of incentives and benefits that motivate each collaborator to work with others to innovate? What kinds of innovation will a balanced collaboration (mutualism) produce relative to a less balanced one (commensalism)? What are the criteria for defining a successful network for collaborative symbiotic innovation and for a successful network orchestrator? What are the best methods of communication among network participants? What can go wrong in a symbiotic collaboration, such as poorly set expectations, environmental challenges, excessive delays, or parasitic partners? Knowledge from future research among academics, adaptive experimentation among practitioners, and reporting of successes and failures can help answer some of these questions and drive the development of significant innovation for the future.

## Notes

1. Symbiosis has been discussed in marketing literature (Adler, 1966, Varadarajan and Rajaratnam, 1986), in organizational literature (Haire, 1959), and in the literature on industrial ecology (Chertow, 2007), but primarily in terms of strategic partnerships, not in terms of innovation.
2. Throughout the chapter, although we often consider innovation in terms of an offering to consumers (product, service, etc.), we also see it as a more generalized opportunity to create something new in any context. New business models, new

forms of communication, new distribution options, new customer relationship management approaches, and new pricing paradigms illustrate the scope of innovation we recognize.

3. The units of gain or loss can be defined by a number of benefits; for example, the reduction of risk in the relationship may be seen as a gain by one participant or the other, or by both.

4. Valuable analyses supporting this discussion of Kodak are provided in Crook (2012) and Scheyder (2012).

5. For a more detailed discussion of adaptive experimentation, see Wind (2011).

## References

Adler, Lee (1966) Symbiotic Marketing, *Harvard Business Review*, 44, 59–71.

Alderson, Wroe (1957) *Marketing Behavior and Executive Action: A Functionalist Approach to Marketing Theory*, Homewood, IL: Richard D. Irwin.

Apple Inc. v. Samsung Electronics Co., Ltd. (2012) Retrieved from Wikipedia: http://en.wikipedia.org/wiki/Apple_Inc._v._Samsung_Electronics_Co.,_Ltd.

Austin, Robert D., Lee Devin, and Erin E. Sullivan (2012) Accidental Innovation: Supporting Valuable Unpredictability in the Creative Process, *Organization Science*, Forthcoming.

Barrett, Paul M. (2012) Apples War on Android, *Business Week*, March 29, Retrieved from: http://www.businessweek.com/articles/2012-03-29/apple-s-war-on-android.

Birkinshaw, Julian and Stuart Crainer (2009) Combine Harvesting, *Lab Notes*, London: London Business School, 12, 15–18.

Brohan, Mark (2010) Nike's Web Sales Flourish in Fiscal 2010, *Internet Retailer*, Retrieved from: http://www.internetretailer.com/2010/06/30/nikes-web-sales-flourish-fiscal-2010.

Business Week (2005) The Power of Us, June 20. Retrieved from http://www.business-week.com/magazine/content/05_25/b3938601.htm.

Cain, Susan (2012) The Rise of the New Groupthink, *New York Times*, January 13, Retrieved from: http://www.nytimes.com/2012/01/15/opinion/sunday/the-rise-of-the-new-groupthink.html?_R=1.

Chesbrough, H. (2003) *Open Innovation: The New Imperative for Creating and Profiting from Technology*, Boston, MA: Harvard Business School Press.

Chertow, Marian R. (2007) "Uncovering" Industrial Symbiosis, *Journal of Industrial Ecology*, 11, 11–30.

Crook, Jordan (2012) What Happened To Kodak's Moment? *TechCrunch*, January 21, Retrieved from: http://techcrunch.com/2012/01/21/what-happened-to-kodaks-moment/.

Dougherty, Deborah and Danielle D. Dunne (2011) Organizing Ecologies of Complex Innovation, *Organization Science*, 22, 1214–1223.

Douglas, Angela E. (2010) *The Symbiotic Habit*, Princeton: Princeton University Press.

Drummond, Mike (2011) Unlocking Open Innovation, *Inventors Digest*, February, Retrieved at www.inventorsdigest.com/archives/5559.

Echeverri, Per and Per Skalén (2011) Co-creation and Co-destruction: A Practice-Theory Based Study of Interactive Value Formation, *Marketing Theory*, 11, 351–373.

Füller, Johann (2010) Refining Virtual Co-Creation from a Consumer Perspective, *California Management Review*, 52, 98–122.

Fung, Victor K., William K. Fung, and Yoram (Jerry) Wind (2007) *Competing in a Flat World: Building Enterprises for a Borderless World*, Upper Saddle River, NJ: Wharton School Publishing.

Gillette, Felix (2011) The Rise and Inglorious Fall of MySpace, *Business Week*, June 22, Cover Story.

Green, Paul E. and Jerry (Yoram) Wind (1975) New Ways to Measure Consumer Judgment, *Harvard Business Review*, 53, 107–117.

Griffin, Abbie and John R. Hauser (1993) The Voice of the Customer, *Marketing Science*, 12, 1–27.

Gupta, Ashok K., S. P. Raj, and David Wilemon (1986) A Model for Studying R&D-Marketing Interface in the Product Innovation Process, *Journal of Marketing*, 50, 7–17.

Haire, Mason (1959) Biological Models and Empirical Histories of the Growth of Organizations, in Mason Haire, (ed.), *Modern Organization Theory*, New York: Wiley, 272–306.

Hawthorne, Fran (2010) A Club to Discuss Discreetly the Issues of Wealth, *New York Times*, October 20, F4; See also: http://www.tiger21.com/default.aspx.

Howe, Jeff (2006) The Rise of Crowdsourcing, *Wired*, February, Retrieved from http://www.wired.com/wired/archive/14.06/crowds.html. See also: http://www.crowd-sourcing.com/.

IBM (2010) *Capitalizing on Complexity*, Somers, NY: IBM Global Business Services. Retrieved from: http://www-935.ibm.com/services/us/ceo/ceostudy2010/index.html

IBM (2011) *From Stretched to Strengthened*, Somers, NY: IBM Global Business Services. Retrieved from: http://www-935.ibm.com/services/us/cmo/cmostudy2011/cmo-registration.html.

Isaacson, Walter (2011) *Steve Jobs*, New York: Simon & Schuster.

Jaruzelski, Barry, John Loehr, and Richard Holman (2011) The Global Innovation 1000: Why Culture Is Key, strategy+business, 65, Retrieved from: http://www.strategy-business.com/article/11404.

Lakhani, Karim R., Lars Bo Jeppesen, Peter A. Lohse, and Jill A. Panetta (2007) The Value of Openness in Scientific Problem Solving, Working Paper, Cambridge, MA: Harvard Business School. Retrieved from http://www.hbs.edu/research/pdf/07-050.pdf.

Laugesen, John and Yufei Yuan (2010) What Factors Contributed to The Success of Apple's iPhone? 2010 Ninth International Conference on Mobile Business, *IEEE*, 91–99.

Magretta, Joan (1998) The Virtual Power of Integration: An Interview with Dell Computer's Michael Dell, *Harvard Business Review*, 66, 73–84.

Nambisan, Satish and Mohanbir Sawhney (2007) *The Global Brain: Your Roadmap for Innovating Faster and Smarter in a Networked World*, Upper Saddle River, NJ: Pearson and Wharton School Publishing.

Page, Scott E. (2007) *The Difference*, Princeton, NJ: Princeton University Press.

Peppers, Don and Martha Rogers (1993) *The One to One Future: Building Relationships One Customer at a Time*, New York: Doubleday.

Pine, B. Joseph (1993) *Mass Customization*, Cambridge, MA: Harvard Business School Press.

Poetz, Marion K. and Martin Schreier (2012) The Value of Crowdsourcing: Can Users Really Compete with Professionals in Generating New Product Ideas? *Journal of Product Innovation Management*, 9, 245–256.

Prahaladad, C. K. and Venkatram Ramaswamy (2000) Co-opting Customer Competence, *Harvard Business Review*, 78, 79–87.

Prahaladad, C. K. and Venkatram Ramaswamy (2004) Co-creating Unique Value with Customers, *Strategy & Leadership*, 32, 4–9.

Prahaladad, C. K. and Gary Hamel (1990) The Core Competence of the Corporation, *Harvard Business Review*, 68, 79–91.

PwC (2012) Delivering Results, Growth, and Value in a Volatile World, PricewaterhouseCoopers 15th Annual Global CEO Survey 2012. Retrieved at pwc.com.

Robertson, Thomas S. (1967) The Process of Innovation and the Diffusion of Innovation, *Journal of Marketing*, 31, 14–19.

Sapp, Jan (2009), *The New Foundations of Evolution: On the Tree of Life*, New York: Oxford University Press.

Scheyder, Ernest (2012), Focus On Past Glory Kept Kodak from Digital Win, *Reuters*, January 19, Retrieved from: http://www.reuters.com/article/2012/01/19/us-kodak-bankruptcy-idUSTRE80I1N020120119.

Schmidt, Gregory (2011), Fans Fill Moleskine's Notebooks With Love, *New York Times*, June 28, p. B3.

Sheth, Jagdish N., Rajendra S. Sisodia, and Arun Sharma (2000) The Antecedents and Consequences of Customer-Centric Marketing, *Journal of the Academy of Marketing Science*, 28, 55–66.

Slotegraaf, Rebecca J and Atuahene-Gima, Kwaku (2011) Product Development Team Stability and New Product Advantage: The Role of Decision-Making Processes, *Journal of Marketing*, 75, 96–108.

Surowiecki, James (2004) *The Wisdom of Crowds: Why the Many are Smarter than the Few and How Collective Wisdom Shapes Business, Economies, Societies, and Nations*, New York: Doubleday.

Terwiesch, Christian and Karl T. Ulrich (2009) *Innovation Tournaments*, Boston: Harvard Business School Publishing.

Varadarajan, P. "Rajan" and Daniel Rajaratnam (1986) Symbiotic Marketing Revisited, *The Journal of Marketing*, 50, 7–17.

Vargo, Stephen. L. and, Robert F. Lusch (2004) Evolving to a New Dominant Logic for Marketing, *Journal of Marketing*, 68, 1–17.

von Hippel, Eric (1986) Lead Users: A Source of Novel Product Concepts, *Management Science*, 32, 791–805.

von Hippel, Eric and Georg von Krogh (2003) Open Source Software and the "Private-Collective" Innovation Model: Issues for Organization Science, *Organization Science*, 14, 209–223.

Wind, Jerry (Yoram) (1981) Marketing and the Other Business Functions, *Research in Marketing*, 5, 237–264.

Wind, Jerry (1982) *Product Policy: Concepts, Methods, and Strategy*, Reading, MA: Addison-Wesley.

Wind, Jerry (2008) A Plan to Invent the Marketing We Need *Today, MIT Sloan Management Review*, 49, 21–28.

Wind, Jerry (2011) Adaptive Experimentation, Wharton@Work, Nano Tools for Leaders, June. Retrieved from: http://executiveeducation.wharton.upenn.edu/wharton-at-work/1106/adaptive-experimentation-1106.cfm.

Wind, Jerry (Yoram), Vijay Mahajan, and Robert Gunther (2001) *Convergence Marketing: Strategies for Reaching the New Hybrid Consumer*, Upper Saddle River, NJ: Prentice Hall/Financial Times.

Wind, Jerry (Yoram) and Arvind Rangaswamy (2001) Customerization: The Next Revolution in Mass Customization, *Journal of Interactive Marketing*, 15,13–32.
Wind, Jerry (Yoram) and Robert J. Thomas (2010) Organizational Buying Behavior in an Interdependent World, *Journal of Global Academy of Marketing Science*, 20, 110–122.
Wolfers, Justin and Eric Zitzewitz (2004) Prediction Markets *Journal of Economic Perspectives*, 18, 107–126.

# 2
# Performance Assessment of Co-creation Initiatives: A Conceptual Framework for Measuring the Value of Idea Contests

*Volker Bilgram*

## 1 Introduction

In 2010, more than 1,200 external users from 80 different countries answered an open call by the BMW Group and submitted 1,072 ideas within 8 weeks (Jawecki et al. 2010). The instrument involving users from outside the company in ideation is called an *idea contest*. The BMW Group's idea contests are just one example of a plethora of similar initiatives recently taken by leading innovative companies. Since the turn of the millennium, many companies such as 3M (von Hippel et al. 1999), Lego (Moon & Sproull 2001), Ducati Motor (Sawhney et al. 2005, 10–12), Procter & Gamble (Sakkab 2002), and Beiersdorf (Bilgram et al. 2011) have undergone a radical change in innovation strategy by opening up their innovation processes and making external stakeholders a part of their innovation endeavors (Bartl 2006; Brem & Voigt 2007). For example, 3M embarked upon an extensive innovation program involving external lead users to generate breakthrough innovations with a revenue potential eight times higher than that of ideas from conventional ideation approaches (von Hippel et al. 1999). The foundations of this phenomenon have been laid by von Hippel's seminal work on the *Customer Active Paradigm* in the late 1970s (von Hippel 1978) and on the sources of innovation outside the company's walls (von Hippel 1988). Von Hippel found that innovations not only originate from the manufacturers' domain, but also to a large extent from users (von Hippel 1988, 2005). Since then, innovation management ushered in a new era and is currently in a state of significant transition. Value creation, the core purpose of companies, is no longer solely accomplished by companies using internal resources, but relies on distributed innovation capabilities outside the company as well. Finding the right *modus innovandi* has become a crucial core competence of organizations today. This means that companies need to decide for every single task of value creation which internal or external resources may be most effectively and efficiently used and combined. Even sophisticated R&D organizations recognize that resources for innovation, e.g., knowledge and creativity, are widely distributed, and they utilize a variety of so called co-creation instruments in order to tap that potential. Idea contests can be considered an instrument in the field of co-creation, which is defined as the "active, creative and social process, based on collaboration between producers and users, that is initiated by the firm to generate value for customers" (Roser et al. 2009, p. 9). Following a community-based innovation approach (Füller et al. 2006), idea contests are social insofar as they support a culture of exchange among participants and a sense of community, and provide users with a shared interest and a collaborative innovation platform on which to create and jointly elaborate new ideas.

## 2 Background

**Idea contests as co-creation instruments**

Particularly with the advent of the Internet and with social applications and networks reaching the mainstream, a proliferation of "outside-in" innovation concepts have emerged, tremendously easing the way for companies to engage and collaborate with users (Dahan & Hauser 2002; Prandelli et al. 2006; Baldwin & von Hippel 2009). Recently, idea contests have gained tremendous popularity within the co-creation domain; they are successfully used by companies such as BMW, DHL, Henkel, Lufthansa or Siemens. Also smaller or publicly less well-known companies are increasingly using this ideation instrument as software-as-a-service (SAAS) offers allow for cost-efficient alternatives to highly customized branded platforms. Despite the recent upswing in popularity, the basic concept of idea contests have a long-standing history, for instance, going back to the Longitude Prize in 1714 by the British government (Hallerstede & Bullinger 2010, p. 1). Today, idea contests are mostly web-based competitions of users who make use of their skills, experiences and creativity to provide ideas or initial solutions for a particular contest challenge formulated by an initiator (Piller & Walcher 2006; Ebner et al. 2009, p. 347; Hallerstede & Bullinger 2010, p. 2). This new breed of idea contests use the Internet to distribute the "call for ideas" among a wide target group and also to provide a community-like experience on the platform. Utilizing the reward structures of tournaments (Morgan & Wang 2010, p. 78) and crowdsourcing principles (Benkler 2006; Howe 2006), a challenge is broadcast to a large general public or targeted group, and the contributions are evaluated by a jury to select and reward the winning ideas. In contrast to traditional cooperation with external partners, idea contests reach out to an undefined crowd of users rather than to a network of suppliers. Thus, idea contests are open to all online users and attract a wide variety of user types. Customers of the company or brand fans contribute their ideas to the task posted by the company as well as consumers (who may be potential future customers), including highly involved consumers such as opinion leaders or leading-edge users. A study of several virtual co-creation projects by Füller (2010a) conceives of a theoretical framework that examines the motives of participation of different user types. Enjoyment, the sense of community, monetary incentives, and firm recognition are among the most prevalent motivations of users participating in co-creation initiatives.

Researchers have predominantly directed their attention to single cases of idea contests to develop an understanding and a taxonomy for this rather new instrument of joint value creation. Ebner et al. (2009) describes the SAPiens contest and conceives of an integrated concept for IT-based idea contests along certain design parameters. Piller and Walcher (2006) conducted action research on the Adidas toolkit for idea competitions, exploring the nature of user contributions along dimensions such as creativity and originality. Research on a more large-scale idea contest reveals the process of implementation for the IBM Innovation Jam (Bjelland & Wood 2008).

## Benefits of co-creation

By opening up their innovation processes, firms are trying to overcome two pivotal challenges, "sticky information" (von Hippel 1994) and the "local search bias" (Lakhani 2006), and thus increase both effectiveness (Ogawa & Piller 2006, p. 66f) and efficiency (Lakhani 2006, p. 33) of new product development. The term "sticky information" is used to describe information that is difficult to obtain, to transfer, and to use in a new location, e.g., tacit or latent "need information" that a company wants to gain from its customers by means of market research (von Hippel 1994, p. 2). The "local search bias" describes the phenomenon that the search for new ideas or technologies to satisfy consumer needs is often local. This means it is conducted in the close vicinity of the searching entity – e.g., inside the company or within a small network of suppliers and cooperative partners. Consequently, ideas are often biased by a reapplication of knowledge, methods, and solutions (Lakhani 2006, p. 2; Birch & Rabinowitzs 1951, 122). Idea contests may provide a remedy for both challenges. On the one hand, idea contests allow the company to gain implicit "need information" residing in the ideas and comments submitted by users. On the other hand, companies search for ideas in the periphery and thus may overcome the local search bias and efficiently gather existent ideas. Idea contests therefore address both types of information, "need information" and "solution information". However, the degree of idea elaboration is still rather crude. Thus, in most cases the submitted ideas only give directions to future solutions and outline potential fields for solutions to answer a consumer need.

Companies consider new *modi innovandi* and open up towards external actors, hoping for a variety of benefits related to these two innovation challenges. Apart from developing customer-centric products with a high fit-to-market, i.e., doing "the right thing," increased efficiency of the innovation process, i.e., reduced time-to-market and cost-to-market, is a key goal (Nambisan & Sawhney 2007; Reichwald & Piller 2009). Although the value of co-creation has not yet been sufficiently assessed, initial results are promising. For instance, projected sales of product ideas which emerged from 3M's lead user program exceeded those from conventional innovation programs by a factor of eight (Lilien et al. 2002). Lakhani et al. (2007, p. 11) demonstrated that scientific problems that are broadcast in an open call to a diverse scientific community have high solution rates. It was found that 29.5 percent of problems that had previously remained unsolved within the company's internal R&D lab could be solved that way (Lakhani et al. 2007, p. 4). At the same time Lakhani (2006, p. 35) states that this form of problem solving is "economically efficient because solvers utilize prior solution information in their submissions." However, Lakhani and colleagues note that they still lack a sufficient empirical basis to compare the performance of this form of distributed innovation with traditional R&D endeavors (Lakhani et al. 2007, p. 11). Matthing et al. (2004) found that customers generate more innovative product ideas than employees. A similar study showed that ideas by customers are more original and valuable than those by developers with a professional background (Kristensson et al. 2004). In addition to the ideas sprung from co-creation with consumers, Blazevic and Lievens (2008, p. 150) point out another benefit of close cooperation with customers. By listening to and interacting with users in a virtual environment, latent needs can be deduced from the conversations and ideas. Gruner and Homburg (2000, p. 11) found support for the idea that a higher intensity of customer interaction in different phases of the innovation process has a positive effect on new product success. However, research on innovation projects partnering with customers, especially in the early days of the Internet, found different results. Campbell and Cooper (1999, p. 512), for instance, found no significant differences between partnerships and in-house innovation projects on any performance metric. They emphasize that "the majority of research studies have restricted their scope to examining the factors associated with the success or failure of collaborative product development rather than examining its performance relative to other options available to firms" (Campbell & Cooper 1999, p. 508).

Besides the innovation-related outcome of co-creation initiatives and idea contests in particular, a variety of indirect benefit dimensions are visible, however, often neglected (Nambisan & Baron 2007). These indirect outcomes primarily refer to the marketing and branding domain, e.g., help to establish close and even passionate consumer-brand relationships and support a positive and innovative image of a brand (Füller 2010b). Hundreds of consumers spend significant amounts of time interacting with companies and their brands on idea contest platforms and thus deep relationships with the brands are developed and tightened. What weighs even more is that consumers share their experiences with a brand and spread positive word-of-mouth in their networks extending the reach of these initiatives (Füller 2010b). The BMW Group idea contest mentioned at the beginning of this chapter illustrates the powerful effects of idea contests in social media. Within only 72 hours after contest launch, more than 3,100 users "liked" the BMW Group contest on Facebook and thus created buzz for the initiative and positive innovative image of the company (Jawecki et al. 2010, p. 6). Irrespective of these insights, "many companies treat their virtual environments strictly as an innovation platform and pay limited attention to other issues," Nambisan and Nambisan (2008, p. 53) note; they claim that "companies that ignore the broader impact of the customer's experience are overlooking an important dimension." Since these side effects are only relevant to the respective internal organizational units, e.g., marketing and branding departments, they are neither sufficiently appreciated nor proactively utilized by companies so far.

Despite the promising advantages of co-creation, the downsides and potential dangers need to be addressed as well. Involving thousands of consumers in new product development by means of an open platform is often tantamount to a great deal of publicity. Consequently, disappointed consumers who, for instance, feel the co-creation contest was not fair and they were not on a par with the company, may cause significant negative publicity (Gebauer et al. 2012). A prominent example of the "dark side" of co-creation is the design contest initiated by German Fast Moving Consumer Goods (FMCG) brand Henkel. The goal of the contest was to have consumers create a new packaging design for the dishwashing liquid Pril. Eventually, the favorite design voted on by users (a "fun" design of a roast chicken, which obviously was not meant to be a serious submission) was not approved by the jury. A chorus of outrage on the platform and social media channels occurred, which also reached traditional mass media. More recently, these so called "shitstorms" (a popular anglicism used in Germany to describe the Internet phenomenon of users expressing their indignation about the behavior of a company in a very harsh way) have occurred more often and have gained tremendous momentum in the social media landscape. The rising phenomenon of empowered consumers rebelling against corporations needs to be considered in the assessment of co-creation initiatives. Apparently, the outcome of co-creation initiatives cannot only be assessed quantitatively (e.g., by the reach of the initiative) but needs to consider quality (e.g., the sentiment of the user feedback) as well.

## Status quo of performance assessment in co-creation

Research on the assessment of co-creation and on metrics capturing the value of co-creation is still in its infancy. With regard to idea contests, research efforts mainly revolve around the question as to how this instrument needs to be designed along a set of parameters (Ebner et al. 2009). The question of the value of online idea contests has so far remained mostly untackled. In practice, prominent metrics applied in the context of idea contests include the number of participants, the number and quality of ideas submitted, and the number and quality of user comments on ideas.[1] A first approach to assessing the performance of idea contests was taken by Blohm et al. (2011, p. 100f), who have adapted the general conception of a balanced scorecard (Kaplan & Norton 1996) to idea contests. Drawing on the example of SAP's SAPiens online community, they divide the measures into four dimensions: innovation, customer, internal, and finance. Additionally, the authors suggest a number of quantitative as well as qualitative measures for each dimension of the idea scorecard – for example, page impressions[2] per member (customer dimension), the number of comments per idea (innovation), the number of recruited new employees (internal), and the efforts for community management (finance). The authors also show how in the innovation dimension the number of submitted ideas can be further specified. For instance, the quality of ideas assessed by a jury of experts or the number of comments is adduced. Additionally, the customer perspective capitalizes on metrics known from website analysis and monitoring (Cothrel 2000, p. 18; Preece 2001, pp. 350–353). For example, in addition to the number of users who registered on the platform, the number of visits, page impressions, and user characteristics serve as proxies for the activity and retention of users. The scorecard also provides cross-category metrics uniting two perspectives, for instance, the number of ideas per registered user (innovation process and customer perspective). In the internal process dimension, Blohm et al. (2011, p. 100) also suggest two often neglected values of idea contests. First, the extensive user conversations and comments on ideas bear a lot of potential for consumer insights and further "need information" beyond the actual idea output. This potential may be unlocked by ethnographic content analyses, for example (Kozinets 2002). Second, the benefit of idea contests for recruitment may be measured by the number of employees/ interns recruited through the platform. To capture the absorption and integration of ideas within the organization, the involvement of SAP employees is measured, e.g., via the metric page impressions of company representatives. Additionally, the authors go beyond the output stage and address outcome measures. For example, the further processing of ideas within the organization, e.g., by allocating budget to advance the idea (Blohm et al. 2011, p. 101), can be considered a very first form of outcome. A further outcome metric, which, however, could not be measured yet is the number of realized ideas. Referring to the input dimension, the authors highlight the costs of rewards, platform implementation, and costs for activation and community management as well as the costs of the expert evaluation jury.

However, the performance assessment of online idea contests is still rather limited to the innovation dimension. Consequently, measures such as the quantity and quality of ideas are already applied in practice. Indirect output dimensions such as effects on brand awareness, brand engagement, and word-of-mouth as seen from social media metrics (Hoffman & Fodor 2010) are still very underrepresented in idea contest assessment. In the last decade, companies have experimented with various forms of co-creation and have gained valuable experience. With increasing diffusion and maturity of co-creation as an approach for distributed innovation management, efforts to capture the value of these instruments, establish a performance assessment system, and provide metrics for performance assessment will have to be enhanced. If co-creation is to leave its reputation as a promising management fashion behind and become an equal innovation management instrument, it cannot shy away from comparing its performance with that of conventional ideation instruments. Measurement of idea contests still lags behind established and more mature forms of R&D performance measurement (Bremser & Barsky 2004; Drongelen & Bilderbeek 1999; Chiesa & Masella 1996; Werner & Souder 1997) or innovation controlling (Littkemann 2005; Bösch 2006).

## 3    The co-creation assessment framework

The research aims to pave the way towards a holistic assessment of idea contests, including relevant benefit dimensions for R&D and innovation management as well as for customer relationship management and marketing. In the following, I formalize the elements relevant to measuring the performance of idea contests. Based on a framework developed for conventional R&D activities, I develop a conceptual framework as a foundation to build on in future research on performance measurement in an open innovation context (see Figure 2.1). Based on the insights derived from the multiple idea contest projects conducted at HYVE,[3] a co-creation enabler specializing in crowdsourcing initiatives, the conceptual framework is substantiated and concretized.

The initial conceptual framework is to serve three major purposes. First, it is to constitute a frame allowing for return on investment considerations, i.e., relating the input to the output of idea contests. Only the combination of both input and output metrics may provide a meaningful measure for efficiency (Griffin 1993, p. 115; Hagedoorn & Cloodt 2003, p. 1368). The underlying economic principle and production theory are traditionally used for economic analysis such as the input–output model by Leontief (1986), but have also found their way in business research in the form of a micro-level input–output model measuring performance of companies

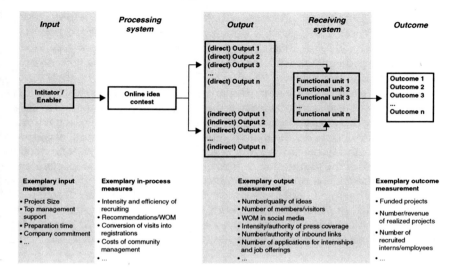

*Figure 2.1* Conceptual framework with exemplary measures
*Source*: Adapted from Brown and Svenson (1988, p. 31).

or innovation management systems (Cordero 1989, p. 187; Diewert 1992). Second, the framework should be designed to link the output of the idea contest to the company. In particular, the framework intends to cast light on the output dimensions from the different functional units of a company directly or indirectly involved in the project. Moreover, as a third requirement, the framework is to map the chronological sequence of actions and their respective output dimensions. As the true value of the output of an idea contest may not be obvious right away, various stages of the processing should be integrated in the framework.

Borrowing from systems theory (Ulrich 1970; Haberfellner 1974) and the conversion process model (Twiss 1980), the framework is based on Brown and Svenson's (1988, p. 31) measurement framework, which focuses on the R&D lab as the relevant system in an input–output model. Similar input–output models are frequently used to map innovation processes, for instance, by Cordero (1989, p. 185f), Hagedoorn and Cloodt (2003, p. 1365f), Griffin (1993, p. 115), Wakasugi and Koyata (1997, p. 384f) or Saren (1984, p. 21). In contrast to hierarchical measurement approaches such as the "Performance Pyramid" (Lynch & Cross 1991), Brown and Svenson's (1988, p. 31) framework is process focused. However, it has several touchpoints with various hierarchy levels (subsystems) in the overall ecosystem of the company. Systems can generally be depicted as input–output models consisting of effects influencing an element (input) and effects emanating from an element (output). Thus, the systems view appears to be capable of mapping innovation processes (Cooper 1990, p. 45; Hauschildt 2004, p. 30; Brockhoff 1999, p. 36). Underpinning the flexibility of systems theory, Ulrich (1970, p. 105f) conceives of systems theory as a very broad and unspecified concept with regard to its purpose as well as its elements, their array, and interrelation. Relating to the analysis of a company's innovation system and co-creation initiatives in particular, Horvath (2008, p. 86) observes that systems theory is drawn upon in literature to provide a holistic view of a corporation, but also to describe the organizational units within a corporation. As the benefit dimensions of idea contests perceived by different functional units are a core question to be answered, system theory provides the foundation to break down the system into its elements to explore these views on the output of co-creation.

In the following I will decompose Brown and Svenson's framework and scrutinize the individual elements, input, processing system, output, receiving system, and outcome. Each element is then adapted to the new context of idea contests and tied to previous research on co-creation.

### Input

In accord with Brown and Svenson (1988), input factors include resources such as the costs of manpower deployed by both the company and the enabler to set up and implement the IT-based online platform of the idea contest. Besides the actual setup of the contest platform, significant input resources are also needed for the activation and management of the community, which accrue during the online phase of the contest. Moreover, the eventual prize awarded to the winner(s) and the efforts of post-processing and assimilation within the organization are to be considered. From the perspective of the users participating in idea contests, input factors range from their time investment, "need information", creativity, and knowledge. These inputs do not incur direct costs for the company conducting the idea contests; however, the incentives, i.e., the rewards, necessary to engage users are likely to be dependent on the required user input and efforts to perform the given task. The input factors have a significant role within the framework as they add a relational perspective to the outputs and outcomes of the idea contest. They reflect the investment required to create a certain result by means of the instrument of an idea contest and thus allow for the measurement of efficiency.

In one of the few studies comparing performance of co-creation initiatives involving users with the performance of conventional innovation projects, Lilien et al. (2002, p. 1052) measure the input of co-creation projects as the costs arising to train employees to apply the lead user method and the person days necessary to conduct the lead user project. Similarly, Lakhani (2006) and Raynor and Panetta (2005) point out two major input factors in their research on problem broadcasting: administrative costs and prize money (Morgan & Wang 2010; Terwiesch & Xu 2008; Clemons et al. 2011).

## Processing system

The idea contest and in particular its online phase constitute the processing system in the framework. Brown and Svenson's (1988) framework was originally tailored to measure the performance of R&D labs as the formerly most important driving force in corporate innovation systems. In the context of open innovation, the framework is adapted to the new instrument for ideation, i.e., the idea contest. In terms of performance measurement, Griffin and Page (1993, p. 295f) distinguish between project, program, and firm level. For the investigation of idea contests, a project-centered view is suggested on account of the still rather experimental stage of idea contests, which quite often are not yet implemented as part of the innovation strategy. Only a few examples are available in which co-creation has been implemented in a more programmatic way or even as a core part of strategy (Bjelland & Wood 2008; Jawecki et al. 2010). According to Patzak and Rattay (2004, p. 17) idea contests can also be considered projects with regard to formal constituent characteristics of a project, such as newness, temporal character, complexity, and goals. The actors who are part of the processing system no longer only stem from inside the company, i.e., functional units. Actors from the company's ecosystem also play a key role in the processing system, e.g., external users and co-creation enablers, providing the organizational structure of the idea contest and the management of the contest. In the processing system, several inputs are necessary to activate, conduct, and internalize the idea contest. Input metrics applied to keep track of the performance during the idea contest aim to monitor two crucial activities: the recruitment of participants to activate the contest and community management during the actual submission phase of the contest. The intensity and efficiency of recruiting activities can be measured by the number of recruitment banners and postings, the resulting number of visitors on the contest website, and the conversion rate of visits into registrations on the platform. The community management helps to stimulate and guide the discussions about ideas submitted to the platform. Communities can be managed by company representatives or by co-creation enablers or both. Thus, the costs of the processing system can be measured by the overall efforts taken for community management during the idea contest.

## Output

In the original framework, outputs of an R&D lab are, for instance, patents, new products, principles, and publications (Brown & Svenson 1988, p. 31). Even though co-creation often takes place at the core of the value chain, i.e., new product development, the outcomes should not be restricted to the ideas and insights gathered by means of idea contests. Instead, remarkable side effects in adjacent domains should be taken into account as well. Nambisan and Nambisan (2008, p. 53) point out that web-enabled co-creation platforms "can offer important (and often hidden) benefits beyond the innovation outcomes" and support Prahalad and Ramaswamy's (2003, p. 15f) proposition that co-creation initiatives in themselves can be of value. Hence, the value of co-creation is suggested to arise from the process of co-creation itself, which induces a flow experience (Csikszentmihalyi 2002; Novak et al. 2000, p. 39) and a feeling of empowerment (Füller et al. 2010). According to Füller (2010b, p. 26), there is support for the idea that an enjoyable and compelling co-creation experience has a positive effect on trust, positive word-of-mouth (WOM), and brand image. The more partici-pants of co-creation initiatives enjoy the collaborative task and commu-nity experience, the greater the positive indirect effects. As a consequence, the right tools in the processing system enabling compelling and enjoy-able co-creation experiences are vital to bring about indirect effects as well (FüNewller et al. 2010). In other words, co-creating with a large number of users online is not exclusively about co-innovating or co-developing but – even if it was not intended in the first place – also appears to yield a great deal of additional effects due to the manifold contact points with consumers and the experience created on co-creation platforms (Bilgram & Jawecki 2011). Although these effects today are often only side effects of an innovation-focused initiative, nonetheless, the value dimensions should be taken into consideration when measuring the value of co-creation.

The multiple benefit dimensions of idea contests are also reflected in prac-tice. Swarovski has conducted a watch design contest to gain product ideas and in-depth information on consumer needs. However, an important goal was also to create a unique brand experience to accompany the launch of a new product brand (Hutter et al. 2010). A further rather unexpected indi-rect effect of the initiative was noticed in the human resources department: two of the participating users were recruited to work for renowned watch manufacturers. Similarly, Blohm et al. (2011, p. 100) describe the benefit of recruiting new employees via idea contests. The dwell time and deep interaction with a brand bears a potential for impactful communication. For example, on average each member of the BMW Group Co-Creation Lab spent 50 minutes on the platform during the first weeks after its launch (Jawecki et al. 2010, p. 8). These are just two examples highlighting how "groundswell initiatives can easily reach across departments, including product development, marketing, customer support and public relations" (Bernoff & Li 2008, p. 41; Nambisan & Baron 2007).

Based on the observation of the phenomenon that idea contests yield multiple value dimensions, the output is therefore differentiated in the framework into direct outputs, i.e., outputs the project owner aimed at, and indirect outputs, i.e., outputs that are side effects of the actual achieve-ment of objectives. In most cases, the direct outputs of idea contests are the ideas or initial solutions submitted by users or the insights that can be derived from the conversations revolving around these ideas. The quantity and quality of ideas is a measure frequently used to capture the innovation output as well as derivatives of these metrics. The quantity and quality of comments on a certain idea, for example, can be proxies for the quality and relevance of an idea. Among others, indirect outputs of idea contests cover marketing and branding-related effects such as increased brand passion or brand evangelism. In addition to these measures, which require survey-based data collection, a few performance indicators can be retrieved from log data, i.e., data automatically collected in the back end of the contest application. For a manager in charge of customer relationship management, the average dwell time of participants on the platform, return rates, and the frequency and reach of sharing contest-related content in social media are meaningful metrics to gauge the success of idea contests. Human resources managers, however, may rather be interested in applications of promising future employees and the consumer insights manager may look into the comments and discussions between participants on the platform.

### Receiving system

The framework encompasses an element called a "receiving system," which describes the various functional units within the organization, such as marketing or engineering, that use the output of the processing system (Brown & Svenson 1988, p. 31). The receiving system thus comprises the multiple recipients inside companies who work with the diverse output of idea contests. Idea contests with external actors provide a bundle of value propositions which are to be realized by the beneficiaries, i.e., the internal functional units of a company. Sawhney et al. (2005) suggested that co-creation needs to be described and analyzed for every specific marketing process and related functional unit inside the company, e.g., customer relationship management, new product development, customer support, and brand building. Even though there may be a main project manager and contact person within the company, other functional units are often involved in the idea contest, or at least they post-process its output. Ebner et al. (2009, p. 348/353) describe the implementation and conduct of the SAPiens contest, in which several internal departments were involved. Besides the SAP R&D department, the human resources department was involved; it aimed to identify the most promising students. Drawing on the experiences from the SAPiens contest, which was conducted as an interdisciplinary project at SAP, Ebner et al. (2009, p. 353) point out that "integrating idea competitions into HR processes of a company seems to be very fruitful." Looking at the receiving system of the output of an idea contest, the framework provides for the fact that "value is always uniquely and phenomenologically determined by the beneficiary" (Vargo & Lusch 2008, p. 7). In order to explore the value of idea contests, it seems to be promising to draw on the body of literature on service-dominant logic and value co-creation (Prahalad & Ramaswamy 2004; Ballantyne & Varey 2006; Vargo & Lusch 2004). Ng et al. (2010) applying service-dominant logic to a business-to-business setting. Hence, it supports my rationale to adopt service-dominant logic to the constellation of firms that have contracted a co-creation enabler or service provider to conduct the initiative and deliver value propositions. Transferring this logic to the framework of this work, the value of idea contests is rather to be considered a "store of potential value" (Ballantyne & Varey 2006, p. 344). Depending on the elements of the receiving system, i.e., the functional units within the company, unique value is realized from the value propositions provided by consumers and the co-creation enabler as main actors during the idea contest. Different functional units and employees are subject to different contingency factors and have different "operant resources" (Vargo & Lusch 2008, p. 6), e.g., knowledge or skills, to realize different value propositions of idea contests and use them to achieve their goals. Further, the theory of differentiation (Blau 1970) helps to underpin the existence of various beneficiaries within companies and their differing goals. It argues that organizational systems have been "differentiated into several distinct subsystems, each subsystem performing a portion of the task" (Lawrence & Lorsch 1967, p. 3f). Evidence was found that each subsystem has a particular goal orientation as it copes with its subenvironment. For instance, marketing managers may be more concerned with customers and competitor actions while managers in charge of production processes would be more focused on the operation of equipment and the management of suppliers (Lawrence & Lorsch 1967, pp. 8f and 21f; Dearborn & Simon 1958, 142f). Thus, the receiving system as an element of the framework appears to be apt to explore the value dimensions of idea contests and to allow for a measurement framework that takes into account the various internal functional units benefiting from co-creation.

## Outcome

As a result of the further processing of the outputs by the elements of the receiving system, value is generated for the company (Brown & Svenson 1988, p. 31). Adapting Brown and Svenson's framework, I differentiate between various stages of further processing. As there is a significant lag between initial ideation and commercialization, I suggest that earlier stages of outcome have to be considered (see stages in the focus in Figure 2.2). For example, it may be seen as an outcome if an idea (output of the idea contest) is promising and thus allocated a budget to be further processed (Blohm et al. 2011, p. 101). Eventually, however, value for the company is not realized before the product resulting from the idea is successfully launched in the market. The outcome is a crucial element within the conceptual framework as it reveals what output is truly valuable to the various functional units. Brown and Svenson (1988, p. 31) also call it the "real value" added to the organization. Hence, the outcome phases can be considered a "moment of truth" when initial output goals are eventually either further processed and developed into outcomes or abandoned. A robust measure for ideas in the early outcome stage is the budget

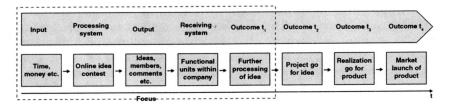

*Figure 2.2*   Different points of outcome measurement in the framework

allocated to them. The budget indicates how ideas are assessed inside the company in comparison to ideas from other sources. Time-to-market of an idea is another important metric, reflecting a potential temporal competitive advantage of co-creation initiatives. The "king" of all metrics, the revenues and profits accrued by selling a product, entails a major challenge. Due to the time lag between ideation and market launch, it is difficult to isolate the credit that can be directly ascribed to the idea behind it.

## 4   Future research directions

In the last decade, open and co-creative innovation approaches have significantly changed the innovation landscape. Understanding consumers not only as buyers but rather as partners in value creation has shaped the way companies consider their core task in today's economic and societal environment. Today firms are on the verge of moving from a piloting stage to establishing continuous open innovation and co-creation programs and making them an integral part of their innovation strategy. This next step in co-creation evolution asks for sophisticated measurement approaches tailored to open innovation processes. Henry Chesbrough, who initiated the research stream on open innovation with his open innovation paradigm, predicts that "finally, out of this new approach will eventually arise new and different metrics for assessing the performance of a firm's innovation process" (Chesbrough 2005, p. 12).

In contrast to traditional R&D, co-creation comprises various parts of the value creation chain within companies. Besides innovation processes, marketing and branding activities may increasingly benefit from co-creation approaches and may add a new dimension to open innovation projects (Vargo & Lusch 2008). Only by closely interweaving innovation, marketing, and branding activities in co-creation, can the full potential of this new paradigm be tapped. Researchers may shed light on the output dimensions of co-creation initiatives to show how these can serve multiple purposes, aiming for innovation as well as for branding effects.

Future research may investigate the multiple benefit dimensions of co-creation initiatives by comparing the performance to alternative approaches.

For instance, the input and output of internal ideation sessions such as brainstormings and project teams may be compared to the performance of idea contests. Furthermore, effects on consumers' relationships towards the brand arising from co-creation could be compared to conventional branding campaigns such as online advertisement. As co-creation aims to combine external and internal innovation processes, research should turn to the question of which parts of value creation are most efficiently achieved inside the company and which outside the company. Answers to this question may help managers to decide which *modus innovandi* is the most effective and efficient approach to tackle single innovation tasks within an innovation project. Whereas internal capabilities may be most promising in one specific part of value creation, external resources may yield better results in other situations. Research may support a balanced combination of internal and external resources in value creation by taking all benefit dimensions into consideration.

## 5   Conclusion

In contrast to existing measurement literature focusing on traditional R&D and innovation measurement (Drongelen et al. 2000; Bremser & Barsky 2004; Bösch 2006; Donnelly 2000) and the measurement of offline co-creation projects (Lilien et al. 2002), the conceptual framework sheds light on an open Internet-enabled innovation instrument like the example of idea contests. The conceptual framework focuses the firm's perspective and helps to gain a multifaceted view of the output of idea contests. It thus provides a holistic view of value dimensions from a multiple internal stakeholder perspective, providing a measurement framework that covers direct (i.e., intended by the initiator) as well as indirect effects (i.e., often unintended outputs) of idea contests for different company departments.

For practice, the research provides the following implications. The analysis of output dimensions captured by the framework creates awareness of the purposes that idea contests may serve. In particular, indirect effects of idea contests may finally get the center of attention. Applying idea contests to achieve non-innovation-related goals, e.g., idea contests for brand engagement and relationship management, is still rather neglected by companies. In addition, a holistic performance measurement may enable R&D to break down the costs of co-creation projects and allocate them to the business units that benefit from the project as well. Integrating the view of other internal functional departments or even involving people in cross-functional teams may help to pursue a specific non-innovation-related goal or emphasize the value dimensions important to the respective stakeholders inside the company during project management. By aligning marketing/branding objectives with innovation goals, companies may discover the full value of co-creation.

# Notes

1. Metrics are often revealed on the online platforms of idea contests. For an example visit a contest by Henkel: http://www.packdesign-contest.com/start.php (last date of retrieval: December 15, 2011).
2. The number of page requests resulting from users clicking on a link.
3. An overview of contests conducted by HYVE can be found at http://www. innovation-community.de (date of last retrieval December 25, 2011).

# References

Baldwin, C. Y., & vonHippel, E. (2009) Modeling a paradigm shift: From producer innovation to user and open collaborative innovation. *MIT Sloan Research Paper*, 4764–09, 1–34.

Ballantyne, D., & Varey, R. J. (2006) Creating value-in-use through marketing interaction: The exchange logic of relating, communicating and knowing. *Marketing Theory*, 6(3), 335–348.

Bartl, M. (2006) *Virtuelle Kundenintegration in die Neuproduktentwicklung*. Doctoral Dissertation, WHU – Otto Beisheim School of Management Vallendar.

Benkler, Y. (2006) *The Wealth of Networks – How Social Production Transforms Markets and Freedom*. New Haven, Conneticut: Yale University Press.

Bernoff, J., & Li, C. (2008) Harnessing the power of the oh-so-social web. *MIT Sloan Management Review*, 49(3), 36–42.

Bilgram, V., Bartl, M., & Biel, S. (2011) Getting closer to the consumer: How Nivea co-creates new products. *Marketing Review St. Gallen*, 1, 34–40.

Bilgram, V., & Jawecki, G. (2011) Erfolgsmessung von Open Innovation Projekten: Über Kennzahlen in Forschung und Praxis. *Controller Magazine*, 4, 60–65.

Birch, H., & Rabinowitzs, H. (1951) The negative effect of previous experience on productive thinking. *Journal of Experimental Psychology*, 41(2), 121–125.

Bjelland, O. M., & Wood, R. C. (2008) An inside view of IBM's "Innovation Jam". *MIT Sloan Management Review*, 50(1), 31–40.

Blau, P. (1970) A formal theory of differentiation in organizations. *American Sociology Review*, 35(2), 201–218.

Blazevic, V., & Lievens, A. (2008) Managing innovation through customer coproduced knowledge in electronic services: An exploratory study. *Journal of the Academy of Marketing Science*, 36(1), 138–151.

Blohm, I., Leimeister, J. M., Rieger, M., & Krcmar, H. (2011) Controlling von Ideencommunities: Entwicklung und Test einer Ideencommunity-Scorecard. *Controlling – Zeitschrift für erfolgsorientierte Unternehmensführung*, 23(2), 96–103.

Bösch, D. (2006) *Controlling im betrieblichen Innovationssystem*. Doctoral dissertation, University of Vienna.

Brem, A., & Voigt, K.-I. (2007) Innovation management in emerging technology ventures: The concept of an integrated idea management. *International Journal of Technology, Policy and Management*, 7(3), 304–321.

Bremser, W. G., & Barsky, N. P. (2004) Utilizing the balanced scorecard for R&D performance measurement. *R&D Management*, 34(3), 229–238.

Brockhoff, K. (1999) *Produktpolitik*. Stuttgart: Lucius & Lucius.

Brown, M. G., & Svenson, R. A. (1988) Measuring R&D productivity. *Research Technology Management*, 41(6), 30–35.

Campbell, A. J., & Cooper, R. G. (1999) Do customer partnerships improve new product success rates? *Industrial Marketing Management*, 28(5), 507–519.

Chesbrough, H. (2005) Open innovation: A new paradigm for understanding industrial innovation. Paper presented at the DRUID 10th Anniversary Summer Conference, Copenhagen.

Chiesa, V., & Masella, C. (1996) Searching for an effective measure of R&D performance. *Management Decision*, 34(7), 49–57.

Clemons, E. K., Pac, M.F., & Savin, S. (2011). Designing innovation tournaments to maximize their value: An economic perspective. *Working Paper*, Wharton University of Pennsylvania, Operations and Information Management Department.

Cooper, R. G. (1990) Stage-gate systems: A new tool for managing new products. *Business Horizons*, 33(3), 44–53.

Cordero, R. (1989) The measurement of innovation performance in the firm: An overview. *Research Policy*, 19(2), 185–192.

Cothrel, J. P. (2000) Measuring the success of an online community. *Strategy & Leadership*, 28(2), 17–21.

Csikszentmihalyi, M. (2002) *Creativity: Flow and the Psychology of Discovery and Invention*. New York, NY: Harper Perennial.

Dahan, E., & Hauser, J. (2002) The virtual customer. *Journal of Product Innovation Management*, 19(5), 332–353.

Dearborn, D. C., & Simon, H. A. (1958) Selective perception: A note on the departmental identifications of executives. *Sociometry*, 21(2), 140–144.

Diewert, W. E. (1992) The measurement of productivity. *Bulletin of Economic Research*, 44(3), 164–198.

Donnelly, G. (2000) A P&L for R&D. *CFO Magazine*, February, 44–50.

Drongelen, I. C. K.-v., & Bilderbeek, J. (1999) R&D performance measurement: More than choosing a set of metrics. *R&D Management*, 29(1), 35–46.

Drongelen, I. C. K.-v., Nixon, B., & Pearson, A. (2000) Performance measurement in industrial R&D. *International Journal of Management Reviews*, 2(2), 111–143.

Ebner, W., Leimeister, J. M., & Krcmar, H. (2009) Community engineering for innovations: The ideas competition as a method to nurture a virtual community for innovations. *R&D Management*, 39(4), 342–356.

Füller, J. (2010a) Refining virtual co-creation from a consumer perspective. *California Management Review*, 52(2), 98–122.

Füller, J. (2010b) Virtual co-creation of new products and its impact on consumers' product and brand relationships. Paper presented at the meeting of the Academy of Management Annual Meeting, Montreal.

Füller, J., Bartl, M., Ernst, H., & Mühlbacher, H. (2006) Community based innovation: How to integrate members of virtual communities into new product development. *Electronic Commerce Research*, 6(2), 57–73.

Füller, J., Mühlbacher, H., Matzler, K., & Jawecki, G. (2010) Consumer empowerment through Internet-based co-creation. *Journal of Management Information Systems*, 26(3), 71–102.

Gebauer, J., Füller, J., & Pezzei, R. (2012) The dark and the bright side of co-creation: Triggers of member behavior in online innovation communities. Accepted for publication in *Journal of Business Research*, http://dx.doi.org/10.1016/j.jbusres.2012.09.013.

Griffin, A. (1993) Metrics for measuring product development cycle time. *Journal of Product Innovation Management*, 10(4), 112–125.

Griffin, A., & Page, A. L. (1993) An interim report on measuring product development success and failure. *Journal of Product Innovation Management*, 10(4), 291–308.

Gruner, K., & Homburg, C. (2000) Does customer interaction enhance new product success. *Journal of Business Research*, 49(1), 1–14.

Haberfellner, R. (1974) *Die Unternehmung als dynamisches System. Der Prozesscharakter der Unternehmungsaktivität.* Zurich: Industrielle Organisation.

Hagedoorn, J., & Cloodt, M. (2003) Measuring innovative performance: Is there an advantage in using multiple indicators?. *Research Policy*, 32(8), 1365–1379.

Hallerstede, S. H., & Bullinger, A. C. (2010) Do you know where you go? A taxonomy of online innovation contests. Paper presented at the XXIth ISPIM Conference, Bilbao.

Hauschildt, J. (2004) *InnovationsManagement.* München: Vahlen.

Hoffman, D. L., & Fodor, M. (2010) Can you measure the ROI of your social media marketing? *MIT Sloan Management Review*, 52(1), 41–49.

Horvath, P. (2008) *Controlling.* Munich: Vahlen.

Howe, J. (2006) The rise of crowdsourcing. *Wired*, 14(6).

Hutter, K., Hautz, J., Füller, J., Matzler, K., & Mayr, A. (2010) Ideenwettbewerbe als innovatives Markenbindungsinstrument. *Marketing Review St. Gallen*, 4, 26–34.

Jawecki, G., Bilgram, V., & Wiegandt, P. (2010) The BMW group co-creation lab: Managing an innovation hub for a panopticon of users. Paper presented at the ESOMAR Innovate Conference, Barcelona.

Kaplan, R. S., & Norton, D. P. (1996) *The Balanced Scorecard.* Boston: Harvard Business School Press.

Kozinets, R. (2002) The field behind the screen: Using netnography for marketing research in online communications. *Journal of Marketing Research*, 39(1), 61–72.

Kristensson, P., Gustafsson, A., & Archer, T. (2004) Harnessing the creative potential among users. *Journal of Product Innovation Management*, 21(1), 4–14.

Lakhani, K. R. (2006) Broadcast search in problem solving: Attracting solutions from the periphery. *Working Paper*, MIT Sloan School of Management.

Lakhani, K. R., Jeppesen, L. B., Lohse, P. A., & Panetta, J. A. (2007) The value of openness in scientific problem solving. *Working Paper 07–050*, Harvard Business School.

Lawrence, P. R., & Lorsch, J. W. (1967) Differentiation and integration in complex organizations. *Administrative Science Quarterly*, 12(1), 1–47.

Leontief, W. W. (1986) *Input-output Economics.* New York: Oxford University Press.

Lilien, G. L., Morrison, P. D., Searls, K., Sonnack, M., & vonHippel, E. (2002) Performance assessment of the lead user idea-generation process for new product development. *Management Science*, 48(8), 1042–1059.

Littkemann, J. (2005) Innovationscontrolling. In S. Albers & O. Gassmann (Eds), *Handbuch Technologie und Innovationsmanagement: Strategie-Umsetzung-Controlling* (pp. 585–602). Wiesbaden: Gabler.

Lynch, R. L., & Cross, K. F. (1991) *Measure Up: The Essential Guide to Measuring Business Performance.* London: Mandarin.

Matthing, J., Sandén, B., & Edvardsson, B. (2004) New service development learning from and with customers. *International Journal of Service Industry Management*, 15(5), 479–498.

Moon, J. Y., & Sproull, L. (2001) Turning love into money: How some firms may profit from voluntary electronic customer communities. *Working Paper*, Stern School of Business.

Morgan, J., & Wang, R. (2010) Tournament for ideas. *California Management Review*, 52(2), 77–97.

Nambisan, S., & Baron, R. A. (2007) Interactions in virtual customer environments: Implications for product support and customer relationship management. *Journal of Interactive Marketing*, 21(2), 42–62.

Nambisan, S., & Nambisan, P. (2008) How to profit from a better "Virtual Customer Environment". *MIT Sloan Management Review*, 49(3), 53–61.

Nambisan, S., & Sawhney, M. (2007) *The Global Brain: Your Roadmap for Innovating Faster and Smarter in a Networked World*. Upper Saddle River: Wharton School Publishing.

Ng, I. C. L., Nudurupati, S. S., & Tasker, P. (2010) Value co-creation in the delivery of outcome based contracts for business-to-business service. *Working Paper, E.P.S.R. Council.*

Novak, T. P., Hoffman, D. L., & Yung, Y.-F. (2000) Measuring the customer experience in online environments: A structural modeling approach. *Marketing Science*, 19(1), 22–42.

Ogawa, S., & Piller, F. (2006) Reducing the risks of new product development. *Sloan Management Review*, 47(2), 65–71.

Patzak, G., & Rattay, G. (2004) *Projekt Management: Leitfaden zum Management von Projekten, Projektportfolios und projektorientierten Unternehmen*. Wien: Linde.

Piller, F., & Walcher, D. (2006) Toolkits for idea competitions: A novel method to integrate users in new product development. *R&D Management*, 36(3), 307–318.

Prahalad, C. K., & Ramaswamy, V. (2003) The new frontier of experience innovation. *MIT Sloan Management Review*, 44(4), 11–18.

Prahalad, C. K., & Ramaswamy, V. (2004) *The Future of Competition: Co-creating Unique Value with Customers*. Boston, MA: Harvard Business School Press.

Prandelli, E., Verona, G., & Raccagni, D. (2006) Diffusion of web-based product innovation. *California Management Review*, 48(4), 109–135.

Preece, J. (2001) Sociability and usability: Twenty years of chatting online. *Behavior and Information Technology Journal*, 20(5), 347–356.

Raynor, M. E., & Panetta, J. A. (2005) A better way to R&D? *Strategy & Innovation*, 3(2), 14–16.

Reichwald, R., & Piller, F. (2009) *Interaktive Wertschöpfung: Open Innovation, Individualisierung und neue Formen der Arbeitsteilung*. Wiesbaden: Gabler.

Roser, T., Samson, A., Humphreys, P., & Cruz-Valdiviesco, E. (2009) *Co-creation: New Pathways to Value*. London, LSE Enterprise.

Sakkab, N. Y. (2002) Connect & develop complements research & development at P&G. *Research Technology Management*, 45(2), 38–45.

Saren, M. A. (1984) A classification and review of models of the intra-firm innovation process. *R&D Management*, 14(1), 11–24.

Sawhney, M., Verona, G., & Prandelli, E. (2005) Collaborating to create: The Internet as a platform for customer engagement in product innovation. *Journal of Interactive Marketing*, 19(4), 4–17.

Terwiesch, C., & Xu, Y. (2008) Innovation contests, open innovation, and multiagent problem solving. *Management Science*, 54(9), 1529–1543.

Twiss, B. C. (1980) *Managing Technological Innovation*. London: Longman.

Ulrich, H. (1970) *Die Unternehmung als produktives soziales System. Grundlage der allgemeinen Unternehmungslehre*. Bern: Haupt.

Vargo, S. L., & Lusch, R. F. (2004) Evolving to a new dominant logic for marketing. *Journal of Marketing*, 68(1), 1–17.

Vargo, S. L., & Lusch, R. F. (2008) Service-dominant logic: Continuing the evolution. *Journal of the Academy of Marketing Science*, 36(1), 1–10.

von Hippel, E. (1978) A customer-active paradigm for industrial product idea generation. *Research Policy*, 7(3), 240–266.

von Hippel, E. (1988) *The Sources of Innovation*. New York: Oxford University Press.

von Hippel, E. (1994) "Sticky information" and the locus of problem solving: Implications for innovation. *Management Science*, 40(4), 429–439.

von Hippel, E. (2005) *Democratizing Innovation*. Cambridge: MIT Press.

von Hippel, E., Thomke, S., & Sonnack, M. (1999) Creating breakthroughs at 3M. *Harvard Business Review*, 77(11), 47–57.

Wakasugi, R., & Koyata, F. (1997) R&D, firm size and innovation outputs: Are Japanese firms efficient in product development? *Journal of Product Innovation Management*, 14(5), 383–392.

Werner, B. M., & Souder, W. E. (1997) Measuring R&D performance: State of the art. *Research Technology Management*, 40(2), 34–42.

# 3
# Measuring the Success of Open Innovation

*Erik Brau, Ronny Reinhardt, and Sebastian Gurtner*

## 1  Introduction and motivation

Managers and researchers agree that innovations are necessary for a firm to endure within a global and competitive environment (Mirow & Linz 2000; Davila et al. 2005). The process of generating innovations, however, has changed significantly. The innovation process is determined by the constantly changing environment, the availability of private venture capital, and increased cooperation between companies and their suppliers (Chesbrough 2003).

The formerly "closed" innovation process, where the creation of ideas and technologies as well as their conversion into marketable products occurred only within the boundary of the firm, is now supplemented by the open innovation strategy. This strategy is a paradigm that describes the opening of a firm's boundaries and the strategic utilization of the external environment to enlarge the innovation potential. Studies show that the trend of opening the firm's boundaries is still continuing, because firms benefit from advantages such as the reduction of development time and expenses (Scinta 2007). Many large and medium-sized firms are still in a transition from closed to open innovation (Lichtenthaler 2008a).

As there is no such thing as a free lunch, companies are faced with internal and external risks and barriers that need consideration while implementing and adopting the new strategy (Gurtner & Dörner 2009). The increasing globalization of stakeholders in the innovation process adds even more complexity to this already complex process, but also facilitates the open innovation paradigm (Gassmann 2006; Gassmann et al. 2010). Traditional management measures and methods are obsolete when applying the open innovation strategy. Managing the open innovation process demands new measures that allow managers to monitor and control the innovation process and the associated risks and opportunities (Chesbrough 2004; Enkel et al. 2009).

Existing research does not yet cover these aspects. Thus, we develop qualitative and quantitative measures and adapt existing methods to manage the inflows and outflows of knowledge and, more generally, to manage the implementation of the open innovation strategy. This study is structured as follows. First, we briefly review the concept of open innovation and describe changes that are required to adapt this strategy. Subsequently, we mention challenges in measuring open innovation and propose key quantitative and qualitative figures that can be used to measure several aspects of open innovation. In Section 4 we adopt the technology roadmap method and the balanced scorecard method as a framework for monitoring and controlling the implementation processes. Finally, we discuss implications for research and management.

## 2　The concept of open innovation

### General aspects of open innovation

Joseph A. Schumpeter was the first to note that innovations are created by companies and thus need to be developed within a closed environment (Schumpeter 1954). This innovation strategy has dominated companies' view on innovations, because it was seen as a prerequisite to persisting on the market. The quote from Henry W. Chesbrough summarizes Schumpeter's understanding of creating innovations in a closed process: "If you want something done right, you've got to do it yourself" (Chesbrough 2003, p. xx). Accordingly, companies primarily relied on their own R&D departments to develop new products and processes. When innovation projects resulted in ideas that did not match the firm's strategy, the idea often remained unused.

Dynamic environmental impacts challenged the former best practice model of closed innovation (Bühner 2004). Increasing competitive pressure, shorter product life cycles, combined with declining R&D budgets and low success rates of innovative products, forced companies to intensify their innovation potential (Lüthje 2003; Reichwald & Piller 2009).

In addition to pressure from the outside on the innovation process, Chesbrough (2003) identified several factors that foster the change from closed to open innovation. One of these factors is the growing mobility of highly experienced and qualified employees, which hinders the preservation of intellectual property within the firm's boundaries. Moreover, the growing presence of private venture capital opens the opportunity for employees to commercialize their own ideas and thus become competitors to their former company (Chesbrough 2003; Gaule 2006). The increased cooperation between firms and their suppliers further drives change (Chesbrough 2003). The democratization of knowledge through the Internet allows firms to acquire valuable information at low cost (Gaule 2006). Consequently, the

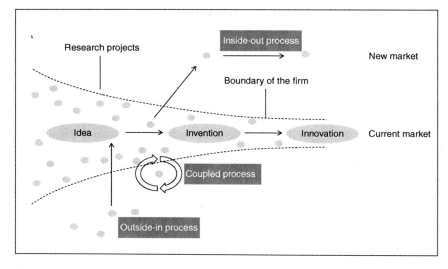

*Figure 3.1* Open innovation processes
Source: Based on Chesbrough (2003) and Gassmann and Enkel (2006).

former strategy of vertical integration becomes obsolete, whereas specialization and modularization as well as outsourcing and networking are rising in importance (Chesbrough 2003; Christensen 2006).

To meet these dynamic requirements, companies need to follow a less linear approach and to look both "inside out" and "outside in" (Figure 3.1). This is called the open innovation paradigm. Chesbrough (2003, p. xxiv) defines open innovation as "a paradigm that assumes that firms can and should use external ideas as well as internal ideas, and internal and external paths to markets, as the firms look to advance their technology." Figure 3.1 summarizes the three possible open innovation processes resulting from this definition.

## Adapting the open innovation strategy

When a company changes its strategy from closed to open innovation, the ability to assimilate and apply external knowledge for internal purposes becomes more and more important. Furthermore, the strategic change needs to be combined with a fundamental organizational and cultural rethinking. Literature points to three major subjects that need to be considered.

### Absorptive capacity

As early as 1990, Cohen and Levinthal (1990) stated that the ability of a firm to recognize, assimilate, and apply external information is critical to its innovative capabilities. They termed this phenomenon "absorptive capacity." The more problem-related knowledge a firm possesses before the acquisition of external information, the simpler and shorter the learning and integration process becomes (Spithoven et al. 2011). In the global economy, there is no single source of information; multinational companies must find good ideas and technologies around the world (Vanhaverbeke 2006).

The concept of absorptive capacity is not limited to the acquisition of external knowledge, but also includes its exploitation. Consequently, firms must not only focus on the environment, but on the intraorganizational transfer of knowledge as well (Lane & Lubatkin 1998).

Absorptive capacity has become a major source of competitive advantage (Zahra & George 2002) and, therefore, its measurement needs to be included in a holistic open innovation strategy. Specific performance measures are needed to plan, monitor, and control absorptive capacity processes. Furthermore, absorptive capacity represents a link between the technology dimension and organizational learning. Consequently, firms need to establish an adequate level of absorptive capacity before they actually benefit from the open innovation strategy.

*Change of the organizational structure*

In addition to recognizing the value of external information, the innovating firm needs to act upon a number of managerial levers. Chiaroni et al. (2011) identified four key levers that have an impact on the implementation of open innovation: networks, organizational structures, evaluation processes, and knowledge management systems.

Switching to an open innovation strategy implies an extensive use of interorganizational relationships. This is the only way firms can absorb knowledge from a variety of external sources and market internal ideas that are not consistent with the current business strategy. A variegated network consisting of universities (Perkmann & Walsh 2007), research institutions, suppliers (Emden et al. 2006), and users (von Hippel 1986) is necessary to successfully realize inbound and outbound processes. Appropriate structures that enable cooperations have to be put in place.

In addition, it is important to develop complementary internal networks. The organizational structure of the firm must be able to access and integrate the acquired knowledge into the companies' innovation process. In the same way, the organizational structure must facilitate external paths to market for internally developed ideas (von Hippel 1986).

*Change in corporate culture*

Corporate culture can be defined as the sum of values, mindsets, and norms that characterize the internal and external appearance of the firm (Krulis-Randa 1984). Researchers and managers agree that corporate

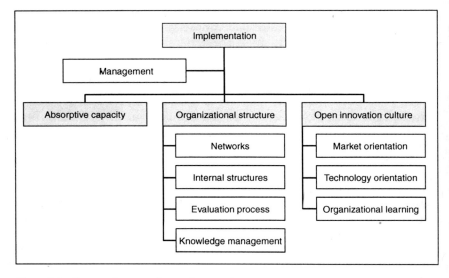

*Figure 3.2*   Success factors of open innovation
*Source*: Based on Herzog (2008).

culture has a strong influence on the success of innovations (Hauschildt 1993). However, maximizing the benefits of external knowledge requires a fundamental shift in corporate culture and the elimination of cultural boundaries.

According to Danneels (2002), product innovation requires competencies related to both technologies and markets. These two key functions include the physical construction of the new product (technology dimension) as well as selling the product to customers (market dimension). Both dimensions are influenced by corporate culture. Values and norms that encourage the openness to new ideas are needed to acquire relevant knowledge (Zhou et al. 2005). To efficiently use and exploit the information gained from the technology or the market dimension, a firm also needs the ability of organizational learning (Probst & Büchel 1998). Moreover, the "not-invented-here" and "only-used-here" syndromes need to be taken into account, because these mindsets can have negative impacts on a successful open innovation strategy (Katz & Allen 1982; Lichtenthaler & Ernst 2006).

Taking everything into consideration, the change from closed to open innovation is not a simple strategic shift; it demands open mindsets, open handling of external ideas and technologies, as well as adaptations of internal and external networks. Following Chesbrough and Crowther (2006), these criteria can be denoted as success factors for an open innovation strategy (see Figure 3.2).

## 3 Measuring open innovation

Knowing about the concept of open innovation and the key success factors to its implementation, the question remains how a firm can ensure that the open innovation strategy positively influences the success of its business.

### Challenges in measuring open innovation

The application of the open innovation approach generates a multitude of ideas. Before these ideas can become inventions and innovations, they need to be evaluated. Assessments of ideas have to take into account both technology and market uncertainties (Chesbrough 2004). Especially when targeting completely new markets, the difficulty of assessment increases (Braunschmidt 2005). Taking into account all of these considerations, a multitude of management measures and instruments can be used. Quantitative methods are applied at a high level of maturity of the idea because predictions for anticipated cash flows and other financial measures become more precise (Vahs & Burmester 2002). Qualitative methods are used earlier in the innovation process. Although qualitative criteria are hard to measure, they should be included in a holistic evaluation (Vahs & Burmester 2002).

To capture the development of innovations, key performance indicators (KPIs) are used to condense corporate information into an informative figure and reveal interrelations within the firm (Botta 1997; Meyer 1994). KPIs can also help to identify vulnerabilities and ways to use scarce resources effectively (Preißler 2008). They are the basis for a structured evaluation by management executives and thus are an essential instrument to monitor and control the profitability of a firm.

### Key figures

In the following sections, we describe categories and key figures for measuring different aspects of open innovation. We conducted an exploratory search in the field of performance measurement methods and indicators. We focused on qualitative and quantitative indicators that are used for R&D, new product development, innovation management, and technology management. In addition, we analyzed the literature on open innovation and extracted relevant indicators.

#### Quantitative key figures

Quantitative key figures are absolute or comparative numbers used for evaluating arithmetically ascertainable data (Gladen 2008). The results of our search suggested that there are five perspectives that can be distinguished: input, output, time, knowledge, and cooperation. The input perspective (i.e., costs of an open innovation strategy) is balanced with the output perspective (i.e., the additional value of an open innovation strategy) allowing the measurement of both sides of the coin. In addition, time is included as a third perspective, measuring various aspects of how quickly a business can achieve progress. Inherent in the open innovation strategy are the perspectives knowledge and cooperation. The knowledge dimension consists of elements from outside-in and inside-out processes as well as from absorptive capacity. The success of cooperation with external partners within the coupled process needs to be measured to complete the overall picture. The five perspectives and associated key figures are shown in Table 3.1.

Output figures are needed to measure the value added by the new strategy. The out-licensing intensity, one of the components of the output perspective, describes the share of the turnover that is generated by externalizing internal knowledge. In contrast, the outside-in intensity comprises the share of turnover that is generated with products that are based on external knowledge. Combining these figures with the return on investment of new products that are based on external ideas and technologies allows us to measure the output of the open innovation strategy.

The input figures comprise expenses needed to adapt the new business strategy. These include the costs to develop an adequate infrastructure to assimilate external knowledge and dispense internal knowledge as well as the costs to identify partners and establish incentives for cooperation (Wecht 2005). The costs of legal safeguarding include expenses for legal protection (e.g., patents) as well as profit-sharing and license agreements (Nagel 1993). The ratio of productivity of external patents indicates whether purchasing a patent is profitable or not. A comparison of output and input key figures provides an indication of whether the costs of an open innovation strategy exceed the benefits (Jeppesen 2005).

Within the time category, the key figure *time-to-market* covers the time required from the first idea to market launch. Combining this figure with the figure *internal cycle time* allows assessing potential time savings. The half-life concept describes the amount of time that is necessary to achieve a defined improvement (Coenenberg et al. 2007). This figure can be considered a meta-figure, which allows planning and measuring the open innovation process and related key figures. For example, the half-life figure could measure how long it takes to reduce the time-to-market.

The knowledge perspective includes key figures that focus on assessing the absorptive capacity of a firm. For example, a large contribution of external ideas to the overall amount of ideas indicates that outbound interfaces help to assimilate knowledge. However, this key figure does not include the quality of the affiliated idea. Further information from other key figures, such as the share of implemented external ideas or the sales of products that are based on external ideas, is necessary. In addition to evaluating the absorptive capacity of a firm, using these measures allows the firm to quantify the applicability of the new knowledge. The internal patent utilization level and the input-output ratio add to the understanding of the processes involved in the knowledge domain. The internal

*Table 3.1*   Key figures for measuring open innovation success

| Category | Key figure |
| --- | --- |
| **Quantitative key figures** | |
| Output | Out-licensing intensity[a] |
| | Outside-in intensity[b] |
| | Return on investment of products based on external ideas and technologies[c] |
| Input | Productivity of ext. patents |
| | Training costs[d] |
| | Costs of legal safeguarding[e] |
| | Adjustment costs[f] |
| Time | Time-to-market[a] |
| | Internal cycle time[a] |
| | Half-life-concept |
| Knowledge | Percentage of external expertise |
| | Percentage of implemented external ideas[g] |
| | Input-/output-ratio[a] |
| | Transfer intensity[a] |
| | Internal patent utilization level |
| Cooperation | Number of partners[h] |
| | External partner turnover rate |
| | Employee acquisition[h] |
| | Outside-in ratio |
| **Qualitative key figures** | |
| Open culture | Motivation of external partners[i] |
| | Openness of employees[j] |
| | Mutability[k] |
| Learning process | Absorptive capacity[l] |
| | Ability to implement |
| | Employee incentives |
| Structure | Interface management[m] |
| | Communication[n] |
| | Overview[o] |

*Sources for quantitative figures:* [a] Chesbrough (2004); [b] Chiaroni, et al. (2011); [c] based on Coenenberg et al. (2007); [d] Kaplan & Norton (1997); [e] Nagel (1993); [f] Reichwald & Piller (2009); [g] Kaplan & Norton (1997); [h] Flores et al. (2009). Sources for qualitative figures: [i] Gronau et al. (2009); [j] Dillon et al. (2005); [k] Robert & Weiss (1990); [l] Cohen & Levinthal (1990); [m] Chiaroni et al. (2011) and Gronau et al. (2009); [n] Kaplan & Norton (1996); [o] Eckelmann (2002).

patent utilization level specifies whether firms use their patent portfolio adequately. A low level of this key figure indicates that the majority of a firm's patents are unused and, thus, the potential for an out-licensing strategy increases (De Backer & Cervantes 2008). The input-output ratio contrasts the share of patents sold with the share of patents received per year. This ratio reveals whether the firm is faced with a one-sided leakage of knowledge (Enkel 2009).

The final category cooperation describes the openness to interact with various firms. The number of partnerships and the share of customer-driven, supplier-driven, or partner-driven projects of all projects represents the degree of openness. These figures vary between industries and a careful interpretation when conducting benchmarks is recommended. The quality of the cooperation can be measured using the rate of partner turnovers (cf. employee turnover; Hafner & Polanski 2008). Managers can use this figure to assess the quality of their own cooperation as well as that of their potential partners. Nevertheless, there are various reasons for resigning from partnerships and, thus, these reasons need to be identified (Hafner & Polanski 2008).

*Qualitative key figures*

Although qualitative criteria are hard to measure, their integration into the assessment of the open innovation strategy is important to allow a holistic evaluation and reveal substantial indications of the success or failure of the strategy. Managers and staff from different organizational levels have to be consulted (Hofstede et al. 1990). Most of the relevant criteria can be measured using metric scales and utility analysis (e.g., Hoffmeister 2008).

The qualitative key figures can be divided into the categories open corporate culture, learning process, and structure (Dillon et al. 2005). The first subdivision, open culture, relates to the not-invented-here and the only-used-here syndrome. Having an open culture minimizes the risk of developing one or both syndromes. Learning processes and knowledge exchange are intertwined processes within the innovation process. Therefore, the dimension "learning process" is an indispensable part of open innovation and needs to be assessed (Gibbert et al. 2002; Sawhney & Prandelli 2000). Whereas an open culture can be viewed as a willingness to innovate, the learning process can be understood as the ability to innovate. Both are essential to an open innovation strategy and have mutual effects on each other (Thom & Etienne 2000). The third element is the internal structure, which describes how internal corporate processes are adjusted to the new strategy.

Examples of qualitative key figures relating to the three domains are shown in Table 3.1. For instance, a firm can assess its corporate culture with the help of case studies, observations, and in-depth interviews. This procedure can be used to identify the employees' openness to novelty as well as the firms' ability to adapt to changes (Herzog 2008). In addition, motivation of external partners to submit ideas and knowledge is an important figure that helps to evaluate the outside-in process's success.

The learning process is a part of the absorptive capacity (Spithoven et al. 2011). In addition to the previously mentioned quantitative key figures, the qualitative key figures "ability to implement" and "absorptive capacity" are used. Firms can use these figures to evaluate the effectiveness of the implementation and diffusion of new ideas, technologies, and products. The figure "employee incentives" describes to what extent staff members receive assistance in continuously seeking improvements, without being restricted to the firm's boundaries (Thom & Etienne 2000).

The third category comprises the corporate structure, including the communication of the strategy inside and outside the company. Firms should be able to assess whether their employees understand and implement the open innovation strategy properly. The second factor of the corporate structure relates to the interfaces' effectiveness of outside-in and inside-out processes (Kaplan & Norton 1996). The third factor is termed "overview" and includes whether the firm is equipped with an organizational structure that allows a general view of know-how, networks, and the innovation process (Eckelmann 2002). The more complete the overview of an organization, the better managers can understand its processes and the more realistic are the goals that can be defined.

It is important to note that key figures vary among industries and that comparisons must be adapted for every industry. Moreover, it is important to avoid isolated key figures and it is important to include input as well as output figures to assure a consistent measure (Muller et al. 2005). An isolated consideration of qualitative criteria leads to misinterpretation, because they are vulnerable to subjective biases (Vahs & Burmester 2002). Furthermore, key figures need to be observed over a period of time (e.g., three to five years) to monitor developments and not avoid being just a snapshot in time (Hafner & Polanski 2008).

Although we have introduced various qualitative and quantitative measures of open innovation, the added value of an open innovation process is still difficult to describe in precise metrics. The fact that existing metrics are firm-specific and the open innovation process requires to measure cooperative innovation processes further adds to the complexity of continuously evaluating the strategy. Untangling external knowledge and internal resources that were used to develop and market a new product is another task that cannot always be measured adequately and often has to remain unmeasured (Enkel 2009).

## 4  Adjusting methods to navigate open innovation strategies

The previous section described key performance indicators that provide information on and allow measuring of separate strategic goals. However, firms often pursue several objectives within the innovation process simultaneously. Hence, in addition to key performance indicators, multidimensional methods are needed to measure and control the open innovation process (Vahs & Burmester 2002).

Technology roadmapping can be an appropriate method to integrate the open innovation strategy into a consistent system. It can assist managers to develop a structured and long-term strategy. Technology roadmapping focuses on visualizing and evaluating essential technology developments and related market trends. However, this method does not provide precise measures and activities to achieve strategic goals (Vahs & Burmester 2002). This gap between strategic and operational tasks could be closed by a balanced scorecard.

A balanced scorecard is designed to measure and navigate corporate success as well as its key factors. The scorecard adopts the strategy from the roadmap and expresses the strategy within four perspectives. The multidimensionality of the scorecard is reflected by the inclusion of financial and nonfinancial aspects of corporate success. All four aspects are interdependent, and their respective key success factors need to be taken into account (Kaplan & Norton 1997; Weber & Schäffer 2000; Baum et al. 2007).

The following section describes modifications of the technology roadmap and the balanced scorecard method to meet the requirements of the open innovation strategy. We term these modified methods "open technology roadmap" and "open innovation scorecard." In addition, we critically assess a simultaneous application of the open technology roadmap and the open innovation scorecard.

## Open technology roadmap

To benefit from advancing technology developments, firms need information about the temporal sequence, required resources and applications of technology developments in the present as well as in the future (Petrick & Echols 2004).

Technology roadmaps are graphical illustrations of all corporate objectives and projects and provide a frame to identify and consolidate market, product, and technology developments within a given period of time (Petrick & Echols 2004; Laube & Abele 2008). Figure 3.3 shows technologies that are the basis of the firm's products, as well as market requirements these products have to fulfill. In addition, market and technological changes are displayed. Anticipated elements are illustrated with dashed ovals or boxes. This visualization simplifies management decisions and the coordination of activities and resources within an increasingly complex environment (Wells et al. 2004). Thus, the value of technology roadmaps lies in collecting information from various sources to formulate short, middle, and long-term strategies. During this process, interdependencies among projects become visible and synergies can be realized (Petrick & Echols 2004; Laube & Abele 2008). Technology roadmaps, however, lack a dynamic approach or an indication of the value of individual nodes or paths through the roadmap. It also remains unknown how to integrate the open innovation approach. Therefore, modifications to the present roadmap are necessary.

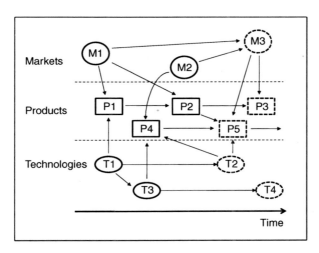

*Figure 3.3*   General technology roadmap
*Source*: Rinne (2004).

First, a change from static to dynamic visualization enables the roadmap to constantly include external as well as internal effects on the firm (Baum et al. 2007). Thus, anticipated effects on existing goals can be identified and appropriate actions can be taken (Laube & Abele 2008). Kafouros et al. (2008) highlight that globally operating firms need not only focus on developing new technologies and products, but also need to invest in the expansion of new markets. Using the open technology roadmap can assist in visualizing interdependencies between product and international market development.

Secondly, metrics need to be introduced to ensure controls that work well. In the approach of the Santa Fe Institute (Laube & Abele 2008), roadmaps are modified by pointing out the relevance of a node by its size (see Figure 3.4). The size of the node can either correspond to the size of the potential market, the creative potential, the expected return on investment of the product, or the importance of the technology for the firm. The value of a node needs to be assessed from the view of the beholder. Accordingly, each firm would assign different values to the respective node (Laube & Abele 2008).

A further extension to the roadmap is the illustration of synergies between technologies. The roadmap is able to visualize the degree of difficulty in substituting other technologies. A similar situation could be projected to the product level. To visualize this, color graduation is used. A light grey indicates an easy change between technologies and dark grey or black reflects a difficult change. The further the color graduations are apart, the less attractive a change would be.

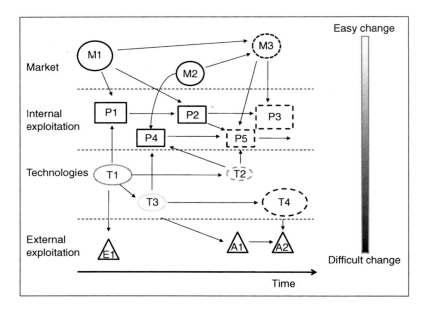

*Figure 3.4*   Open technology roadmap

*Source*: Based on Lichtenthaler (2008b) with a few modifications (e.g. a new level "markets" and third dimension "difficulty of change").

Figure 3.4 shows the proposed modifications to the roadmap. In the example, a change from T1 to T2 is simple, whereas a change from T3 to T4 causes high costs. However, T4 has the highest potential for success (represented by its size) and, thus, is much more promising than T2. The technology roadmap visualizes the tradeoff between switching costs and technology impact (color graduation vs. size).

The increasing significance of the inside-out process demands the integration of this process into the current roadmap and, therefore, into the company-specific technology planning processes (Kostoff & Schaller 2001; Gassmann & Enkel 2006). The goal is to map possible flows of internally generated knowledge to the environment.

Exploitation of external technology, as illustrated in Figure 3.4, forms an additional level. Besides the established internal application of technologies (T1) to develop marketable products (P1), it is now possible to visualize the external exploitation of technologies (E1). The informative content of this roadmap is not just limited to external commercialization. It is also expendable to depict cooperation and alliances. A1, for example, represents a strategic alliance, which includes the technologies T1 and T3 and further on leads into an alliance A2, which comprises T4 (Lichtenthaler 2008b). Thus, the coupled process can be partially displayed. The roadmap only displays the firm's own technologies, but does not display technologies of alliance partners, thus ignoring the outside-in process.

## Open innovation scorecard

The balanced scorecard method is a strategic management system. It consists of multiple perspectives and therefore eliminates a weakness of traditional indicator systems, which usually focus on a single measure (Kudernatsch & Arminger 2001). The balanced scorecard method adds a customer, internal business process, and learning and growth perspective to the existing financial perspective. Depending on the strategy, various indicators are assigned to each of the four perspectives. In turn, goals, measures, and activities are assigned to each indicator (Horváth & Kaufmann 1998; Baum et al. 2007). To avoid an isolated view of the indicators and perspectives, balanced scorecard users have to capture the internal structure using cause-and-effect relationships. These causal relationships visualize the impact of activities on strategic goals.

In the following, we propose adaptions to the balanced scorecard method to use it in open innovation projects. Usually the structure of a balanced scorecard is based on the assumption of closed innovation, because there is no externally oriented perspective, except for the customer perspective (Flores et al. 2009; Neely et al. 1995). To better support the implementation of a business unit's strategy, we propose adaptations to and the creation of perspectives, as have Kaplan and Norton (1997).

The basic structure of the open innovation scorecard is based on Flores et al.'s (2009) balanced scorecard for research and collaboration. Table 3.2 shows five perspectives, which are organized hierarchically.

The internal business structure defines the capabilities of a firm and acts as a driving force of intellectual capital. Concerning the internal perspective, managers should focus on creating processes and networks that allow the acquisition of knowledge as well as generating appropriate interfaces that allow the firm to offer internal knowledge to the market (Hansen & Nohria 2004; Chiaroni et al. 2011). These internal processes influence the intellectual capital perspective. According to Brooking (1996), intellectual capital consists of market-based capital (e.g., licensing agreements), human capital (e.g., absorptive capacity), legally protected capital (e.g., intellectual property), and infrastructure capital (e.g., corporate culture). The third perspective, collaboration, emphasizes a joint idea generation and technology development among different partners such as universities, customers and suppliers (Flores et al. 2009). The innovation category then elucidates if and how the other perspectives contribute to new technologies and marketable new products. The last perspective, finance, eventually links the open innovation strategy to financial performance indicators. Various types of indicators, such as turnover and cost indicators, have to be included so that managers can have a complete picture of the open innovation strategy. Some

66    *Erik Brau et al.*

examples of possible indicators for all five perspectives are shown in Table 3.2. As these indicators are examples only, they cannot be globally applied to every company in every situation. Depending on the specific characteristic and the specific open innovation strategy, managers need to adapt indicators and create entirely new ones that reflect the firm's strategy.

After selecting appropriate indicators, managers need to define causal relationships among all five perspectives (Kaplan & Norton 1997). An example of causal relationships is shown in Figure 3.5.

*Table 3.2*  Open innovation scorecard perspectives and indicators

| Perspective | Main question | Indicators |
|---|---|---|
| Internal business structure | How to create processes that allow external knowledge acquisition as well as internal knowledge exploitation? | Capabilities to absorb external know-how[a] <br> Commercialization capabilities[a] <br> Internal communication |
| Intellectual capital | How good is the ability to create, acquire, share, and disseminate knowledge? | Openness of employees[c] <br> Acquisition of employees resulting from collaborations[d] <br> Absorptive capacity <br> Training and education of employees[e] <br> Motivation of external partners[f] |
| Collaboration | How open is the business to collaboration with different partners? | Collaboration with universities (per year)[d] <br> Percentage of projects driven by customers, suppliers, and other partners <br> Partner turnover rate |
| Innovation | How does opening the innovation process improve the innovation capability of a firm? | Time-to-market for products based on external ideas[g] <br> Percentage of realized external ideas[h] <br> External patents usage rate[g] <br> Customer satisfaction[i] |
| Finance | How does the open innovation strategy influence the financial performance? | Percentage of turnover generated with products based on external ideas[g] <br> Cost of legal security[i] <br> Profitability of external resources <br> External licensing intensity[g] |

*Sources*: [a] Chiaroni et al. (2011); [b] Eckelmann (2002, 2009); [c] Dillon et al. (2005); [d] Flores et al. (2009); [e] Kaplan and Norton (1997); [f] Gronau et al. (2009); [g] Chesbrough (2004); [h] Kaplan and Norton (1997); [i] Baum et al. (2007).

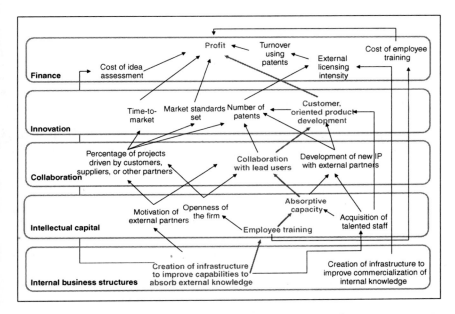

*Figure 3.5* Causal relationships in an open innovation scorecard

In the example, the starting point is the adaption of the internal business structure to improve the absorption of external know-how. Following this goal, the second objective includes employee training and awareness-raising for the new innovation paradigm. The next step comprises recognizing, absorbing, and utilizing external knowledge (i.e., absorptive capacity). For example, collaborating with lead users can improve the absorptive capacity. Combining the first three perspectives allows the firm to develop products that are more customer-oriented, implying reduced market risks (Reichwald & Piller 2009). In turn, lower risks and more customer-oriented products will have an impact on profits.

This example shows that the open innovation scorecard is a flexible tool, which can be used in different settings. However, the causal relationships have to be critically reflected because there is no empirical evidence for the theorized relationships (Kudernatsch & Arminger 2001; Morganski 2001). Managers also have to be aware of the danger of incomplete scorecards. Incomplete scorecards can lead to neglecting relevant dimensions and missing essential feedback loops and temporal effects (Baum et al. 2007). On the other hand, increasing the number of indicators also increases complexity, limiting the advantages of the open innovation scorecard.

## Combining the open technology roadmap with the open innovation scorecard

The open technology roadmap supports the development of a long-term strategic perspective and precedes the open innovation scorecard (Vinkemeier 2008). Open technology roadmaps can illustrate internally used and externally commercialized technologies as well as know-how utilized in collaborations. On the basis of internal and external knowledge, market and technology trends can be captured and assessed. Various potential development paths can be systematically displayed. Using this visual aid, managers can better communicate the relevance of certain strategies and developments to the company.

Bridging the gap from the strategic and long-term perspective employed by the open technology roadmap, managers can use the open innovation scorecard for the operational and mid-term perspective. The open innovation scorecard uses the strategies displayed in the open technology roadmap as a basis and provides measurable goals using specific open innovation indicators. The open innovation scorecard usually contains several feedback and feed-forward loops that require a constant measurement and assessment of the current strategy and tactic (Baum et al. 2007). The scorecard can be a helpful framework for this purpose.

However, using both an open technology roadmap and an open innovation scorecard significantly increases effort and costs. Consequently, managers need to evaluate whether using both instruments adequately fits the firm's strategy.

## 5   Implications

Both research and practice demand metrics for measuring open innovation strategies (Chesbrough 2004; Enkel et al. 2009). We proposed various metrics, from single indicators to combining the open technology roadmap with the open innovation scorecard. Using these measures and techniques allows a firm to empirically assess the implementation process of the open innovation strategy.

It allows managers to monitor and assess their company's specific strategy, and it can help researchers to compare metrics of different implementation strategies. However, it is important to educate and train staff members to gain an understanding of open innovation. Solely using the measures to implement the open innovation strategy will not be successful, because it usually requires a change in culture.

Open innovation strategies can have positive impacts on innovative performance (Laursen & Salter 2006) and on new product ideas (Poetz & Schreier 2012), to name just two examples. However, more is not always better, and some open innovation strategies can be costly (Laursen & Salter 2006). Therefore, the implementation of the new paradigm needs to be monitored and controlled to be able to come to conclusions concerning the success of the implementation. If no metrics or insufficient metrics are used, it will remain an open question which actions have to be taken to increase the chances of successfully employing the open innovation strategy.

Innovation always requires communication (Allen 1977); open innovation requires even more communication. Using the proposed measures as a framework to communicate the open innovation strategy to various parts of the company in different countries can foster the success of its implementation.

We also believe that incorporating open innovation measures into compensation systems for managers will foster the implementation process. If the compensation system is created using key figures from multiple levels of the open innovation scorecard, managers will need to cooperate across business units to achieve their goals, thus fostering the implementation of the open innovation strategy. In addition, a staff position could be responsible for monitoring the measures and could regularly report to the executive in charge.

Moreover, different open innovation scorecards can be used in different regions to reflect local differences. For example, there are differences in knowledge capabilities because of different regional innovation systems and clusters (Vanhaverbeke 2006), country differences (Simard & West 2006), and the availability of lead users (Gassmann et al. 2010).

The open technology roadmap and the open innovation scorecard should also be easily understood by various stakeholders of different companies. Multinational firms are especially in need of appropriate tools to communicate strategic directions. Both the open technology roadmap and the open innovation scorecard are applicable in many countries, because they are visually oriented. However, there is still a research gap in analyzing cross-cultural issues in open innovation (Lichtenthaler 2011).

The proposed metrics could also be used for benchmarking (c.f. Ragatz et al. 2002). Two or more companies that do not compete in the same market but are comparable in other measures (e.g., size, location) could compare indicators, roadmaps, and scorecards to facilitate learning and exchanging best practices for open innovation. Managers could also apply the benchmarking method to different strategic business units of the same company. If they are located in different regions, the proposed measures can help detect differences and reveal opportunities. For example, firms could open several test branches in close proximity to elite universities to establish knowledge spillover effects (Simard & West 2006) and later evaluate which locations worked well.

## 6  Limitations and further research

We presented the results of a conceptual study on measuring the implementation of an open innovation strategy. We conducted an exploratory search of existing measures and methods and proposed adaptions in order to meet the requirements of the open innovation concept. However, as this study is of a conceptual nature, there are other measures that could possibly be useful that we did not include in our overview. In addition, we did not yet empirically test the effects of our concepts. Further research could assess the applicability of our propositions.

## References

Allen, T. J. (1977) *Managing the Flow of Technology. Technology Transfer and the Dissemination of Technology Information within the R&D Organization*. Cambridge, MA: MIT Press.

Baum, H.-G., Coenenberg, A. G., & Günther, T. (2007) *Strategisches Controlling*. Stuttgart: Schäffer-Poeschel.

Botta, V. (1997) *Kennzahlensysteme als Führungsinstrumente: Planung, Steuerung und Kontrolle der Rentabilität im Unternehmen*. Berlin: Erich Schmidt.

Braunschmidt, I. (2005) *Technologieinduzierte Innovationen: Wege des innerbetrieblichen Technologie-Transfers in innovative Anwendungen*. Wiesbaden: Gabler.

Brooking, A. (1996) *Intellectual Capital: Core Asset for the Third Millennium Enterprise*. London: Itp Life Long Learning.

Bühner, R. (2004) *Betriebswirtschaftliche Organisationslehre*. München: Oldenbourg.

Chesbrough, H. W. (2003) *Open Innovation: The New Imperative for Creating and Profiting from Technology*. Boston, Mass: Harvard Business School Press.

Chesbrough, H. W (2004) Managing open innovation. *Research Technology Management*, 47(1), 23–26.

Chesbrough, H. W., & Crowther, A. K. (2006) Beyond high tech: Early adopters of open innovation in other industries. *R&D Management*, 36(3), 229–236.

Chiaroni, D., Chiesa, V., & Frattini, F. (2011) The open innovation journey: How firms dynamically implement the emerging innovation management paradigm. *Technovation*, 31(1), 34–43.

Christensen, J. F. (2006) Whither core competency for large corporations. In H. W. Chesbrough, W. Vanhaverbeke & J. West (Eds), *Open Innovation: Researching a New Paradigm* (pp. 35–61). Oxford: Oxford University Press.

Coenenberg, A. G., Fischer, T. M., & Günther, T. (2007) *Kostenrechnung und Kostenanalyse*. Stuttgart: Schäffer-Poeschel.

Cohen, W. M., & Levinthal, D. A. (1990) Absorptive capacity: A new perspective on learning and innovation. *Administrative Science Quarterly*, 35(1), 128–152.

Danneels, E. (2002) The dynamics of product innovation and firm competences. *Strategic Management Journal*, 23(12), 1095–1121.

Davila, T., Epstein, M. J., & Shelton, R. (2005) *Making Innovation Work: How to Manage It, Measure It, and Profit from It*. Philadelphia, Pa: Wharton School Publishing.

De Backer, K., & Cervantes, M. (2008) *Open Innovation in Global Networks*. Paris: Organisation for Economic Co-operation and Development.

Dillon, T., Lee, R., & Matheson, D. (2005) Value innovation: Passport to wealth creation. *Research Technology Management*, 48(2), 22–36.

Eckelmann, O. (2002) *Die Innovation Scorecard als Instrument des Innovations- und Technologiemanagements: Möglichkeiten und Grenzen*. Diplomarbeit. European Business School. Wiesbaden.

Emden, Z., Calantone, R. J., & Droge, C. (2006) Collaborating for new product development: Selecting the partner with maximum potential to create value. *Journal of Product Innovation Management*, 23(4), 330–341.

Enkel, E. (2009) Chancen von open innovation. In A. Zerfass & Möslein K. M. (Eds), *Kommunikation als Erfolgsfaktor im Innovationsmanagement. Strategien im Zeitalter der open innovation* (pp. 178–192). Wiesbaden: Gabler.

Enkel, E., Gassmann, O., & Chesbrough, H. W. (2009) Open R&D and open innovation: Exploring the phenomenon. *R&D Management*, 39(4), 311–316.

Flores, M., Al-Ashaab, A., & Magyar, A. (2009) A balanced scorecard for open innovation: Measuring the impact of industry-university collaboration. In L. M. Camarinha-Matos, I. Paraskakis & H. Afsarmanesh (Eds), *Leveraging Knowledge for Innovation in Collaborative Networks* (pp. 23–32). Berlin: Springer.

Gassmann, O. (2006) Opening up the innovation process: Towards an agenda. *R&D Management*, 36(3), 223–228.

Gassmann, O., & Enkel, E. (2006) Open innovation: Die öffnung des innovationsprozesses erhöht das innovations potenzial. *Zeitschrift Führung und Organisation*, 75(3), 132–138.

Gassmann, O., Enkel, E., & Chesbrough, H. W. (2010) The future of open innovation. *R&D Management*, 40(3), 213–221.

Gaule, A. (2006) *Open Innovation in Action: How to be Strategic in the Search for New Sources of Value*. London: H-I Network.

Gibbert, M., Leibold, M., & Probst, G. (2002) Five styles of customer knowledge management, and how smart companies use them to create value. *European Management Journal*, 20(5), 459–469.

Gladen, W. (2008) *Performance Measurement: Controlling mit Kennzahlen*. Wiesbaden: Gabler.

Gronau, N., Reger, G., Adelhelm, S., Bahrs, J., & Vladova, G. (2009) Planung und Steuerung von offenen Innovationsprozessen in Life Sciences KMUs. *Industrie Management*, 25(1), 9–12.

Gurtner, S., & Dörner, N. (2009) From roles to skills – key persons in the innovation process. *International Journal of Technology Marketing*, 4(2/3), 185–198.

Hafner, R., & Polanski, A. (2008) *Kennzahlen-Handbuch für das Personalwesen*. Zürich: Praxium.

Hansen, M. T., & Nohria, N. (2004) How to build collaborative advantage. *MIT Sloan Management Review*, 46(1), 22–30.

Hauschildt, J. (1993) Innovationsmanagement: Determinanten des Innovationserfolges. In J. Hauschildt & O. Grün (Eds), *Ergebnisse empirischer betriebswirtschaftlicher Forschung: zu einer Realtheorie der Unternehmung* (pp. 295–326). Stuttgart: Schäffer-Poeschel.

Herzog, P. (2008) *Open and Closed Innovation: Different Cultures for Different Strategies*. Wiesbaden: Gabler.

Hoffmeister, W. (2008) *Investitionsrechnung und Nutzwertanalyse*. Berlin: Berliner Wissenschafts-Verlag.

Hofstede, G., Neuijen, B., Ohayv, D. D., & Sanders, G. (1990) Measuring organizational cultures: A qualitative and quantitative study across twenty cases. *Administrative Science Quarterly*, 35(2), 286–316.

Horváth, P., & Kaufmann, L. (1998) Balanced scorecard – Ein Werkzeug zur Umsetzung von Strategien. *Harvard Business Manager*, 20(5), 39–50.

Jeppesen, L. B. (2005) User toolkits for innovation: Consumers support each other. *Journal of Product Innovation Management*, 22(4), 347–362.

72    *Erik Brau et al.*

Kafouros, M. I., Buckley, P. J., Sharp, J. A., & Wang, C. (2008) The role of internationalization in explaining innovation performance. *Technovation*, 28(1–2), 63–74.

Kaplan, R. S., & Norton, D. P. (1996) *The Balanced Scorecard: Translating Strategy into Action*. Boston, Mass: Harvard Business School Press.

Kaplan, R. S., & Norton, D. P. (1997) *Balanced Scorecard: Strategien erfolgreich umsetzen*. Stuttgart: Schäffer-Poeschel.

Katz, R., & Allen, T. J. (1982) Investigating the Not Invented Here (NIH) syndrome: A look at the performance, tenure, and communication patterns of 50 R&D Project Groups. *R&D Management*, 12(1), 7–20.

Kostoff, R. N., & Schaller, R. R. (2001) Science and technology roadmaps. *IEEE Transactions on Engineering Management*, 48(2), 132–143.

Krulis-Randa, J. S. (1984) Reflexionen über die Unternehmenskultur und über ihre Bedeutung für den Erfolg schweizerischer Unternehmen. *Die Unternehmung*, 38(4), 358–372.

Kudernatsch, D., & Arminger, G. (2001) *Operationalisierung und empirische Überprüfung der Balanced Scorecard*. Wiesbaden: Dt. Univ.-Verl.

Lane, P. J., & Lubatkin, M. (1998) Relative absorptive capacity and interorganizational learning. *Strategic Management Journal*, 19(5), 461–477.

Laube, P. J., & Abele, T. (2008) Technologie-Roadmapping zur Planung und Steuerung der betrieblichen Forschung. In M. G. Möhrle & R. Isenmann (Eds), *Technologie-Roadmapping. Zukunftsstrategien für Technologieunternehmen* (pp. 353–386). Berlin: Springer.

Laursen, K., & Salter, A. (2006) Open for innovation: The role of openness in explaining innovation performance among U.K. manufacturing firms. *Strategic Management Journal*, 27(2), 131–150.

Lichtenthaler, U. (2008a) Open innovation in practice: An analysis of strategic approaches to technology transactions. *IEEE Transactions on Engineering Management*, 55(1), 148–157.

Lichtenthaler, U. (2008b) Integrated roadmaps for open innovation. *Research Technology Management*, 51(3), 45–49.

Lichtenthaler, U. (2011) Open innovation: Past research, current debates, and future directions. *Academy of Management Perspectives*, 25(1), 75–93.

Lichtenthaler, U., & Ernst, H. (2006) Attitudes to externally organising knowledge management tasks: a review, reconsideration and extension of the NIH syndrome. *R&D Management*, 36(4), 367–386.

Lüthje, C. (2003) Kundenorientierung als Erfolgsfaktor im Innovationsprozess. In C. Herstatt & B. Verworn (Eds), *Management der frühen Innovationsphasen* (pp. 35–56). Wiesbaden: Gabler.

Meyer, C. (1994) *Betriebswirtschaftliche Kennzahlen und Kennzahlen-Systeme*. Stuttgart: Wissenschaft und Praxis.

Mirow, M., & Linz, C. (2000) Planung und Organisation von Innovationen aus systemtheoretischer Perspektive. In G. E. Häfliger & J. D. Meier (Eds), *Aktuelle Tendenzen im Innovationsmanagement* (pp. 249–268). Heidelberg: Physica.

Morganski, B. (2001) *Balanced Scorecard*. München: Vahlen.

Muller, A., Välikangas, L., & Merlyn, P. (2005) Metrics for innovation: Guidelines for developing a customized suite of innovation metrics. *Strategy & Leadership*, 33(1), 37–45.

Nagel, R. (1993) *Lead-User-Innovationen: Entwicklungskooperationen am Beispiel der Industrie elektronischer Leiterplatten*. Wiesbaden: Dt. Univ.-Verlag.

Neely, A., Gregory, M., & Platts, K. (1995) Performance measurement system design: A literature review and research agenda. *International Journal of Operations & Production Management*, 15(4), 80–116.

Perkmann, M., & Walsh, K. (2007) University–industry relationships and open innovation: Towards a research agenda. *International Journal of Management Reviews*, 9(4), 259–280.

Petrick, I. J., & Echols, A. E. (2004) Technology roadmapping in review: A tool for making sustainable new product development decisions. *Technological Forecasting and Social Change*, 71(1–2), 81–100.

Poetz, M. K., & Schreier, M. (2012) The value of crowdsourcing: Can users really compete with professionals in generating new product ideas? *Journal of Product Innovation Management*, 29(2), 245–256.

Preißler, P. R. (2008) *Betriebswirtschaftliche Kennzahlen: Formeln, Aussagekraft, Sollwerte, Ermittlungsintervalle*. München: Oldenbourg.

Probst, G., & Büchel, B. (1998) *Organisationales Lernen. Wettbewerbsvorteil der Zukunft* (2nd edn). Wiesbaden: Gabler.

Ragatz, G., Handfield, R., & Peterson, K. (2002) Benefits associated with supplier integration into new product development under conditions of technology uncertainty. *Journal of Business Research*, 55(5), 389–400.

Reichwald, R., & Piller, F. T. (2009) *Interaktive Wertschöpfung: Open Innovation, Individualisierung und neue Formen der Arbeitsteilung*. Wiesbaden: Gabler.

Rinne, M. (2004) Technology roadmaps: Infrastructure for innovation. *Technological Forecasting and Social Change*, 71(1–2), 67–80.

Robert, M., & Weiss, A. (1990) *Die permanente Innovation: Anleitung für die Unternehmenspraxis*. Frankfurt, Main: Campus-Verlag.

Sawhney, M., & Prandelli, E. (2000) Communities of creation: Managing distributed innovation in turbulent markets. *California Management Review*, 42(4), 24–54.

Schumpeter, J. A. (1954) *Economic Doctrine and Method*. New York: Oxford University Press.

Scinta, J. (2007) Industrial research institute's R&D trends forecast for 2007. *Research Technology Management*, 50(1), 17–20.

Simard, C., & West, J. (2006) Knowledge networks and the geographic locus of innovation. In H. W. Chesbrough, W. Vanhaverbeke & J. West (Eds), *Open Innovation: Researching a New Paradigm* (pp. 220–240). Oxford: Oxford University Press.

Spithoven, A., Clarysse, B., & Knockaert, M. (2011) Building absorptive capacity to organise inbound open innovation in traditional industries. *Technovation*, 31(1), 10–21.

Thom, N., & Etienne, M. (2000) Organisatorische und personelle Ansatzpunkte zur Förderung eines Innovationsklimas im Unternehmen. In G. E. Häfliger & J. D. Meier (Eds), *Aktuelle Tendenzen im Innovationsmanagement* (pp. 269–281). Heidelberg: Physica.

Vahs, D., & Burmester, R. (2002) *Innovationsmanagement. Von der Produktidee zur erfolgreichen Vermarktung*. Stuttgart: Schäffer-Poeschel.

Vanhaverbeke, W. (2006) The interorganizational context of open innovation. In H. W. Chesbrough, W. Vanhaverbeke & J. West (Eds), *Open Innovation. Researching a New Paradigm* (pp. 205–219). Oxford: Oxford University Press.

Vinkemeier, R. (2008) Gesamtkonzept zur langfristigen Steuerung von Innovationen. In M. G. Möhrle & R. Isenmann (Eds), *Technologie-Roadmapping. Zukunftsstrategien für Technologieunternehmen* (pp. 279–295). Berlin: Springer.

VonHippel, E. (1986) Lead users: A source of novel product concepts. *Management Science*, 32(7), 791–805.

Weber, J., & Schäffer, U. (2000) *Balanced Scorecard & Controlling: Implementierung – Nutzen für Manager und Controller – Erfahrungen in deutschen Unternehmen.* Wiesbaden: Gabler.

Wecht, C. H. (2005) *Frühe aktive Kundenintegration in den Innovationsprozess.* Dissertation. Universität St. Gallen.

Wells, R., Phaal, R., Farrukh, C., & Probert, D. (2004) Technology roadmapping for a service organization. *Research Technology Management*, 47(2), 46–51.

Zahra, S. A., & George, G. (2002) Absorptive capacity: A review, reconceptualization, and extension. *Academy of Management Review*, 27(2), 185–203.

Zhou, K. Z., Yim, C. K., & Tse, D. K. (2005) The effects of strategic orientations on technology- and market-based breakthrough innovations. *Journal of Marketing*, 69(2), 42–60.

# 4
# Can SMEs in Traditional Industries Be Creative?

*Jon Mikel Zabala-Iturriagagoitia*

## 1  Introduction

Innovations are new creations of economic significance, mainly carried out by firms. They can be new products, processes, business models, or marketing-oriented innovations. According to their character, they may be incremental or radical. But how do we manage innovation? Is it possible at all to do so? This chapter will try to shed some light on the management of innovation, clarifying certain terms that are usually used interchangeably in managerial contexts as well as in research environments.

Innovation management (IM) has been a growing field of study for the past four decades (Phaal et al. 2006). An increasing literature has been developed over the years, contributing to the advance of IM as a field of study (Van de Ven 1986; Currie 1999; Libutti 2000; Kärkkäinen et al. 2001; Tidd 2001; Hidalgo & Albors 2008). However, not many contributions looking at firms as the main unit of analysis have been undertaken to date (Mol & Birkinshaw 2009).

This chapter intends to present an overview of IM tools through a thorough review of IM-related articles. The chapter will combine a theoretical overview about IM with a more user-oriented description of the routines and tools available to firms for the daily practice of IM. We aim to determine the characteristics of IM tools and what makes them different from general management tools. In particular, we will focus on creativity processes and how they can be exploited in firms involved in traditional industries, particularly those regarded as small and medium-sized enterprises (SMEs). We would like to demonstrate the potential for SMEs to internally manage their innovation processes and implement and exploit creativity processes. We will illustrate this by focusing on SMEs operating in the machine tool industry.

Section 2 of the chapter will introduce the three managerial levels that need to be considered in the management of innovation activities – the strategic, the operational, and the instrumental levels. Section 3 will provide an overview of the most common IM tools described in the literature. Section 4 will then focus on creativity as a potential IM tool, presenting its main characteristics, rationales, and phases. Section 5 concludes by discussing some of the key aspects required for a sucessful use of IM tools and the main differences between general-purpose and IM techniques.

## 2  Innovation and its management

Innovation management (IM) is a multidisciplinary topic as it involves disciplines such as science, engineering, economics, strategic management, sociology, and psychology (García & Calantone 2002; Phaal et al. 2006). For many practitioners, the management of innovation is equivalent to the management of research and development (R&D) (Brady et al. 1997). Innovation is about creating products, processes, and organizational models that have a direct impact on the performance of the company, either in terms of increases in turnover (product innovations) or improvements in efficiency (process and organizational innovations). R&D is just one of the multiple determinants of innovation activities, but definitely not the only one (Edquist 2005). Accordingly, if innovation is not only driven by R&D-related factors, the management of innovation can not only refer to the management of R&D (Cetindamar et al. 2009).

IM is not the same as technology management either (Cetindamar et al. 2009). IM is about managing both "hard" and "soft" factors (Phaal et al. 2004). Technology, R&D, market commercialization, etc. all refer to so-called "hard" factors – that is, determinants that are usually embedded in companies' activities and hence are part of the agendas of their general managers. However, innovation also involves "soft" elements, e.g., learning, development of skills, acquisition of capabilities, knowledge sharing, which are of intangible nature and much more difficult to quantify and qualify (Brady et al. 1997). In this sense it needs to be stated that not all determinants of innovation management are internal to the company. The management of innovation also implies taking into consideration other external aspects such as the customers' needs, the technology life cycle, competitors, organizations embedded in the innovation system with which the company may share some interests and hence would be interested in cooperating with, standards, regulations, etc. (Van de Ven 1986; Tuominen et al. 1999; Tidd 2001; Martins & Terblanche 2003; Mol & Birkinshaw 2009).

IM needs to be understood as a core process of the organization, which requires a continuous and systematic application and which will eventually lead to a certain level of restructuring and reorganization (Leonard-Barton 1988). It could be said that the main purpose of IM is to introduce and stimulate change in organizations in order to create new opportunities or exploit existing opportunities (Mol & Birkinshaw 2009). It is a long-term race whose key feature is about being systematic (Mogee 1993; Burgelman et al. 2001; Bessant et al. 2005; Hamel 2006). Therefore, managers should not expect returns on investment as soon as the company has started to establish certain routines and use certain tools. IM is in this regard defined as the invention and implementation of a management practice, process, structure, or technique that intends to enhance firm performance (Birkinshaw et al. 2008, p. 825; Mol & Birkinshaw 2009, p. 1269).

For an effective management of innovation, we need to take into consideration the links the strategic level of the company has with its internal organization. In fact, a strategy is only of value if mechanisms for its implementation and renewal are in place (Gregory 1995). In line with Skilbeck and Cruickshank (1997), Papinniemi (1999), and Martins and Terblanche (2003), Phaal et al. (2004) conclude that the management of innovation is a multilevel process that requires three levels of activity: strategic, operational, and instrumental. Following this, we will divide the remainder of this section into three subsections. The first will deal with the strategic dimension, where decisions about the corporate strategy, the organizational culture, the values and the objectives of the company are made. The operational level will provide firms with the routines that can be established over time so that companies can pursue their strategic goals (Cetindamar et al. 2009). Finally, the instrumental level will provide a set of tools in order to aid companies bringing the previous routines into practice.

### The strategic level

The strategic level is primarily concerned with the corporate goals of the company (Phaal et al. 2004). During the formulation of a strategy, every company faces similar challenges when trying to answer the following questions: What can we do and why? What do we want to do and why? What are we going to do and why? And finally, how are we going to do it?

Innovation increasingly plays an important role in providing an answer to the previous questions. Developing a strategy is complicated due, among other reasons, to the existence of changing customer needs, preferences, technologies, and a dynamic environment. Thus, the corporate strategy needs to be aligned with the innovation strategy the company wants to pursue. As discussed above, the management of innovation depends on both the internal and the external environments. The internal factors that have a direct influence on the setting up of a corporate strategy are the availability of core competences in the firm, either technological (hard) or intellectual (soft), its organizational culture, the size of the firm, the mechanisms that support learning activities, the products/services offered, the technologies required, and the networks the company is involved in. The external factors include, among others, the type of industry/sector the company belongs to, the national/regional innovation system the company is integrated in, the technological trajectory of the technologies the company is interested in, their relationships with customers and suppliers, the alliances with competitors, and institutional support. Hence, due to the mix of internal and external components, being systematic in the analysis of these factors becomes a key aspect when formulating a corporate strategy.

From here it can be concluded the key role played by strategic management, whereas the management of innovation is focused on appropriately adapting, integrating, and reconfiguring of internal and external organizational skills, resources, and competencies that the firm faces due to the changing environment (Cetindamar et al. 2009).

## The operational level

From the previous description, two of the main features of IM can be revealed. First, managing innovation is about establishing and maintaining links between the strategy of the company and the available resources. Second, management implies the need to develop certain routines (Phaal et al. 2004) since a firm's internal knowledge is incorporated in them. These routines contain and transmit the way in which tasks are performed in the organization (Nieto 2003) – in other words, the way things are done in the company.

The goal of the routines is to represent the processes within a company by which its core competences are being developed (Bessant et al. 2005). One needs to note that the term "activity" is often used interchangeably with "process" or "routine," being understood as the approach to achieving a managerial objective (Cetindamar et al. 2009, p. 238) and as the procedure firms adopt in order to apply a particular tool in a systematic way (Hamel 2006).

In general terms, we can identify five routines as the determining factors for the successful management of innovation in firms (Cotec 1998; Tidd et al. 2001; Phaal et al. 2004): scanning, focusing, resourcing, implementing, and learning. These five routines do not have to follow a particular order. Also, companies do not need to have all the routines in place in order to manage their innovation processes. However, this is recommended in order to have a comprehensive and systemic approach to the management of innovation.

Scanning aims at detecting signals of change. Some of its activities imply examining the external environment for technological, market, regulatory, and other opportunities or collecting and filtering signals from competitors or potential partners. The second routine, focusing, aims to make the company aware of its current technological base and the fit with its overall business strategy. In addition, it sets the basis for the diagnosis of the company's knowledge base so those areas in which further competences are needed for the survival of the company can be identified. Resourcing implies finding all the possible combinations of new or existing assets of the company that may offer a solution a particular problem. This can imply the development of an innovation through in-house R&D activities, the use of technology transfer activities to absorb/gain competences in those technologies the company has identified as critical, the provision of the conditions for creativity, etc. The implementing routine requires a close interaction between marketing and technical-related activities in order to develop an idea into a marketable product or service. Finally, the learning routine aims to analyze the failures and successes had by the company as input for its future innovation processes.

### The instrumental level

The last level is labeled *instrumental* as several tools are made available in it. These tools aim at bringing organizations into action, so they can develop their own competences, technologies, ideas, and concepts and bring them into the market in the form of innovations. Additionally, they allow innovation to be a continuous and systematic process that can be sustained over time.

Many of these tools are structured in very well-delimited stages or steps. The reason for delimiting the tools to certain stages is simplicity. By making IM tools instrumental (i.e., delimited in very simple stages), their users do not need to concentrate on the content of the tasks, since this will be mastered and their content easily processed (Van de Ven 1986). Instead, the users of the tools will focus on the purpose of the task (e.g., scenario analysis, technology watch, auditing). Section 3 will provide a much more detailed description of the tools included in this instrumental level.

## 3 Tools for innovation management

Brady et al. (1997, p. 418) define a management tool as a "document, framework, procedure, system or method that enables a company to achieve or clarify an objective." But what about IM tools in particular? To what extent do they differ from general management tools? The literature in this regard is quite misleading. The definitions that address what is meant by IM tools are in fact very similar to if not the same as those referring to general management tools (Brady et al. 1997; Phaal et al. 2006; Hidalgo & Albors 2008; Mol & Birkinshaw 2009).

A large number of tools have been developed during last couple of decades in order to understand the practical and conceptual issues associated with the management of innovation (Cotec 1998; Tidd et al. 2001; Phaal et al. 2006).[1] Some of these tools include diagnostic audit methodologies (Chiesa et al. 1996; Hallgren 2009), methods for supporting creativity (Amabile 1995, 1996), foresight approaches (Chakravarti et al. 1998; Major et al. 2001), knowledge management (Nevo & Chan 2007), crowdsourcing (Lee et al. 2012), intellectual capital (Rivette & Kline 2000), the lead-user approach (Herstatt & Von Hippel 1992; Urban & Von Hippel 1988), technology outlook (Escorsa Castells et al. 2000; Veugelers et al. 2010), project management, team-building, and open innovation (Chesbrough 2003; Dahlander & Gann 2010). Each of these tools has a structure that allows it to be applied in different organizational contexts.

The European Commission (Brown 1997) launched a project for the review of existing practices concerning the application of innovation management tools. The study includes cases from around 20 European companies, illustrating the way they exploit those methodologies and the main results they achieved. However, not many studies deal with the application of IM tools in SMEs. This is the main reason why we focus on the adoption of certain tools (creativity, in the particular case of this chapter) and their implementation and use in the context of SMEs.

A number of research programs have resulted in the publication of practical guidelines supporting the application of IM tools (Brady 1995; Cotec 1998). The last of these initiatives is the one undertaken by the European Commission with the development of the IMP3ROVE catalogue. With this act, the commission wanted to provide firms with a tool to benchmark their own innovation management processes against those of other companies.[2]

Tools are not meant to be used in isolation; we need to have a systemic view of how IM is managed within firms, and this implies having a clear mindset about which tools are useful for which goals (routines). Table 4.1 includes some of the IM tools that are most extensively applied in organizational contexts, according to the existing literature to date on IM. The list could be much more extensive, including new approaches such as crowdsourcing and open innovation, which are receiving a great deal of attention by the research community in recent years.[3] In general terms, the IM tools included in Table 4.1 are of quite a different nature. For example, the lead user method is conducted as a project and revolves around the right recruitment of participants for a creativity workshop, i.e., selecting creative people with trend-leading needs. However, knowledge management is not project-based; it spans all value creation phases and thus can be regarded a much more general concept/approach that supports innovation activities in a continuous manner. As the reader can observe, most tools can be used for most of the routines outlined in the previous section, as there is no one-to-one correlation between an IM tool and the routine it is intended to contribute to (Kärkkäinen et al. 2001).

We again need to highlight here that this set of tools are not meant to be mechanistic, nor do we mean them to be prescriptive or applicable in all industrial contexts. As we will see in the next section, the application of all these tools requires some development and customization by the company (Phaal et al. 2006). In particular, this becomes true in SME environments, where the requirements of firms, even those operating within the same sector, substantially differ.

*Table 4.1* Relation between innovation management (IM) routines and some potential IM tools

| IM Tools / Routines | Audits | Knowledge Mgmt. | Lead user | Technology watch | Technology foresight | Creativity |
|---|---|---|---|---|---|---|
| Scan | ▓ | | ▓ | ▓ | ▓ | |
| Focus | | | ▓ | ▓ | ▓ | ▓ |
| Resource | | ▓ | ▓ | | | ▓ |
| Implement | | ▓ | ▓ | | | |
| Learn | ▓ | ▓ | ▓ | ▓ | | ▓ |

*Source*: Author's own elaboration.

*Note*: The gray cells mean that a particular tool (for example audits) can serve the purposes of those routines (such as scan and learn in the case of audits).

## 4 Creative processes and their role in fostering innovation

The concepts of creativity and innovation are very often used interchangeably in the literature (Martins & Terblanche 2003) and in organizational contexts. Creativity can be defined in general terms as the action of developing new thoughts and readapting previous knowledge in order to develop new theories, paradigms, and knowledge (Boone & Hollingsworth 1990). This definition views creativity as a personal characteristic, a state of mind, focusing on the role of the individual. However, within organizational environments creativity is understood as "the generation of new and useful/ valuable ideas for products, services, processes and procedures by individuals or groups" (Martins & Terblanche 2003, p. 67). This latter definition puts more emphasis on the creative process and its exploitation in firm contexts, which is the current focus of the chapter.[4] Innovation and creativity are strongly related to each other, but still are far from being the same. From our point of view, innovation, or its stimulation, could be understood as the main intended outcome of the creative process, but not as its major output, which would be the generation of mere ideas that satisfy a particular goal and its contiguous requirements. In this regard, there are some authors that understand creativity as a tool for problem-solving, which is closer to organizational approaches to innovation (Guilford 1979). Obviously that is a plausible (although narrow) way of understanding the creative process. We do not support that statement here, since we believe creativity can be used and promoted in many different and proactive environments and situations, and not just as a life jacket, in an old and reactive fashion.

So what can companies expect with the application of a creative process? According to Kärkkäinen et al. (2001, p. 171), the tools that effectively support the development of creativity also: (i) help to raise problems and development needs in a positive manner, while not allowing direct criticism; (ii) help to democratically bring out all participants' opinions, committing them better to achieving results, and (iii) encourage the generation of new and wild ideas, helping to discover new and hidden needs. Creativity can be used for both radical and incremental innovations. As discussed by Sternberg et al. (2004, p. 148), creativity through "forward incrementation" is probably the kind that is most easily recognized and appreciated, as it extends existing notions and does not threaten the assumptions of such notions (i.e., incremental innovation). However, we also believe that creativity can be used to nurture more radical ideas. From our point of view, the type of results that may emerge from the creative process will depend in part upon the company managers and the way innovation is understood by managing bodies, but also in part upon the organizational environment (Sternberg et al. 2004, p. 152).[5]

Hidalgo and Albors (2008) show how lateral thinking (or creativity) and the Delphi Method are the less exploited IM tools among European firms. In fact, most of the cases one can find in the literature refer to multinational companies in the high-tech sector, a sector embedded in the emergent phases of its technology life cycle, which hence is much more open to the development of new ideas and innovations of a radical character. However, not many examples of its application in SMEs operating in traditional industries are found (Dou & Dou 1999). This is one of the goals of this chapter. By focusing on an industry segment like machine tools, we want to show that all sectors, regardless of where in the technology life cycle they may be, are prone to the implementation of creativity processes.

## Methodology

The research outlined below is based on in-the-field experiences in several machine tool companies in the Basque Country (Spain). The machine tool sector is regarded as a medium-tech industry, despite its high investments in R&D activities. This is a traditional industry whose technology life cycle is at its maturity stage. The Basque Country is the leading region for this type of technology in Spain, which is illustrated by the investments in R&D activities, which amounted to €23 million in 2009 (Eustat 2011).

The creative process outlined in this chapter (see Figure 4.1) was applied in several Basque machine tool SMEs. To develop a creativity process that could be adopted by these companies, different sources were used to gather information, such as literature review, interviews, document analysis and questionnaires, on-the-spot trial-and-error processes, and application in several small projects of fine-tuning of the process. This implied that the creativity process was made as suitable as possible to the demands of these local SMEs. The action research approach followed lead to more than 30 creativity sessions held during 2002, with more than 20 people involved in them and around 500 ideas obtained. Each creative process followed all the steps detailed in Figure 4.1. The average time frame for the development of the whole creative process was 3 to 4 months. As a result of the application of the creative process, new ideas, products, services, and technologies that could match existing or future market opportunities in this industry segment were identified. This created further technological projects and cooperation agreements for these companies valued at €1.5 million.

This case study approach is considered a suitable research methodology, as it not only enables a situation to be described, but also allows a theory to be developed, constructed, and tested (Cabello Medina et al. 2005).

## Activities

As stated above, in this chapter we do not follow a person-centered view or attributional approach in the way creativity is understood, but we rather adopt (following Amabile 1995) systems thinking by focusing on creativity as a tool/process that can be systematically managed within organizations. The aim of this section is to develop a set of generic stages or activities into which the creative process can be divided.

The creative process refers to the sequence of thoughts and actions that lead to novel, adaptive productions (Lubart 2001, p. 295). The first approaches to the study of the creative process (Wallas 1926; Guilford 1950) divided it into four steps: preparation, incubation, illumination, and verification. Amabile (1996) then incorporated new insights by describing the creative process as consisting of the following phases: problem or task identification, preparation, response generation, and response validation and communication.

The nature of the process we outline below, shown in Figure 4.1, is closer to that of Amabile, but adapted to the particularities of manufacturing firms in the traditional industry described above. Thus, the development of the creativity process is based on previous scholarly contributions and my own involvement in the development and application of the process in small machine-tool companies in the Basque Country.[6]

As Amabile (1995, p. 425) suggested, "We must consider...simultaneously...[the] complex interactions and feedback cycles throughout the creative process," and this is precisely one of the goals of this chapter: to provide a systematic development of the creative process, adapted to the demands of industrial organizations.

We have divided our creative process into several well-delimited stages for the sake of clarity in its application to industrial organizations, which are not usually that keen on discussing the semantic differences in connotation among analogous constructs. A very similar approach to the systematic understanding of the creativity process we propose in this chapter is also shared by Alves et al. (2007), who divide the creative process into new idea generation, idea classification and selection, and new (product) concept generation and development. As the reader may note, the creativity process outlined in Figure 4.1 shows the very short durations for the individual phases, e.g., 2 to 3 hours for idea classification. However, the realization of the whole creativity process took several months in all cases. The five stages

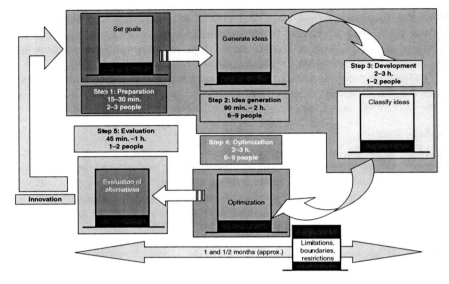

*Figure 4.1*   The creativity process
Source: Author's own elaboration.

detailed below took place at different moments in time (which include time slacks among them), there were time-consuming managerial processes (e.g., workshop preparation, circulating relevant documents) that needed to be in place before each stage could be completed. In the following subsections we will dig into the major characteristics and requirements for each of these stages.

### Setting goals

Setting the goal is regarded as the first step in the creative process, so its future success will depend to a great extent on the right formulation of the goal to accomplish.

Usually, the people involved in this goal-setting stage are rather limited. In fact, it is mostly company managers in charge of the innovation strategy who are more active in this stage. However, it is also worth noting that, particularly in SMEs, where decision-making structures are much more decentralized, the degree of involvement of the staff is much higher than in larger corporations. This implies that the initiative to carry out the creative process can also start from an employee of the company who detects a problem or a potential opportunity that could benefit from new ideas and concepts. In many cases, the definition of the goal needs to be complemented by some mandatory or desirable requirements (or constraints) the company would like to meet. Our experience has shown that for the sake of clarification, sheets like the one proposed in Figure 4.2 are of great help for company managers in order to explicitly state which are the constraints or limitations the creative process will have. In this regard, we agree with Crawford (1984), who stated that mandatory attributes should be kept to a minimum.

However, these requirements are not meant to be circulated or made public to the participants involved in the remaining stages of the creative process. Experience has taught us that the broader the goal could be, the more divergent and creative the ideas obtained were. In turn, when mandatory requisites were also known by the participants in the idea-generation stage, their ideas were more constrained, as they were focused on meeting the previous requirements instead of trying to develop as many ideas as possible. Thus, we strongly advocate for keeping these requirements (and also the document shown in Figure 4.2) for the final evaluation stage.

*Generation of ideas*

Usually creativity is regarded as idea generation. But, as observed, the creative process includes many other stages besides the generation of ideas. One of the most important demands for this stage lies in the suspension of total

| What? | Which is the object/product with the potencial opportunity identified? | | |
| | What does the opportunity consist about? | | |
| | | Mandatory requisites | Desirable requisites |
| Where? | Where would be the new opportunity applied? | | |
| When? | Under which conditions will the new opportunity work? | | |
| How? | How do we know if the new opportunity succeeds? | | |
| How much? | How much would the project cost? | | |

*Figure 4.2* Goal-setting and requirement-setting document
*Source*: Author's own elaboration.

judgment over the ideas that may emerge from the participants. In other words, it is crucial to avoid premature judgment. The ideas proposed in this stage, no matter how foolish or without value one may consider them be, should be taken as the starting point for further development rather than as finished solutions that need to be accepted or rejected (Nolan 2003). In fact, one has to bear in mind that even if some of these ideas may not be evaluated positively at the end of the creative process (stage 5 in Figure 4.1, evaluation of alternatives), they still may influence or induce another person to go in a certain divergent direction. Accordingly, at this stage there is no risk of rejection in expressing any idea. So one could state that the rationale for this second phase is "quantity is quality."

However, no matter how positive the environment may be, participants usually encounter several difficulties in turning to a creative mode or, in other words, to start thinking laterally and suggest creative ideas. After some efforts in trying to find the reasons for that behavior, we ended up suggesting a warm-up period. The goal of this warm-up period is to propose some creative games to the participants that are not related to the purpose of the session but will make them think in a divergent way. We found that after this warm-up, the people involved in this stage of the creative process were much more relaxed, cheerful, and open-minded, which had a direct impact on the type of ideas obtained.

We also observed that the physical atmosphere (e.g., having music adapted to the goal, being in an open-air environment, having a set of pictures to stimulate people's thoughts in case they ran out of ideas) is of vital relevance. In this regard, some of the companies where the process was applied decided to implement a "creativity room" (Kristensen 2004; Haner 2005). This is in line with the findings of Amabile (1995, p. 425), who also commented that "too few of our theories give serious attention to the social environment, too few of our research paradigms rigorously examine social factors at any level, and too few of our creativity-enhancement programs go beyond the training of creative-thinking techniques or the discussion of creativity styles to the explicit consideration of how work environments can impact the creativity of individuals and groups." These are precisely some weaknesses we tried to avoid in our research.

This idea-generation stage involved 6 to 9 people for about 90 minutes. The teams were intended to be as heterogeneous as possible, so we tried to mix the capabilities and skills of the participants. Several creativity techniques were used in these sessions, some of them individual oriented and some others group oriented (McAdam & McClelland 2002).[7] We observed that the best results were obtained with group-oriented creativity techniques. However, from our point of view, the most successful way to stimulate the thoughts of the participants in the session was by circulating their ideas among themselves. That way, one person could not only read the ideas had by another individual in the session, but could also be provoked by them, and hence be more eager to develop a new lateral thinking.

*Classification of ideas or hexagonal thinking*

The main goal of this stage is to classify the ideas generated in the previous step into homogeneous groups. Accordingly, some groups including similar ideas or common topics were identified. The session usually took place 2 to 3 days after the previous idea-generation session, so that the ideas were still fresh in the mind of the coordinator. As one can see in Figure 4.1, the number of people involved in this labor-intensive stage was rather limited (1 or 2 people). With it, we aimed to let the staff contribute to the further development of their company with their ideas, and not to be bothered by them. This is the reason why the coordinator or animator of the creative process was usually the one in charge of this activity. The time required for it varied depending on the amount of ideas to be grouped, but as a general rule it took between 2 and 3 hours.

This necessarily leads us to introduce the figure of the animator. The relevance of this role has also been observed by other scholars (Kärkkäinen et al. 2001; Kapier & Nilsson 2006). The main role of the animator is to create a collective orientation towards creative thinking. Small companies in traditional sectors undergo a great resistance to change, and the introduction of any type of IM tools to them usually leads to opposition by the staff members. Accordingly, we consider that in order to make the creativity process work effectively from the very beginning, appointing an animator who has a certain responsibility and charm within the company was regarded as key.

As noted, the goal of this stage was to classify the creative ideas into more coherent and homogeneous groups, shown as same-colored hexagons in Figure 4.3A. Each of these groups would then be given a certain name that characterized the ideas included in it (Fig. 43B). The main rationale for pursuing this classification was to find what we labeled as "creative gaps." We defined these creative gaps as those spaces where new ideas were still needed in order to complement the ideas already generated in stage 2. The reason why we called this classification stage "hexagonal thinking" is related to the hexagons we used in order to label each of these groups of ideas (Figure 4.3B). We considered the hexagon to be the geometrical figure that could best illustrate the need to complement a certain idea with other ideas, and the different sides from which the same idea could be understood. Thus, by using hexagonal thinking and identifying the creative gaps among the ideas already obtained in the previous step, we could move on to the next stage.

## Optimization of ideas

As Sternberg et al. (2004, p. 145) declared, people typically want others to love their ideas, but immediate universal applause for an idea usually indicates that it is not particularly creative. This is the reason why we believe that the strengths of the ideas gathered in the previous stages can be combined,

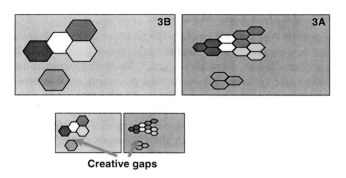

**Creative gaps**

*Figure 4.3*   Hexagonal thinking and creative gaps

in order to produce new alternatives that can better satisfy the proposed targets and meet the specific requisites.

Thus, the main goal of this session is to optimize the ideas from the two previous stages (idea generation and idea classification). This stage is intended to complement those ideas whose characteristics make them attractive with other ideas that may reinforce the weaknesses or shortcomings the original idea may also have. It is a matter of saying "yes-and" rather than "yes-but."

In order to carry out this optimization, it is we recommend analyzing the strengths and weaknesses of each of the ideas, so that a new idea can be generated by combining the strengths of already existing ones. Since the ideas being analyzed were generated in stage 2, the participants in this session should be the same as those who participated in the idea-generation stage. Accordingly, this optimization stage can be understood as a new idea-generation step. As to the length of this session, in our case we considered the optimization stage to be finished when no creative gaps could be found, i.e., when all ideas were connected among themselves.

*Evaluation of alternatives*

The main purpose of this final stage is to assess the potential of all the ideas generated during the process. This is the time when the requirements set in the first stage (goal setting) will be needed in order to assess the plausibility of the ideas generated during the process. Accordingly, the people involved in this final stage should also be those that set the general goals for the entire process as well as the mandatory and desirable requirements in stage 1.[8]

This session aims to analyze and evaluate the ideas, so that projects with potential application for the company are finally detected. It is at this stage when all the ideas will need to meet all the requirements set in Figure 4.2.[9] From our point of view, it is of vital importance to make the evaluation process as transparent as possible, in order to provide creative people with arguments about why their ideas were or were not chosen. We believe this transparency also brings credibility to company managers and decision-makers (e.g., stakeholders). Accordingly, the outcome of this session will be an action plan with concrete project proposals, their funding schemes, and a potential time schedule for their development that should be circulated within the company.

## 5  Conclusions and discussion

Some scholars argue that the business community has not been served by IM literature, mostly due to tunnel vision in which every discipline focuses on its own contribution instead of searching for cross-disciplinarity (Currie 1999). In our opinion, the IM literature has mostly focused on theoretical discussion rather than on the practice (implementation) of these tools. In this chapter we have precisely aimed to offer the reader a user-oriented perspective in which the theoretical foundations of creativity (as one of the tools for IM) were not at the core of the chapter, but the focus was on their implementation, particularly in the case of traditional SMEs.

IM tools can be defined as the range of tools, techniques, and method-ologies that support the process of innovation in firms and help them in a systematic way to meet new market challenges (Hidalgo & Albors 2008, p. 125). We have seen how there are multiple tools that can assist managers in the challenge of managing innovation in companies. However, this does not mean that IM can simply be understood as the use of some techniques included in a toolbox. Managing innovation is much more complex than that, and it implies the commitment of the managers, the alignment of corporate and innovation strategies, and the consideration of both internal and external factors affecting the company. Tools are just an aid in order to face this challenging task.

Needless to say, the tools need to be used in a continuous manner and not by leapfrogging. Innovation must be a continuous process within the company, just like manufacturing or the provision of services to customers. Otherwise, by the time companies notice that innovations are needed in products or processes, it will be too late to start implementing and exploiting these IM tools.

IM tools need time to be established within the company. Firms have constraints and are change averse, so employees, and particularly managers, will need time to adapt to them. The implementation of IM tools will imply changes in organizational routines and the adaptation (or the adoption) of certain processes concerning issues such us communication, interaction and learning. Managers in particular need to be aware of these structural changes prior to the application of the tools in the company (Sternberg et al. 2004). We cannot assume though that all IM tools are applicable to all firms and that using them will automatically lead to competitive advantages.

From our point of view, IM tools provide the guidelines, stages, or a process that could be followed to make them applicable to SMEs. They should not be understood as being prescriptive and normative. We thus advocate for the adaptation of IM tools to the particularities of each firm and to the local environment in which the company is rooted (Bessant et al. 2005).

From the literature, it can be concluded that most general-purpose management techniques are oriented to improving business performance (measuring performance in terms of efficiency gains, quality, customer satisfaction, financial revenues, introduction of new products or services, etc.). So what features make IM tools different from general management practices? This is in fact a tricky question without a clear answer in the literature. We see some degree of confusion between general management tools and actual IM tools (Hidalgo & Albors 2008; Kärkkäinen et al. 2001). Clearly, both sets of tools should be oriented to provide some competitive advantage to the company, so they have some aspects in common. However, from our point of view, the characteristics that distinguish IM tools from general-purpose tools are that IM tools are intended for use with all types of knowledge and skills within an organization, not only in an attempt to improve and develop products and processes, but also in order to change the business environment (i.e., the innovation system) in which the firm is embedded (Cetindamar et al. 2009). We could say that IM and hence IM tools are not only aiming at technological change, but to produce a cultural change supporting innovation.

We would also like to raise the readers' attention as to the need to complement the set of tools described in Section 3. In fact, firm managers should not expect that the application of one of the tools described above will convert a non-innovative company into an innovative one. It is the continuous and systematic application of the routines and processes (scanning, focusing, resourcing, implementing, and learning) that will lead to change and innovative development. In this regard, we here advocate for the search for complementarity among these tools (Bessant et al. 2005; Prajogo & Sohal 2006).

Finally, experience shows that if IM tools are to provide benefits to firms in general, and to SMEs in particular, the tools need to be simple and quickly exploitable by the employees and managers in the company. As we have seen, most tools are structured into well-delimited stages. One of these examples is the one provided by creativity processes, which constituted the main focus of the chapter. However, these steps do not mean this is the "best" and only way of approaching the tool. The several stages into which creativity has been divided should not thus be understood as a prescriptive design. As has been emphasized throughout the chapter, there are too many factors (internal, external, and institutional) we cannot control and which mean that a tool that works in one company may probably fail in another. As a result, best practices are not "best" examples when it comes to IM tools

(Tidd 2001). These are just practices, or case studies, that prove how IM tools are an important asset to improve the innovative performance of the company, a practice companies can learn from.

## Notes

1. For a full list of IM tools and their associated methodologies, see Hidalgo and Albors (2008, p. 118).
2. https://www.improve-innovation.eu/
3. The chapter by Volker Bilgram in this book highlights the relevance of co-creation initiatives such as idea contests and crowdsourcing in supporting innovation activities.
4. From now on, and for the purpose of this chapter, we will use the terms "creativity" and "creative process" interchangeably.
5. This organizational culture is defined as the "deeply seated (often conscious) values and beliefs shared by personnel in an organization" (Martins and Terblanche 2003, p. 65). In order to see the influence that organizational culture can influence on creativity, see Martins and Terblanche (2003, p. 70, figure 2).
6. It is also worth mentioning that there have been some scholars who have discussed the factors that affect team creativity and who have developed diagnostic tools for the assessment of the creative performance of organizations (Torrance 1966; Rickards & Bessant 1980).
7. For a plausible list of creativity techniques, see Geschka (1983).
8. In case the reader is interested in identifying a set of possible measures that could help organizations evaluate the degree of performance of their creativity process, I recommend reading the chapter by Volker Bilgram in this book. In it, Bilgram suggests a number of measures that are used for assessment of the value of online idea contests, which are also directly applicable in the general creativity process discussed in this chapter.
9. For example, ideas can be divided into: ideas to be materialized immediately, ideas to be materialized later (the managers will need to decide when), ideas to be improved or reconsidered, and rejected ideas (also providing the arguments for rejecting them).

## References

Alves, J., Marques, M. J., Saur, I., & Marques, P. (2007) Creativity and innovation through multidisciplinary and multisectoral cooperation. *Creativity and Innovation Management*, 16(1), 27–34.

Amabile, T. M. (1995) Attributions of creativity: What are the consequences? *Creativity Research Journal*, 8(4), 423–426.

Amabile, T. M. (1996) *Creativity in Context.* Boulder: CO, Westview.

Bessant, J., Lamming, R., Noke, H., & Phillips, W. (2005) Managing innovation beyond the steady state. *Technovation*, 25, 1366–1376.

Birkinshaw, J., Hamel, G., & Mol, M. J. (2008) Management innovation. *Academy of Management Review*, 33(4), 825–845.

Boone, L. W., & Hollingsworth, T. A. (1990) Creative thinking in business organizations. *Review of Business*, 12, 3–12.

Brady, T. (1995) Tools, management of innovation and complex product systems. *Working paper prepared for CENTRIM/SPRU/OU Project on Complex Product Systems.* EPSRC Technology Management Initiative, GR/K/31756.

Brady, T. B., Rush, H., Hobday, M., Davies, A., Probert, D., & Banerjee, S. (1997) Tools for technology management: An academic perspective. *Technovation*, 17(8), 417–426.

Brown, D. (1997) *Innovation Management Tools: A Review of Selected Methodologies.* European Commission. EUR 17018. EIMS Project N 94/135.

Burgelman, R. A., Maidique, A., & Wheelwright, S. C. (2001) *Strategic Management of Technology and Innovation.* New York, NY: McGraw-Hill.

Cabello Medina, C., Carmona Lavado, A., & Valle Cabrera, R. (2005) Characteristics of innovative companies: A case study of companies in different sectors. *Creativity and Innovation Management*, 14(3), 272–287.

Cetindamar, D., Phaal, R., & Probert, D. (2009) Understanding technology management as a dynamic capability: A framework for technology management activities. *Technovation*, 29, 237–246.

Chakravarti, A. K., Vasanta, B., Krishnan, A. S. A., & Dubash, R. K. (1998) Modified Delphi methodology for technology forecasting case study of electronics and information technology in India. *Technological Forecasting and Social Change*, 58, 155–165.

Chesbrough, H. W. (2003) The era of open innovation. *MIT Sloan Management Review*, 35–51.

Chiesa, V., Coughlan, P., & Voss, C. (1996) Development of a technical innovation audit. *Journal of Product Innovation Management*, 13(2), 105–136.

Cotec (1998) *TEMAGUIDE: A Guide to Technology Management and Innovation in Companies.* European Commission.

Crawford, C. M. (1984) Protocol: New tool for product innovation. *Journal of Product Innovation Management*, 2, 85–91.

Currie, W. L. (1999) Revisiting management innovation and change programmes: Strategic vision or tunnel vision? *Omega, The International Journal of Management Science*, 27, 647–660.

Dahlander, L., & Gann, D. M. (2010) How open is innovation? *Research Policy*, 39, 699–709.

Dou, H., & Dou, J. M. (1999) Innovation management technology: Experimental approach for small firms in a deprived environment. *International Journal of Information Management*, 19, 401–412.

Edquist, C. (2005) Systems of innovation: Perspectives and challenges. In J. Fagerberg, D. Mowery & R. R. Nelson (Eds), *The Oxford Handbook of Innovation* (pp. 181–208). Oxford: Oxford University Press.

Escorsa Castells, P., Rodríguez Salvador, M., & Maspons Bosch, R. (2000) Technology mapping, business strategy and market opportunities. *Competitive Intelligence Review*, 11(1), 46–57.

Eustat (2011) Retrieved December 2011, from http://www.eustat.es.

García, R., & Calantone, R. (2002) A critical look at technological innovation typology and innovativeness terminology: A literature review. *The Journal of Product Innovation Management*, 19, 110–132.

Geschka, H. (1983) Creativity techniques in product planning and development: A view from West Germany. *R&D Management*, 13(3), 169–183.

Gregory, M. J. (1995) Technology management: A process approach. *Proceedings of the Institution of Mechanical Engineers*, 209, 347–356.

Guilford, J. P. (1950) Creativity. *American Psychologist*, 5, 444–454.
Guilford, J. P. (1979) Some incubated thoughts on incubation. *Journal of Creative Behavior*, 13, 1–8.
Hallgren, E. W. (2009) How to use an innovation audit as a learning tool: A case study of enhancing high-involvement innovation. *Creativity and Innovation Management*, 18(1), 48–58.
Hamel, G. (2006) The why, what and how of management innovation. *Harvard Business Review*, February, 1–13.
Haner, U. E. (2005) Spaces for creativity and innovation in two established organizations. *Creativity and Innovation Management*, 14(3), 288–298.
Herstatt, C., and VonHippel, E. (1992) From experience: Developing new product concepts via the lead user method: A case study in a "low-tech" field. *Journal of Product Innovation Management*, 9, 213–221.
Hidalgo, A., & Albors, J. (2008) Innovation management techniques and tools: A review from theory and practice. *R&D Management*, 38(2), 113–127.
Kapier, N., & Nilsson, M. (2006) The development of creative capabilities in and out of creative organizations: Three case studies. *Creativity and Innovation Management*, 15(3), 268–278.
Kärkkäinen, H., Piippo, P., & Tuominen, M. (2001) Ten tools for customer-driven product development in industrial companies. *International Journal of Production Economics*, 69, 161–176.
Kristensen, T. (2004) The physical context of creativity. *Creativity and Innovation Management*, 13(2), 89–96.
Lee, S. M., Olson, D. L., & Trimi, S. (2012) Innovative collaboration for value creation. *Organizational Dynamics*, 41, 7–12.
Leonard-Barton, D. (1988) Implementation as a mutual adaptation of technology and organization. *Research Policy*, 17(5), 251–267.
Libutti, L. (2000) Building competitive skills in small and medium-sized enterprises through innovation management techniques: Overview of an Italian experience. *Journal of Information Science*, 26(6), 413–419.
Lubart, T. I. (2001) Models of the creative process: Past, present and future. *Creativity Research Journal*, 13(3–4), 295–308.
Major, E., Asch, D., & Cordey-Hayes, M. (2001) Foresight as a core competence. *Futures*, 33, 91–107.
Martins, E. C., & Terblanche, F. (2003) Building organisational culture that stimulates creativity and innovation. *European Journal of Innovation Management*, 6(1), 64–74.
McAdam, R., & McClelland, J. (2002) Individual and team-based idea generation within innovation management: Organisational and research agendas. *European Journal of Innovation Management*, 5(2), 86–97.
Mogee, M. (1993) Educating innovation managers: Strategic issues for business and higher education. *IEEE Transactions on Engineering Management*, 40, 410–417.
Mol, M. J., & Birkinshaw, J. (2009). The sources of management innovation: When firms introduce new management practices. *Journal of Business Research*, 62, 1269–1280.
Nevo, D., & Chan, Y. E. (2007) A Delphi study of knowledge management systems: Scope and requirements. *Information & Management*, 44, 583–597.
Nieto, M. (2003) From R&D management to knowledge management. An overview of studies of innovation management. *Technological Forecasting and Social Change*, 70, 135–161.

Nolan, V. (2003) Whatever happened to synectics. *Creativity and Innovation Management*, 12(1), 24–27.

Papinniemi, J. (1999) Creating a model of process innovation for reengineering of business and manufacturing. *International Journal of Production Economics*, 60–61, 95–101.

Phaal, R., Farrukh, C. J. P., & Probert, D. R. (2004) A framework for supporting the management of technological knowledge. *International Journal of Technology Management*, 27(1), 1–15.

Phaal, R., Farrukh, C. J. P., & Probert, D. R. (2006) Technology management tools: Concept, development and application. *Technovation*, 26, 336–344.

Prajogo, D. I., & Sohal, A. S. (2006) The integration of TQM and technology/R&D management in determining quality and innovation performance. *OMEGA, The International Journal of Management Science*, 34, 296–312.

Rickards, T., & Bessant, J. (1980) The creative audit: Introduction of a new research measure during programmes for facilitating organisation change. *R&D Management*, 10(2), 67–75.

Rivette, K. G. & Kline, D. (2000) Discovering new value in intellectual property. *Harvard Business Review*, January–February, 1–12.

Skilbeck, J. N., & Cruickshank, C. M. (1997) *A Framework for Evaluating Technology Management Processes*. PICMET 1997, Portland.

Sternberg, R. J., Kaufman, J. C., & Pretz, J. E. (2004) A propulsion model of creative leadership. *Creativity and Innovation Management*, 13(3), 145–153.

Tidd, J. (2001) Innovation management in context: Environment, organization and performance. *International Journal of Management Reviews*, 3(3), 169–183.

Tidd, J., Bessant, J., & Pavitt, K. (2001) *Managing Innovation: Integrating Technological, Market and Organizational Change* (2nd edn). Wiley: Chichester.

Torrance, E. P. (1966) *Torrance Tests of Creative Thinking*. Princeton, NJ: Personnel Press.

Tuominen, M., Piippo, P., Ichimura, T., & Matsumoto, Y. (1999) An analysis of innovation management systems characteristics. *International Journal of Production Economics*, 60–61, 135–143.

Urban, G. L., & Von Hippel, E. (1988) Lead user analyses for the development of new industrial products. *Management Science*, 34(5), 569–582.

Van de Ven, A. H. (1986) Central problems in the management of innovation. *Management Science*, 32(5), 590–607.

Veugelers, M., Bury, J., & Viaene, S. (2010) Linking technology intelligence to open innovation. *Technological Forecasting & Social Change*, 77, 335–343.

Wallas, G. (1926) *The Art of Thought*. New York, NY: Harcourt Brace.

# 5
# Scenario-based Learning for Innovation Communication and Management

*Nicole Pfefferman and Henning Breuer*

## 1 Introduction

Imagination and communication are critical factors for innovation but also are potential bottlenecks. For organizations in telecommunication working with mid- and long-term investments in particular, insufficient information and persistence in mindsets can lead to fundamental threats. Like other high-investment infrastructure companies, telecommunications providers work on 5- to 10-year plans in major parts of their business. Long investment life cycles (10 years and more for infrastructures, 1 to 5 years for services) for high-volume investments in durable infrastructures like cables, satellites, and the energy provision to run them pair up with market driving power, even within related service fields. Being well-prepared for future developments is essential for all actors in the telecommunications industry. In spite of this fact, the telecom industry's track record in antici-pating future developments is poor. The next big thing has been regularly out of sight – not only regarding technological topics like ATM, ISDN, 3G, but especially for usability and user-driven developments like the World Wide Web, SMS, and Web 2.0. Oftentimes outdated mindsets and tunnel vision when imagining the future persist. Tunnel vision for imagination and fragmentation of communication systems may turn out to be bottle-necks for innovation.

Moreover, the understanding of value creation through innovation – in terms of ideas, concepts, prototypes, practices, objects, programs/initia-tives, models, designs, etc. that are perceived as new by any stakeholder group (Rogers 2003) – has shifted from a closed to an open innovation view (e.g., Lichtenthaler & Lichtenthaler 2009; Herzog 2008; Chesbrough 2003). "Successful companies will be those that transform information into value-creating knowledge, and ... use this knowledge to innovate and capture additional profit," Davenport et al. said (2006, p. 17). The construct "open innovation" can be understood as "...the use of purposive inflows and outflows of knowledge to accelerate internal innovation, and expand the markets for external use of innovation" (Chesbrough 2006, p. 1). In particular, for frugal innovation, the ability to manage open innovation within and across organizations in different relational arrangements will be crucial for organizations, and especially for telecommunication providers (Löscher 2012). This implies that the ability to manage communication processes, tools, and activities for innovation at all levels will also be increasingly important and that systematically managing communications for innovation is essential for any organization involved in open innovation.

In this context, scenarios, as a strategic tool in innovation management, are a means to sensitize people to future opportunities, future-attentive thinking, and communication about threats and potentials for innovation. However, the key problem is that scenarios and communication activities for innovation have to be coordinated across different corporate units – for instance, marketing, research and development (R&D), corporate communications, and human resource management – and across different time zones, cultures, management standards and policies, and political/legal contexts.

Consequently, complex network structures in open innovation projects, dynamic collaborative arrangements, and the fragmentation of communication systems lead to the need to develop a new framework for scenario-based learning (SCEBAL), linked to communication for innovation.

While numerous methodologies to develop and model scenarios exist, their value for innovation management has not yet been fully perceived. In addition, implementation in daily practice and application as a strategic tool in innovation management is missing. Regarding research in communications related to innovation, scientists mainly concentrate on marketing activities for sales markets (e.g., Trommsdorff & Steinhoff 2007; Crosby & Johnson 2006); communicative perspective in diffusion research, focusing on social processes (e.g., Mazzarol 2011; Peres et al. 2010); as well as building stakeholder relationships and introducing novelties and the organizations behind them (i.e., communication *of* innovations) (e.g., Zerfaß 2009; Zerfaß et al. 2004). But from a strategic innovation management view, research in managing SCEBAL architectures linked to communication for innovation is still deficient.

Therefore, our chapter focuses on the key question of how SCEBAL can be applied as a daily practice to foster future-attentive thinking and effective communication for innovation. Our chapter draws on scenario methods and innovation communication from a management perspective. The objective is to enhance SCEBAL in innovation management and to inspire debates by providing new perspectives in scenario planning and innovation communication, as well as creating a practice-oriented integrated management framework for applying SCEBAL as a daily practice for future-attentive thinking in order to effectively support communication for innovation – exemplified by a case study about a telecommunications provider.

This chapter is structured as follows: After a brief literature review and definition of concepts, we present the framework of SCEBAL for future-attentive thinking and effective communication for innovation. Then we present a case study in telecommunications, to discuss and to study the first managerial implications and limitations. Finally, we conclude and give an outlook on future directions.

## 2 Related literature

### Scenario-based learning

Depending on the project goals and setup of the process, scenarios may serve different purposes. In order to gain new perspectives and opportunities for innovation and change, organizations can apply scenario techniques (Ramirez et al. 2010). Scenarios aim at (1) generating new ideas and filtering ideas and projects, and (2) challenging paradigms/assumptions and creating new perspectives based on learning and on driving change (Lindgren & Bandhold 2009). For instance, scenario workshops "are powerful instruments in the process of challenging existing paradigms and creating shared perspectives on the future" (Lindgren & Bandhold 2009, 27). In this chapter we will focus on the potential of scenarios to foster future-attentive thinking i.e., on scenario learning as a prerequisite for change and innovation, with the focus on new business.

Scenarios can be understood as means of organizational learning and innovation.

Scenario-based learning (SCEBAL) is defined as an architecture to cultivate "mindfulness for innovation" (Breuer & Gebauer 2011) in dynamic business environments. In this context the term "architecture" stresses the need to understand, review, analyze, and manage alternative scenarios, and to establish such practices as a key capability for innovation in organizations.

Scenarios enable us to anticipate and structure discussion about the shape of things to come. Since the Royal Shell Dutch Group conducted the first systematic scenario studies in the 1970s, based on the work of Kahn and Wiener (1967), numerous scenario processes have been conducted and several scenarios have been published. The Shell approach gained impetus with the oil price shock in 1973 when Shell was better prepared than its competitors. Ever since scenarios became known as a valuable approach to address and prepare for upcoming uncertainties as those related to the dynamic developments in the IT (information technology) and telecommunications industries and its environments.

By modeling scenarios, researchers and consultants point out alternative and logically consistent development possibilities in the face of abounding uncertainties. As well-informed projections of uncertainty factors into a dated future, scenarios form an internally consistent image that can be plausibly derived from the present state of affairs. Often they are presented as stories around constructed plots. Different scenario approaches are discussed in Mietzner and Reger (2005) and in Steinmüller (1997).

Working with scenarios and thinking of alternative futures prepares and informs decisions about how to strengthen desirable developments (Breuer et al. 2012). Unlike traditional forecasting and even Delphi Studies, scenarios do not try to predict the future. Instead, they fuel strategic conversations (van der Heijden 2005) and challenge conventional assumptions. They prevent linear extrapolation and foster thinking in alternatives. According to Peter Schwartz, one of the founders of the approach, scenarios are vehicles for helping people to learn. Unlike simulations, they identify patterns and clusters among possible futures and include subjective interpretations (Schoemaker 1995, p. 27). Scenario planning then aims at changing mindsets about external factors before the formulation of specific strategies. "A constant stream of rich, diverse and thought provoking information" (Schwartz 1991) is needed to foster organizational learning. Principles and best practices of scenario planning have been described (van der Merwe 2008). While numerous methodologies to develop and model scenarios exist, at least two major weaknesses persist: their insufficient integration in daily practices and their scientific elaboration and evaluation (Schoemaker 1993). The essential question in our view is, How can we increase the impact of scenarios to foster innovation and to support communication for innovation on a steady basis?

To foster innovation, many organizations started installing practices to encourage and detect surprising deviances like failures, new ideas, and spontaneous changes. Scenarios are one means to sensitize organizations to alternative future directions, and they may facilitate the recognition of new business opportunities and development. Different types of scenarios have been analyzed and applied in corporations, but usually their usage has been limited to exercises done once or a few times. Just recently, SCEBAL architectures have been proposed in order to cultivate "mindfulness for innovation" (Breuer & Gebauer 2011) in dynamic business environments.

However, although scenarios may be understood as learning experiences, few organizations move on from a single exercise to a continuous endeavor to challenge existing assumptions and re-create shared visions. This chapter promotes the understanding of "future-attentive thinking" and scenarios as a daily practice tool in innovation management.

### Communications: enabler of innovation

Communication of innovation has become an emerging issue in academia and in the business world (e.g., Zerfaß & Möslein 2009; Mohr et al. 2009;

Zerfaß & Ernst 2008). It is of particular interest to the business world, because of the increasing demand for innovation, knowledge-empowered stakeholders, the ubiquitous availability of information, the breadth of enterprises' innovation portfolios, and new business models in new and different markets (Davenport et al. 2006).

Four main streams of research can be identified in the communication of innovation: (1) communicative perspective on marketing diffusion; (2) innovation marketing; (3) innovation communication from idea to launch; and (4) innovation communication as a strategic capability. Regarding the first stream, several innovation diffusion models have been introduced in the marketing diffusion literature "to understand the spread of innovations by modeling their entire life cycle from the perspective of communications and consumer interactions," but managerial implications are missing (Peres et al. 2010, p. 91; Mahajan et al. 2000). The second stream, innovation marketing, embodies all consumer market-related activities of innovation management (Steinhoff & Trommsdorff 2011). This research area conceives of marketing as an essential part in the innovation process (Crosby & Johnson 2006) and focuses on strategic and operational activities (e.g., Trommsdorff & Steinhoff 2007), such as the commercialization of radical innovations, technologies, and services (e.g., Mohr et al. 2009; Sandberg 2008; Sowter 2000). The third stream, innovation communication research (Zerfaß et al. 2004; Mast & Zerfaß 2005; Mast et al. 2005), is concerned with communication of novelties and the organization behind them from idea to launch as a constitutive element in innovation management (Zerfaß 2009) and as a part of corporate communication (e.g., Fink 2009; Zerfaß 2009). The fourth stream, innovation communication from a strategic management perspective, is understood as a strategic capability/ ability to plan, execute, coordinate, and evaluate stakeholder innovation dialog and to reconfigure the firm's resources and capabilities for innovation (Pfeffermann 2012b; Pfeffermann 2011a/b). This research area focuses on the management of innovation communication on the intraorganizational and interorganizational level in different relational arrangements to create impact and to exchange with stakeholders (Pfeffermann 2012b, 2011b), and thus to facilitate building and strengthening competitive advantages in the long run. In this context, *scenario planning represents a strategic tool of innovation communication* to create new mindsets and foster awareness of innovations in communication markets (Pfeffermann 2011a).

In fact, in the innovation economy, communication represents a valuable impact factor for organizations to positively influence (1) information processing for consumers about the characteristics and advantages of an innovation throughout the adoption process (Hofbauer et al. 2009); (2) innovation diffusion in social systems (e.g., Mazzarol 2011; Martilla 1971); (3) building stakeholder relationships in innovation processes from idea to launch (Zerfaß et al. 2004); and (4) the management of stakeholder dialogue

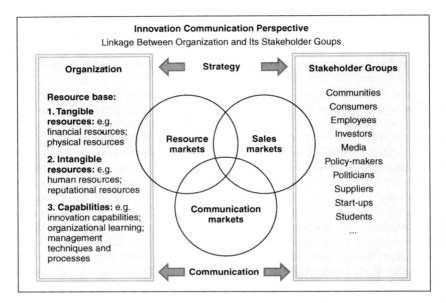

*Figure 5.1* · Innovation communication from a strategic management perspective
*Source*: Pfeffermann (2012a).

and building, re-configuring and extending resources and capabilities for the sustainable creation, dissemination, and success of the innovation as well as for strengthening the firm's long-term reputation and its reinvention in a global environment (Pfeffermann 2012b; Pfeffermann 2011b). The new perspective is illustrated in Figure 5.1.

We define communication for innovation as follows (adapted from Pfeffermann 2012b):

> Innovation communication can be understood as a strategic capability in terms of (1) managing stakeholder dialogue – from idea to launch – in open innovative networks to address communications markets, resource markets, and sales markets and (2) building, reconfiguring, and extending useful resources and capabilities for sustainable creation, dissemination, and success of novelties and emerging issues as well as for the organization's long-term strengthened reputation and reinvention in a global environment.

## 3   3C SCEBAL integrated management framework

Relating to the three markets – resource, sales, and communication markets – (see Figure 5.1 and the literature review), three basic interrelations between the concepts "scenario-based learning" and "innovation communication" are as follows:

1. *Resource markets.* As a prerequisite for change scenarios, resource markets are means for organizational learning, and SCEBAL can support information transmission and reflection to map out pathways for new ways of thinking as a key organizational capability in the twenty-first century.
2. *Communication markets.* Scenario planning represents a strategic tool of innovation communication to create new mindsets and foster awareness of innovation in communication markets, such as framing of key innovative themes and agenda setting.
3. *Sales markets.* Scenarios may function as an inspiration/imaginative tool and a filter for new ideas in innovation management for a smart dialogue between an organization and stakeholder groups about innovation, such as co-creation of new products and services to address demands in sales markets.

Hence, the role of scenarios can be described as an enabler of future-attentive thinking and effective communication for innovation. *However, complex network structures in open innovation projects, dynamic collaborative arrangements, and the fragmentation of communication systems lead to the need to develop a new management framework for applying SCEBAL for future-attentive thinking and to support effective communication for innovation.*

This section describes a practice-oriented integrated management framework (3C SCEBAL framework) using the three key elements for SCEBAL, as illustrated in Figure 5.2: (1) Communications management; (2) Contextualizing innovation; and (3) Common ground among stakeholders.

### Communication management

Scenarios are means of organizational learning and are necessary to map out paths to new ways of thinking and to ensure a shared meaning for future directions, which in turn represents an enabler of effective communication for innovation. What are the communication tools used in scenario planning?

In the 1950s, Kahn and others described scenarios as stories that seemed as if they were written by people in the future. Other formats of (visual) storytelling have been tried out since then, but have only rarely been documented or analyzed. Alternative visualizations that have been documented include, for instance, a world map of islands – each representing one scenario, and caricatures of succinct features of scenarios (Fink & Siebe 2006). Still, even though several authors have stressed the importance of having suitable ways of communicating scenarios, the potentials, strengths, and weaknesses of each (visual) approach have never been explored in a systematic way. Such

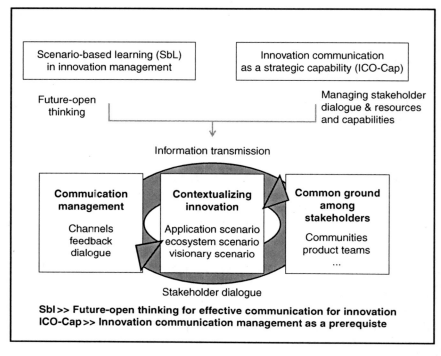

*Figure 5.2*   The 3C SCEBAL integrated management framework
*Source*: Created by the authors.

research would need to focus on the different communicative purposes of scenarios (e.g., to initiate public discussion, trigger organizational learning, or inform strategic decisions). The research would continue work on instructional media (Niegemann et al. 2008; Weidenmann 1994), but would refocus on animating "as-if" worlds of the future. Advanced communication channels (animated illustrations, movies, storytelling, etc.) have been set up. Animation is the most advanced in terms of an interpretable storyline, visualization, and specification, since examples are actually shown, working in hypothetical future situations. They are best suited to gathering attention and initiating discussion, but do not qualify well as a shared reference for critical discussions and strategic planning. New media channels for rich feedback, such as social media tools like blogs and Twitter, might enrich communication and the organizational learning intended around scenarios.

But the most important point in using different communication channels is to find a way to systematically manage all communication activities relating to scenarios at a specific period of time and over time – for instance, communication as a suitable feedback channel with the topics and presentations of different scenarios. In this context, efficient management of innovation communication (Pfeffermann 2012a/b; Pfeffermann 2011b) can be a prerequisite for communication management for SCEBAL.

## Contextualizing innovation

One element of the integrated management framework is contextualizing innovation, which means identifying, analyzing, and communicating different types of scenarios in the context of the innovation. Three types of scenarios are application scenarios, ecosystem scenarios, and visionary scenarios.

*Application scenarios*: application or usage scenarios are product-related and refer to imaginary products in a future context of use. As speculative design, they serve as a cognitive reference or "just-enough prototype" to think through and generate alternative setups of new technology, to detail ideas, and to explore the design space at an early stage of development (Breuer & Matsumoto 2010). According to Alexander (2004), application or usage scenarios describe sequences of actions at the starting point of system design on various levels of granularity. Considerations of desirability and feasibility may be combined to provide orientation. To use an application scenario, previous work is required to explore the design space of mobile learning and to get different notions about the context involved (Breuer & Matsumoto 2011).

*Ecosystem scenarios*: Ecosystem scenarios are also called exploratory scenarios in the related literature (e.g., Kosow & Gaßner 2008). An exploratory approach tries to identify influencing factors, key uncertainties, and future possibilities (e.g., Fink & Siebe 2006; Fink et al. 2000).While exploratory scenarios traditionally focus on external developments, their combination with organizational requirements, called a "future scorecard," is also proposed in literature (Fink et al. 2005). It combines the external, market-based, and the internal, resource-based view of the organization to create a strategic early-warning system. "Explorations" of future uncertainties and consistent scenarios depend on present projections and constructions of meaning that depend on the organization's own position and desires.

*Visionary scenarios*: A preferred vision of the future allows backcasting to derive necessary measures to be taken today and alternative implementation strategies. A normative vision enables strategists to reconsider and focus on sensing and seizing developments and to direct in-depth exploration of critical topics (Kosow & Gaßner 2008). Blind spots and search fields for innovation can be identified, and strengths and weaknesses of the present organization may be analyzed in the context of each scenario (Breuer & Gebauer 2011).

## Common ground among stakeholders

Innovation is a people business. Thus, communication requires convincing and motivating people and cooperating with them, and interacting with different stakeholder groups in different relational arrangements for intensive, positive, and reflective dialogue. Scenarios support the creation of a common ground among stakeholders. A shared future orientation provides a basis for stakeholders to make a collaborative innovation because "successful collaborative innovation can... flourish if the communication culture and climate is open and shared between the partners" (Donaldson et al. 2011, p. 226).

A major reason for using scenarios in innovation management is their potential to make innovation accessible to a wider audience and allow for a narrative evaluation. Not only may the narrative structure of most scenarios help to lower the cognitive distance, the potential scope of scenarios also may reach a broader audience, far beyond the product development teams of SCEBAL, and may affect the creation of innovation. It has been said that scenario workshops "are powerful instruments in the process of challenging existing paradigms and creating shared perspectives on the future" (Lindgren & Bandhold 2009, p. 27).

To sum up, based on the literature review, scenario techniques as a daily practice in innovation management support future-attentive thinking, and scenario planning represents a strategic tool for innovation communication to adjust mindsets, overcome resistance, go beyond boundaries, and foster awareness of innovations. SCEBAL can facilitate effective communication for innovation. We assume that a common ground among stakeholders; differentiation of scenarios in an innovation context; and management of communication processes, channels, and activities are key impact factors for SCEBAL, described as key elements of the 3C SCEBAL Integrated Management Framework, as shown in Figure 5.2.

## 4  Case: convergence 2020

### Methodology

Innovation and communication scientists need data that are rich in content and leads to information fundamental to communication processes in order to obtain evidence for interpretation (Wert-Gray et al. 1991). Qualitative research fulfills these requirements by observing and investigating contextualized situations (Jones et al. 2004). Moreover, it provides an explanation in depth of what exists, and why it exists (Ritchie & Lewis 2003). Thus, by applying qualitative research, the world "out there" is observed and data are gathered to investigate the field of interest.

Among the various qualitative methods available, the case-study method (Yin 2003, 2009) seems to fulfill the requirements of analyzing organizational communication processes and phenomena in innovation management research (e.g., Meissner & Sprenger 2010; Stake 2005; Brown & Eisenhardt 1995). Case studies provide useful information to document individualized outcomes and capture specific difficulties / limitations of a situation in depth (Patton 1987). Focusing on a specific single case includes describing the functional body of a system (case), which is an exploratory approach of qualitative research (Stake 2005). Certain features within and outside a system are mainly considered, in order to understand the nature of the phenomenon.

This chapter uses a single case as an example to provide useful information to document individualized outcomes and to capture specific difficulties / limitations for the field of interest. The report focuses on three subordinate parts: (1) an introduction to learn about the organizational environment; (2) a project description to learn about the context; and (3) information about individualized outcomes that focus on the field of interest. The project description finally leads to managerial implications and limitations regarding the single case/example and the SCEBAL Integrated Management Framework (see Figure 5.2).

## 5 The example: The Convergence 2020 Project at Telekom Innovation Laboratories

### Telekom Innovation Laboratories

Telekom Innovation Laboratories (T-Labs) is the central research and innovation (R&I) unit of Deutsche Telekom. Experts from the Telekom Group work together with scientists to develop novel services and solutions for Telekom customers. Establishing new business units and spin-off companies is another way of utilizing research results.

For Deutsche Telekom, innovations are key to opening new fields of growth. Telekom Innovation Laboratories, founded in 2005 as a research and development unit, has since established itself as the primary point of contact for all innovation topics in the Telekom Group. The link between Telekom's R&I activities with Technische Universität Berlin (TU), the Berlin University of the Arts (UdK), and a variety of other universities, institutes, and industry partners worldwide creates close ties between the worlds of science and business.

The open innovation model is a guiding principle for the work of T-Labs. This encompasses the inclusion of customers, users, and domain experts in the innovation process, enabled by advanced methodology. T-Labs has its headquarters in Berlin; here and at other locations – in Darmstadt and Bonn in Germany and in Silicon Valley in the United States – more than 360 Telekom experts and scientists from a wide variety of disciplines from all over the world work on developing novel services and solutions for Telekom customers. T-Labs also cooperates closely with Ben-Gurion University in Israel.

The R&I work at T-Labs is concentrated primarily on topics and new technologies that can be brought to market within a time horizon of up to five years. It pursues two major strategic directions. First, it supports the development of applications and services, such as in the key areas of interactive high-end media, future communication, and information relevance. Second, its ideas and developments help secure the future infrastructure and complex services in the areas of cross-domain middleware, global cloud, large-scale infrastructure optimization, and broadband access evolution.

The objective of the work of T-Labs is to transfer and realize innovations in collaboration with the Telekom Group's business units. The R&I projects are coordinated closely with the existing portfolio, the product roadmap, and the network expansion strategy, and can be realized within a period of up to one-and-a-half years. Other projects are dedicated to longer-term innovation topics, which will result in strategic innovations in a window of up to five years. In addition, T-Labs works on topics that do not fit into the defined core areas or that demonstrate disruptive potential. This last area includes multidisciplinary topics and candidates for start-ups and investments.

The results of the R&I projects form the foundation of a multitude of current and future Deutsche Telekom products and services. They include solutions for payment with mobile devices, features for Internet protocol television (IPTV), and methods for more effectively utilizing fiber-optic cables and radio channels. Another method of exploiting results is the founding of new start-up companies such as Zimory, and QiSec.

### The Convergence 2020 Project

Scenario projects can be conducted for different purposes and with different focuses. Our project example, the Convergence 2020 project, focuses on future-attentive thinking and organizational learning. The project addresses innovation topics and strategies in telecommunication up to the year 2020. While scenarios had been developed as tools and triggers for strategic management discussion in a previous project (Breuer, Grabowski & Arnold 2011), the Convergence Project was designed in concordance with the SCEBAL approach from the beginning. Thus, scenarios were understood as a medium for collaborative learning, first within the project team and second for related teams and subordinate team members. In order to support scenario-based learning within this project, a series of workshops was designed and conducted. The first goal was to sensitize people to alternative developments and to foster future-attentive thinking among the participants and related stakeholders. The second goal was to identify not only possible but also desirable scenarios for a convergent telecommunications market in 2020 and to derive appropriate strategic measures and courses of action. The third goal was to create a "big picture" and positive vision for convergent telecommunications in 2020, for orientation and for a joint effort based on a shared understanding of critical issues.

The project started with a discussion of the strategic decisions to be addressed and a definition of the scenario field, followed by an exploration of future customer needs and behavior in a converging technological environment, and also of alternative technological potentials, limitations, and developments. In order to focus on learning, a purely qualitative scenario modeling approach was chosen (Breuer et al. 2012). A moderated discourse within the project team was initiated in order to identify the influencing factors and key uncertainties within the scenario field. Following the exploratory scenarios and a creative elaboration of potential upcoming business opportunities, the most desirable possibilities helped to form a positive vision for telecommunications in 2020. Transferring the results back into current business ecosystems, a roadmap of required activities and strategic milestones was derived. While the full methodology and tools used cannot be described here, an example illustrates the kind of content that was discussed: The conflict between usage-on-demand as a growing trend in society and property based on ownership was identified as one key uncertainty for the near future, and a strong usage orientation became a key characteristic of one of the exploratory scenarios. Considering that hardware and software companies may keep up the walled garden of their ecosystems, opportunities for new business may emerge, such as bridging artificial borders between ecosystems and media formats.

## Scenario-based learning and communications in the Convergence Project

In order to sensitize people to alternatives and therefore foster learning with "attentiveness to innovation" (Breuer & Gebauer 2011), several techniques were applied, most importantly a recursive shift in perspectives, throughout the process. This shift in perspectives was reinforced through the integration of changing stakeholders and external participants and through the careful preparation and utilization of different instructions, media, and communication formats.

Participants outside the scenario team contributed user perspectives and expert knowledge (e.g., on future technology and cultural trends) from within and outside the company (for example, trend researchers and futures researchers). User perspectives were integrated in the form of videographic footage that was created by users in the format of cultural probes on topics related to the core factors. Their videos and quotes contributed authentic insights, served as empathy tools for the group activities, and delivered footage to illustrate aspects of the discussions. Experts were invited to participate in the controversial negotiation of influencing factors. Disagreement on the development of these factors served as one indicator of their uncertainty and thus their potential impact. Instructions and templates included the creation of future headlines (designed like newspaper headlines of today), idea napkins as templates for homework, and live exercises. Media formats included graphic recordings by professional illustrators and realistic narratives that enabled participants to think through the consequences of potential developments at refined levels of detail. Finally, an ideation session in a future gallery (where new technology that is still under development is shown in tangible ways) augmented by advanced telecommunication technology proved to be a powerful medium of learning and interaction.

These shifts in perspective not only triggered creative thinking, but also yielded a rich repository of results to create and stage suitable communication media and acts for a variety of purposes, such as offsite team events and management reports. Results were documented in a comprehensive box of cards for individual adaptation, and in an animated movie that summed up the results in an abstract fashion. Participants noticed that the initial agreement on and definition of central concepts and factors ensured highly efficient and effective communications beyond the immediate scope of the project. They also reported high satisfaction with the results and the process and high motivation throughout. As a major reason for this, they listed the collaborative reflection of their strategic options, the big-picture vision that resulted, and last but not least, the learning process they experienced. According to the team leader, the learning experience was the most valuable asset for follow-up work and dissemination of results.

In addition to the core team of about a dozen participants (and four moderators), numerous stakeholders had to be informed after the modeling and transfer workshops had been finished. These communicative follow-up activities included a moderated walk through an exhibition of key results and visualizations as a physical learning architecture of the process and its results with about 160 team members that had not been immediately involved, informing of collaboration partners within and outside the company, and the presentation of status updates and success stories to the top management.

### Managerial implications and limitations

Although the dissemination of project results continues, a feedback round with the project participants has already yielded a high appreciation of the workshops, their results, and the learning experience they provided so far. The handover of the project results was also used as a handover of the responsibility to disseminate the insights and scenarios to stakeholders inside and outside the organization. On the downside, like people doing qualitative research in the social sciences, the project members had to make a big effort to prepare, document, and analyze the activities in transparent ways.

While the multimodal processing of materials and results helped to increase empathy (e.g., regarding authentic user reports), in most reported cases, scenarios have been communicated in a representational rather than in an operational manner with lasting impact, or they have aimed at the generation of declarative rather than procedural knowledge. Activating events and live interventions are even more engaging than prefabricated media communicating insights, as the team leader also reported at the end of the project.

Only a few systematic applications of scenario processes in organizations have been reported in literature. Given the long time frames being addressed, proving the value contribution of scenarios remains a great challenge. Additional scientific research on the benefit of scenarios (see Schoemaker 1993), their comparative performance, and the different communications media and formats is needed. The main limitations we identified are divided into content-related limitations and methodological limitations.

*Content-related limitations*:

- Regarding scenario techniques for different purposes and focuses, this chapter only focuses on future-attentive thinking; it is not a comprehensive approach for all scenario techniques.
- The three key elements of the 3C SCEBAL integrated management framework are identified as *the* critical factors for SCEBAL for future-attentive thinking and effective communication for innovation, based on practice and a literature review in the field of interest. Hence, this framework is based on scenario planning and innovation communication from a management view in the open innovation era.

*Methodological limitations*:

- This chapter presented a single case/project example, which means it is limited to an individualized context in telecommunications with a specific objective for SCEBAL.
- Exploratory analysis has lead to a better understanding of the SCEBAL framework, but it is limited in its evidence, and subsequent research will be required in order to provide conclusive evidence.

## 6  Conclusions and outlook

For telecommunications providers working with mid-term and long-term investments in particular, it is essential to be well-prepared for future developments. Among others, imagination and communication are two critical factors for innovation, but they also are potential bottlenecks. In the innovation economy, the ability to manage communication processes, channels, and activities for innovation at all levels is important for any organization involved in open innovation.

Scenario planning helps an organization to visualize future developments and identify new paths for mid-term and long-term investments. Moreover, scenario planning represents a strategic tool for innovation communication to create new mindsets and foster awareness of innovations. Hence, SCEBAL represents an enabler for future-attentive thinking and communication for innovation.

This chapter focused on the key question of how SCEBAL can be applied for future-attentive thinking and for effective communication for innovation. It draws on scenario planning and innovation communication from a strategic management perspective. The answer is the 3C SCEBAL Integrated Management Framework with its three key elements – communication management, contextualizing innovation, and common ground among stakeholders – which are used to support future-attentive thinking and effective communication for innovation through information transmission and intensive, reflective, and smart dialogue between an organization and its stakeholders.

The discussion of the Convergence 2020 Project, focusing on future-attentive thinking and organizational learning, showed that an organization applying SCEBAL mainly coped with challenges in communications management as one of the key elements of the 3C SCEBAL integrated management framework; for instance, to put forth a large effort to prepare, document, and analyze the activities in transparent ways. Hence, if SCEBAL aims at positively influencing effective communications for innovation, then planning, execution, monitoring, and evaluation of a broad range of communication activities are crucial. This chapter recommends applying efficient management of innovation communication, as illustrated in Figure 5.2.

Hence, as a result, we see that efficiency in the management of innovation communication is a prerequisite for communication management in scenario planning, which in turn may lead to future-attentive thinking and effective communication for innovation.

Regarding the identified limitations, future innovation and research agendas might focus on the interrelations of the key elements for different types of organizations and the relational arrangements to identify drivers and barriers to effective communication for innovation in the long run. Furthermore, we suggest investigating multiple case study research to analyze in depth the interrelations between the key elements and intensive, reflective dialogue between an organization and its stakeholders, related to different types of scenarios and communication formats.

# References

Alexander, I. (2004) Scenarios: An introduction. In I. Alexander & N. Maiden (Eds), *Stories, Use Cases: Through the Systems Development Life-Cycle*. John Wiley.

Breuer, H. & Gebauer, A. (2011) Mindfulness for innovation. Future scenarios and high reliability organizing preparing for unforseeable. *SKM Conference for Competence-based Strategic Management*, pp. 1–18. Linz, Austria. (Online: www. bovacon. com/papers/2011_Mindfulness.pdf, April 18, 2012).

Breuer, H., Grabowski, H., & Arnold, H. (2011) The shape of things to come: Scenarios and visual stories for telecommunication in 2020. In *Proceedings of the IADIS International Conference on Telecommunications, Networks, and Systems* (pp. 107–114), Rome, Italy.

Breuer, H. & Matsumoto, M. (2010) Scenarios for mobile learning across contexts. *ACE Asian Conference on Education Proceedings* (ISSN 2185–6133) (pp. 1935–1945), Osaka, Japan.

Breuer, H. & Matsumoto, M. (2011) Ubiquitous learning environments: Notions of context, application scenarios and reflective stories. In T. Bastiaens & M. Ebner (Eds), *Proceedings of World Conference on Educational Multimedia, Hypermedia and Telecommunications* (pp. 2366–2374). Chesapeake, VA: AACE. Retrieved from http://www.editlib.org/p/38189.

Breuer, H., Schulz, J. & Leihener, J. (2012) Learning from the future – modeling scenarios based on normativity, performativity and transparency. *Proceedings of XXIII ISPIM Conference – Action for Innovation: Innovating from Experience*, Barcelona, Spain (Online: http://ssrn.com/abstract=2125750, August 16, 2012).

Brown, S. & Eisenhardt, K. (1995) Product development: Past research, present findings, and future directions. *Academy of Management Review*, 20(2), 343–378.

Chesbrough, H. W. (2003) The era of open innovation. *MIT Sloan management review*, 44(3), 35–41.

Chesbrough, H. W. (2006) Open innovation: A new paradigm for understanding industrial innovation. In H. W. Chesbrough, J. West, & W. Vanhaverbeke (Eds), *Open Innovation: Researching a New Paradigm* (pp. 1–14). Oxford: Oxford University Press.

Crosby, L. A. & Johnson, S. L. (2006) Customer-centric innovation. *Marketing Management*, 15(2), 12–13.

Davenport, T. H., Leibold, M., & Voelpel, S. (Eds) (2006) *Strategic Management in the Innovation Economy: Strategy Approaches and Tools for Dynamic Innovation Capabilities*. Erlangen: Publics Publishing and Wiley-VCH.

Donaldson, B., O-Toole, T., & Holden, M. (2011) A relational communication strategy for successful collaborative innovation in business-to-business markets. In M. Hülsmann, & N. Pfeffermann (Eds), *Strategies and Communications for Innovations* (pp. 2093–2228). Berlin: Springer.

Fink, A., Marr, B., Siebe, A., & Kuhle, J.-P. (2005) The future scorecard: Combining external and internal scenarios to create strategic foresight. *Management Decision*, 43(3), 360–381.

Fink, A., Schalke, O. & Siebe, A. (2000) Wie Sie mit Szenarien die Zukunft vorausdenken. *Harvard Business Manager*, 2, 34–47.

Fink, A. & Siebe, A. (2006) *Handbuch Zukunftsmanagement. Werkzeuge der strategischen Planung und Früherkennung*. Frankfurt: Campus.

Fink, S. (2009) Strategische Kommunikation für Technologie und Innovationen – Konzeption und Umsetzung. In A. Zerfaß & K. M. Möslein (Eds), *Kommunikation als Erfolgsfaktor im Innovationsmanagement: Strategien im Zeitalter der Open Innovation* (pp. 209–226). Wiesbaden, Germany: Gabler.

Herzog, P. (2008) *Open and Closed Innovation: Different Cultures for Different Strategies.* Wiesbaden: Gabler.

Hofbauer, G., Körner, R., Nikolaus, U., & Poost, A. (2009) *Marketing von Innovationen. Strategien und Mechanismen zur Durchsetzung von Innovationen.* Stuttgart, Germany: Kohlhammer.

Hülsmann, M. & Pfeffermann, N. (2011) *Strategies and Communications for Innovations.* Berlin, Germany: Springer-Verlag.

Jones, E., Watson. B., Gardner, J., & Gallois, C. (2004) Organizational communication: Challenges for the new century. *Journal of Communication*, 12, 722–750.

Kahn, H. & Wiener, A. J. (1967) The next thirty-three years: A framework for speculation. *Daedalus*, 96(3), 705–732.

Kosow, H. & Gaßner, R. (2008) *Methods of Future and Scenario Analysis Overview, Assessment, and Selection Criteria. Studies 39.* German Development Institute: Bonn.

Lichtenthaler, U., & Lichtenthaler, E. (2009) A capability-based framework for open innovation: Complementing absorptive capacity. *Journal of Management Studies*, 46(8), 1315–1338.

Lindgren, M. & Bandhold, H. (2009) *Scenario Planning: The Link between Future and Strategy* (2nd edn). Basingstoke: Palgrave Macmillan.

Löscher, P. (2012) Less is more. *The Economist*, Special Issue "The World in 2012", p. 124

Mahajan, V., Muller, E., & Wind, Y. (2000) New-product diffusion models: From theory to practice. In V. Mahajan, E. Muller, & Y. Wind (Eds), *New-product Diffusion Models* (pp. 3–27). Düsseldorf: Springer.

Martilla, J. A. (1971) Word-of-mouth communication in the industrial adoption process. *Journal of Marketing Research*, 8(2), 173–178.

Mast, C., Huck, S., & Zerfass, A. (2005) Innovation communication: Outline of the concept and empirical findings from Germany. *Innovation Journalism*, 2(7), 1–14.

Mast, C. & Zerfaß, A. (2005) *Neue Ideen erfolgreich durchsetzen. Das Handbuch der Innovationskommunikation.* Frankfurt am Main: F.A.Z.-Institut für Management-, Markt und Medieninformation.

Mazzarol, T. (2011) The role of word of mouth in the diffusion of innovation. In M. Hülsmann, & N. Pfeffermann (Eds), *Strategies and Communications for Innovations* (pp. 117–132). Berlin, Germany: Springer-Verlag.

Meissner, J. O. & Sprenger, M. (2010) Mixing methods in innovation research: Studying the process-culture-link in innovation management. *FQS*, 11(3), Art. 13. (ISSN 1438–5627) Online: http://www.qualitative-research.net/index.php/fqs/article/view/1560 [1 February, 2012]

Mietzner, D. & Reger, G. (2005) Advantages and disadvantages of scenario approaches for strategic foresight. *International Journal of Technology, Intelligence and Planning*, 1(2), 220–239.

Mohr, J., Sengupta, S., & Slater, S. (2009) *Marketing of High-Technology Products and Innovations* (3rd edn). New Jersey: Pearson.

Niegemann, H. M., Domagk, S., Hessel, S., Hein, A., Hupfer, M., & Zobel, A. (2008) *Kompendium multimediales Lernen.* Heidelberg: Springer.

Patton, M. Q. (1987) *How to Use Qualitative Methods in Evaluation.* Thousand Oaks: Sage.

Peres, R., Muller, E., & Mahajan, V. (2010) Innovation diffusion and new product growth models: A critical review and research directions. *International Journal of Research in Marketing*, 63, 849–855.

Pfeffermann, N. (2011a) The scent of innovation – towards an integrated management concept for visual and scent communication of innovation. In M. Hülsmann, & N. Pfeffermann (Eds), *Strategies and Communications for Innovations* (pp. 163–181). Berlin, Germany: Springer-Verlag.

Pfeffermann, N. (2011b) Innovation communication as a cross-functional dynamic capability: Strategies for organizations and networks. In M. Hülsmann, & N. Pfeffermann (Eds), *Strategies and Communications for Innovations* (pp. 257–289). Berlin, Germany: Springer-Verlag.

Pfeffermann, N. (2012a) Managing Communication for Frugal Innovation. Available at: http://blog.iseic-consulting.com/2012/03/15/frugal-innovation/ [date: 15 March, 2012].

Pfeffermann, N. (2012b) Innovation Communication – Enabler of Innovation as Strategic Change. Available at http://www.ispim.org/iwsep12_np.php [date: September 01, 2012].

Ramirez, R., Selsky, & van der Heijden (2010) *Business Planning for Turbulent Times: New Methods for Applying Scenarios* (2nd edn). London, Washington: Earthscan Publications.

Ritchie, J., & Lewis, J. (2003) *Qualitative Research Practice: A Guide for Social Science Students and Researchers*. Thousand Oaks, CA: Sage.

Rogers, E. M. (2003) *Diffusion of Innovations* (5th edn). New York: The Free Press.

Sandberg, B. (2008) *Managing and Marketing Radical Innovations*. London [u.a.]: Routledge.

Schoemaker, P. J. H. (1993) Multiple scenario developing: Its conceptual and behavioral basis. *Strategic Management Journal*, 14(3), 193–213.

Schoemaker, P. J. H. (1995) Scenario planning: A tool for strategic thinking. *Sloan Management Review*, 36(2), 25–40.

Schwartz, P. (1991) *The Art of the Long View: Planning for the Future in an Uncertain World*. New York: Doubleday.

Sowter, C. V. (2000) *Marketing High Technology Services*. Aldershot [u.a.]: Gower.

Stake, R. E. (2005) Qualitative case studies. In N. K. Denzin, & Y. Lincoln, *The Sage Handbook of Qualitative Research* (pp. 443–466) (3rd edn). Thousand Oaks: Sage.

Steinhoff, F. & Trommsdorff, V. (2011). Innovation marketing – an introduction. In M. Hülsmann, & N. Pfeffermann (Eds), *Strategies and Communications for Innovations* (pp. 105–116). Berlin, Germany: Springer-Verlag.

Steinmüller, K. (1997) Grundlagen und Methoden der Zukunftsforschung. Szenarien, Delphie, Technikvorschau. *Werkstattbericht 21. Sekretariat für Zukunftsforschung.* Gelsenkirchen.

Trommsdorff, V., & Steinhoff, F. (2007) *Innovationsmarketing*. München: Franz Vahlen.

Van der Heijden, K. (2005) *Scenarios: The Art of Strategic Conversation* (2nd edn). Chichester, New York: John Wiley.

Van der Merwe, L. (2008) Scenario-based strategy in practice: A framework. *Advances in Developing Human Resources*, 10(2), 216–239.

Weidenmann, B. (Ed.) (1994) *Wissenserwerb mit Bildern*. Bern: Huber.

Wert-Gray, S., Center, C., Brashers, D., & Meyers, R. (1991) Research topics and methodological orientations in organizational communication: A decade in review. *Communication Studies*, 42, 141–154.

Yin, R. K. (2003) *Case Study Research: Design and Methods.* (3rd edn). Newbury Park, CA: Sage.

Yin, R. K. (2009) *Case Study Research: Design and Methods* (4th edn). Newbury Park, CA: Sage.

Zerfaß, A. (2009) Kommunikation als konstitutives Element im Innovations management – Soziologische und kommunikationswissenschaftliche Grundlagen der Open Innovation. In A. Zerfaß & K. M. Möslein (Eds), *Kommunikation als Erfolgsfaktor im Innovationsmanagement: Strategien im Zeitalter der Open Innovation* (pp. 23–56). Wiesbaden, Germany: Gabler.

Zerfaß, A., & Ernst, N. (2008) *Kommunikation als Erfolgsfaktor im Innovationsmanagement. Ergebnisse einer Studie in deutschen Zukunftstechnologie-Branchen.* Leipzig: Universität Leipzig.

Zerfaß, A. & Möslein, K. M. (2009) *Kommunikation als Erfolgsfaktor im Innovations management: Strategien im Zeitalter der Open Innovation.* Wiesbaden, Germany: Gabler.

Zerfaß, A., Sandhu, S., & Huck, S. (2004) Innovationskommunikation – Strategisches Handlungsfeld für Corporate Communications. In G. Bentele, M. Piwinger, & G. Schönborn (Eds), *Kommunikationsmanagement. Strategien, Wissen, Lösungen* (August 2006; 1.24, 1–30). München: Luchterhand.

# 6
# Social Network Analysis: An Important Tool for Innovation Management

*Gerhard Drexler and Bernard Janse*

## 1 Introduction

Innovation can be defined as the creative application of technologies, processes, and ideas to some useful purpose. It is increasingly becoming a highly valued commodity, viewed as key to competitiveness and economic growth. Within organizations, there is continuous pressure to identify those areas that present the greatest opportunity for innovation and to develop process models that will accelerate the pace of innovation (Horn 2005). The management of technological innovation includes the organization and direction of resources toward effectively (1) creating new knowledge; (2) generating technical ideas aimed at new and enhanced products, manufacturing processes, and services; (3) developing those ideas into working prototypes; and (4) transferring them into manufacturing, distribution, and use (Roberts 2007).

In the past decades, firm-level models of innovation have become increasingly sophisticated. Rothwell (1992) summarized the evolution of innovation models from the 1950s to the 1990s in five successive generations, pointing out that the key features of fifth-generation models include systems integration, continuous innovation, and extensive networking.

As innovations originate from ideas that are the results of the creative or rational thinking process, one basic task of the innovation process is to foster the generation of ideas. Traditionally, the focus was to generate ideas internally. Recently, open innovation models suggest that companies seek people of genius from both inside and outside the firm to provide fuel for their innovation process (Chesbrough 2003). Thus, it is now recognized that novel ideas may be generated by employees, customers, suppliers, universities, and other groups of stakeholders. Many studies have focused on the challenges and experiences that firms face in dealing with open innovation approaches (Elmquist et al. 2009; Antikainen et al. 2010). Unfortunately, there have only been a few studies that specifically illuminate the processes of identifying and selecting appropriate external partners (Philbin 2008).

In a highly connected world, innovation networks are a logical result of the increasing complexity of innovative products and services. These networks represent an organizational solution for product and service innovation, since they integrate different organizational skills, focused on a common goal (Pyka & Küppers 2002). Social network research is not new, but it has gained renewed interest from different disciplines due to its practical implications within organizations. In order to shed light on mechanisms of networked innovation, the purpose of this chapter is to motivate innovation managers to adopt network analysis as a tool for scrutinizing and improving their innovation processes.

The main objective of this work is to describe some important theoretical and operational aspects of intraorganizational and interorganizational networks as the basis for idea generation and collaborative activities. A review of extant literature and a demonstration of the operational utilization of network analysis highlight the opportunities provided by internal and external networks. The social networking theory around how interactions that were identified using these tools were initiated, maintained, and fed will form the basis of future work.

## 2   Background

Interorganizational collaboration has been recognized as important in supplementing the internal innovative activities of organizations (Hagedoorn 2002). Firms that engage in interorganizational collaboration in the framework of their innovation strategy tend to be more effective in terms of innovative performance (Faems et al. 2005). "Collaboration" is a broadly used term, and research on collaboration can be found in numerous domains, each focusing on slightly different aspects. Depending on the focus, the definition and intended meanings of collaboration differ. Teasley and Roschelle (1993) define collaboration as "...a coordinated, interactive activity that is the result of a continued attempt to construct and maintain a shared conception of a problem."

A variety of approaches can be used to explain why organizations form partnerships and alliances and how they choose their partners. The main benefits of research and development (R&D) collaborations include risk sharing, exploitation of economies of scale and scope, reduced duplication of research efforts, access to complementary assets, and reduction of time to market. Firms undertaking R&D collaboration acquire new capabilities and improve their ability to monitor, absorb, and exploit external knowledge (Cohen & Levinthal 1990). Recent papers on collaboration strategies shed some more light on various reasons for engaging in collaboration. For example, Bjerregaard (2009) revealed that informal interpersonal ties between industry and university researchers were in some cases utilized as vehicles for establishing new contacts and partner selection; in other cases a more explorative contact-making approach was pursued. Abramovsky et al. (2009) reported evidence of a positive relationship between the extent to which firms were able to benefit from external information flows and the fact that firms that place a high value on external information flows were more likely to collaborate with firms outside their own industry or with research institutes. This enabled them to access a broader range of knowledge.

Although interorganizational collaboration may take on many different forms, studies show that a high number of them are focused on technological issues. Depending on the type and focus of different fields of research, the definitions and intended meanings of collaboration differ. Collaboration involves a form of partnership, alliance, or network aimed at a mutually beneficial, clearly defined outcome. Member entities may include firms, universities, and other organizations. Major objectives of firms to join in collaborative R&D are to foster the establishment of new relationships; to get access to complementary resources and skills, technological learning, and cost sharing; and to keep up with major technical developments (Caloghirou et al. 2003).

Ahuja (2000) views networked innovation in terms of an interfirm collaborative linkage, i.e., as a voluntary arrangement between independent organizations to share resources. Besides joint research ventures, innovation-related interactions in R&D among firms can take a number of other forms, ranging from legal ownership agreements to informal know-how trading. A major process that accompanies interfirm relations is the knowledge flow that occurs between the actors (Ozman 2009). Fetterhoff and Voelkel (2006) argue that as the majority of firms are not used to promoting external innovations, a number of issues have to be taken into consideration by these firms' innovation managers: (1) seeking opportunities, (2) evaluating the market potential and inventiveness of a given opportunity, (3) recruiting potential partners by building a convincing argument, (4) capturing value through commercialization, and (5) extending the innovation offering to outside partners. This implies that the so-called "open innovation" (Chesbrough 2003) is not a simple outsourcing of R&D activity, but rather that it is an integration of internal and external competences, cooperative research, and development. Thus, for Companies to take advantage of open innovation, they must develop their internal knowledge of how to increase their absorptive capacity.

Networks of organizations have been called different things in the literature. An example is the groups of firms that are bound together in some formal or informal way in Japan that are called the *keiretsu* (Kutschker & Schmid 2008). Other authors talk about partnerships, strategic alliances, interorganizational relationships, coalitions, and cooperative arrangements. The role of networks in disseminating information and ideas; providing access to resources, capabilities and markets; and allowing the combination of different pieces of knowledge, has become important for innovation and economic competitiveness. During the past decade, many governments have made a concerted effort to promote cooperative research. This is particularly true within the European Union (Caloghirou et al. 2002). Dundon (2002) emphasized that "the evolution of organizations toward a networked style represents an overall shift in organizational life from centralized command and control to a more participatory, shared leadership style, in which skill, expertise, and ideas rule." She points out some ideas of how one can network and involve others. Examples include network within your organization, network to learn from others, network to share research and development insights and responsibilities, and network to provide a hub or meeting place for others. Based on the dimensions of institutionalization and proximity, Verburg and Andriessen (2011) discern four basic types of knowledge networks: strategic networks, informal networks, question-and-answer networks, and online strategic networks. This variety highlights the different ways in which creating and sharing knowledge can be organized.

In his seminal paper, Coleman (1988) introduced and illustrated the concept of social capital. Adler and Kwon (2002) provide a comprehensive list of definitions of social capital and develop a common conceptual framework that identifies the sources, benefits, risks, and contingencies of social capital. Networking and social capital are, therefore, highly interrelated; networking can lead to the initiation of relationships and the subsequent development of social capital, which in turn can be used to initiate further relationships and social capital. As trust is a major aspect of social capital, alliance trust is said to be a precursor of internal and external social capital between partners (Suseno & Ratten 2007). For an overview of social capital and its three core dimensions, see Lee (2009).

Innovation managers in all kinds of organizations are increasingly challenged to create and exploit networks. Basic knowledge of network theory and practical implications of networking, form the foundation of a number of managerial tasks, and managers should be aware of the importance of this kind of knowledge. As it would be beyond the scope of this chapter to introduce the science of social network analysis, we will only provide a brief overview of this topic as an introduction to the three case studies described later.

A social network is a network of people that are connected via some form of a relationship. In other words, a network is a set of actors connected by a set of ties. The actors (often called nodes) can be a single person, a team, an organization, a concept, etc. Ties connect pairs of actors and can be directed or undirected (interaction takes place in only one direction from one actor to the other or takes place in both directions), dichotomous (present or absent, as in whether two people are friends or not), or valued (measured on a scale, as in strength of friendship). A set of ties of a given type (such as friendship ties) constitutes a binary social relation, and each type of relation defines a different network (Borgatti & Foster 2003).

Networking is defined as the sum of interactions of an actor in a network, comprising all activities concerning the management of the existing relationships, the management of the position occupied in the surrounding network, and the strategies of how to network (Håkansson & Ford 2002). The resources available in the network include technical, organizational, and knowledge resources and are accessed through relationships between key actors. Of particular importance is that resources are provided by actors within the network through their performance of key roles. Importantly, while several actors may have the resources needed, it is through interaction and execution of specific roles that such resources are accessed. We might categorize these roles as task-oriented roles (articulating, funding, developing, prototyping, and producing) or network-oriented roles (connecting, integrating, and endorsing). All of these roles are not necessarily enacted for each kind of activity, and several roles may be played by individual actors (Heikkinen et al. 2007). Another important aspect of social network analysis is the assumption that relationships are important among interacting units, and that relations defined by linkages among units are a fundamental component of network theories. For example, Zhou et al. (2010) suggest that managers can enhance knowledge transfer among co-workers by enhancing their ties to each other.

The strength of linkages between members of a network can be weaker or stronger, depending on the intensity of interactions between the actors. For example, interpersonal ties that are built through frequent communication can lead to more effective interactions (Uzzi 1997). A number of papers deal with different modes of interaction between members of a network, explained by the characteristics of the ties between them. Table 6.1 summarizes relevant papers about the strengths and characteristics of ties.

Weak ties are based on loose emotional tendencies and are maintained via infrequent communication. This kind of relationship provides access

*Table 6.1*  Papers about characteristics of network ties

| Author | Ties |
| --- | --- |
| Granovetter (1973) | Weak vs. Strong |
| March (1991) | Exploration vs. exploitation |
| Ahuja (2000) | Direct vs. indirect |
| Putnam (2000) | Bonding vs. bridging |
| Hardy et al. (2005) | Generalized vs. particularized |
| Simard & West (2006) | Formal vs. informal |
| Simard & West (2006) | Deep vs. wide |

to novel information by bridging otherwise disconnected groups and individuals. In contrast, strong ties involve a strong degree of trust and are characterized by frequent contacts over a longer period. Strong ties more likely provide redundant information, because they occur among a small group of people in which almost everyone knows what the others know (Granovetter 1973).

Social network analysis methods have developed over the past 50 years as an integral part of advances in social theory, empirical research, and formal mathematics and statistics (Wasserman & Faust 1994). The majority of network analysis studies use either whole network or egocentric designs. Whole-network studies examine sets of interrelated objects or actors. Egocentric studies focus on a focal actor or object and his (or its) relationship to his (its) locality within the social network (Marsden 2005).

There are a variety of ways in which network data can be gathered, e.g., questionnaires, interviews observations, and experiments. There are also a variety of methods and tools for social network analysis. Social network analysis software facilitates quantitative or qualitative analysis of social networks by describing features of a network, either through numerical or visual representation. An overview of software for social network analysis is provided by Huisman and van Duijn (2005). Some of the most common software tools are, for example:

- UCINET (http://www.analytictech.com/ucinet/)
- igraph (http://igraph.sourceforge.net/)
- Pajek (http://vlado.fmf.uni-lj.si/pub/networks/pajek/)

One-mode network analyses involve measurements on a single set of actors, and their relations are usually viewed as representing relational contents of many different types. Two-mode network analysis involves measurements on two sets of actors, or on a set of actors and a set of events. The latter represents an affiliation network, which arises when one set of actors is measured with respect to attendance at a set of events or activities (Wasserman & Faust 1994).

Networks depicted in this chapter are visualized by the use of Pajek, freeware provided by the University of Ljubljana.

## 3   Social networks and idea generation

Creativity is the ability to produce work that is both novel and appropriate (Sternberg & Lubart 1996). Recent findings have shown that creativity is not only limited to individuals, but that it is also a social phenomenon. By interacting and communicating with others, individuals get access to novel perspectives and unique knowledge, and they can get support for their ideas

(Perry-Smith and Shalley 2003; Hargadon and Bechky 2006). Support from a network of social relationships can contribute to a person's creativity (Madjar et al. 2002) and can facilitate the creation and implementation of creative ideas (Axtell et al. 2000). Previous studies in this area have focused on the effect of network structure on creativity and innovation. Perry-Smith (2006) showed that research scientists with many weak ties were rated as more creative than those with few weak ties, and Obstfeld (2005) stated that engineers with dense social networks were more engaged in developing new products or processes. These studies examined the relationship of network structure and creativity, but research has generally neglected the anteced-ents of network structure (Brass et al. 2004).

Granovetter (1973) suggested that weak network ties are more likely to be the source of novel information and therefore should increase the probability of stimulating creativity. Communication about ideas can serve different functions during the creative process. In communicating ideas, knowledge is shared, new insights are stimulated, and ideas can be evaluated according to standards valid in the social context. These are cognitive processes stimu-lated by communication. Perry-Smith and Shalley (2003) suggest that weak ties favor individual creativity because they have structural properties that facilitate access to diverse knowledge, reinforce creative-related skills, and encourage autonomous thinking.

Tsai (2001) reported a positive influence of the degree of centrality (the number of neighbors to which an actor is connected) on innovation perform-ance. Informal networks of people that share expertise and knowledge in creative ways that foster new approaches to problems have been recognized as being important for innovation (Wenger & Snyder 2000). Recent research provides evidence for a positive correlation between network centrality of employees and the creation of high-quality ideas (Björk & Magnusson 2009; Drexler 2010). Employee's engagement in knowledge-sharing is highest when network centrality, motivation, and ability are all high (Reinholt et al. 2011).

In the first example presented in this study, we wanted to confirm that members of a firm would be more likely to generate high-quality ideas if they were more densely networked with their colleagues. Network analysis was conducted on a competence center (department) of an international corporation. Ideas were collected by an idea management tool and graded according to quality criteria – e.g., amortization, intangible benefits, and risk. The actors and their relationships were identified by mapping of e-mail communication of all employees (actors) of the entity over a period of several weeks. Networks were plotted using the Pajek network analysis and visualization tool.

Figure 6.1 depicts the whole network. All 93 actors were connected to others within the collective by their exchange of e-mails, but the number of colleagues they were connected with, varied a lot.

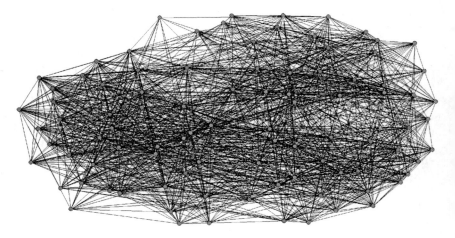

*Figure 6.1*    Graphic representation of the intraorganizational network

Notes: Number in brackets is degree centrality of the node; "v" is the firm's internal code number of the node.

The simplest indicator of interconnectedness is the number of neighbors to which an actor (node) is connected. This is termed the "degree centrality" and can be calculated by comparing the actual number of connections of an actor in a network with the maximum possible number (Wasserman & Faust 1994). The degree centrality of a node can be expressed as follows:

$$C_D(n_i) = d(n_i)/(g-1)$$

where $C_D(n_i)$ is the degree centrality of node $n$, $d(n_i)$ is the number of connections of node $n_i$ (d is the number of connections, $n_i$ is the identification number of the node), and $g$ is the number of nodes in a network. The degree centrality of the actors in the network members in Figure 6.1 ranged between .08 (a person with ties to 8 percent of the network actors) and .67 (a person with ties to 67 percent of the actors). The number in the brackets refer to the degree centrality of the node, "v" indicates the firm's specific code of the actors.

Comparing the group of actors that submitted high-quality ideas with the group of actors that submitted low-quality ideas revealed that actors with a high degree centrality in the network submitted ideas that tended to be of higher quality when compared to the other group. In summary, a $\chi^2$ test revealed a strongly significant interrelationship between the quality of an idea and the degree centrality of its provider.

Despite the fact that these results may not solely be due to social interactions and networking, our data would suggest that it is important for organizations to enable and empower people to socially interact and share their problems and ideas. Our data are supported by the literature and we recommend that innovation managers consider the following:

- Informal networks between employees could be an efficient tool in improving the quality of ideas generated.
- Managers should enable and encourage all actors in their network to establish and maintain informal communications in order to share information and ideas.
- There are many opportunities in the office that can stimulate informal communication, e.g., access to Internet and Intranet, e-mails, coffee corners, and department printers. Managers should take care not to inadvertently restrict informal communication by changes in policies or structural modifications; rather, they should consider actively reviewing their physical and electronic assets to find ways to maximize informal communication.

## 4   External events and social networks

In the second example, we evaluated the dynamics of forming external networks for the purpose of opportunity identification and idea sourcing. Opportunity identification is the process whereby organizations identify opportunities to create better, or totally new products, processes, or services (Koen et al. 2001). Idea sourcing can be fueled by internal idea generation, as well as by global idea sourcing.

It is recognized that only a small percentage of initial ideas result in a commercial success (Kelley 2001). In most innovation programs there is a balance between (1) increasing the total number of ideas and (2) increasing the quality of those ideas. Both approaches can be fostered by collaboration, but the evaluation of many ideas consumes more resources than selection from a small pool of good ideas. Thus, practical management of the ideation process should focus on high-quality ideas from internal or external sources (Tidd et al. 2005).

In addition to tacit knowledge (e.g., literature, patents, and surveys), informal contacts at meetings, trade fairs, and congresses are for most industries very important channels to access external knowledge. Interaction with external partners stimulates a firm's innovativeness, because it makes a far more diversified range of knowledge sources accessible than in the case of intrabusiness interaction. As a consequence, firms cooperating with universities and research organizations increase their ability to realize more radical innovations and to introduce new products to the market. It is, therefore, reasonable that firms increase the interaction between their employees and actors in the science arena (Kaufmann & Tödtling 2001).

In our study, we evaluated the creation of linkages between attendees at a number of public workshops during 2010. Data were provided by the list of attendees at quarterly events of the Platform for Innovation Management (www.pfi.or.at), an Austrian nonprofit organization. These events are aimed at exchanging knowledge about innovation methods and technology knowledge among companies, research institutes, and universities. Nine workshops were monitored over a period of 27 months and a two-mode network analysis was calculated and plotted in order to visualize the creation of connections between different parties.

Initially we analyzed four of the nine workshops; the resulting data is presented in Figure 6.2 (workshops are shown by big bubbles and their attendees by small bubbles).

People from many organizations only attended one workshop (depicted on the perimeter of the figure). Attendees to two workshops are located between the big bubbles of the respective workshops, and attendees to three or four workshops are located in the center of the graph.

We then constructed the two-mode network for all nine workshops. Figure 6.3 provides a comprehensive picture of all nine workshops and their attendees. The nine big bubbles close to the center of the figure symbolize the nine workshops. The outer perimeter of the figure shows people of organizations that attended only one workshop, the next perimeter in shows those attending two, etc. The closer to the center, the more workshops the

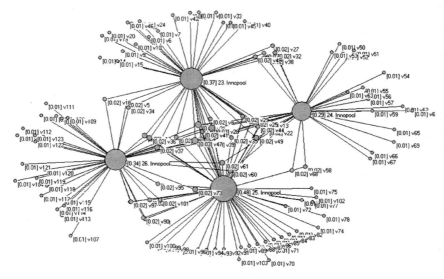

*Figure 6.2*   Two-mode network of four workshops and their attendees

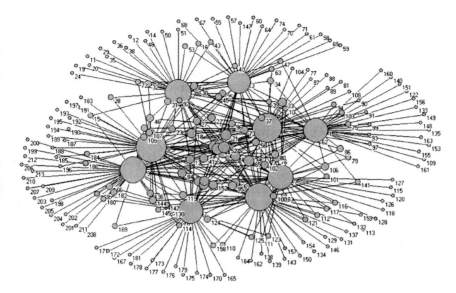

*Figure 6.3* Two-mode network of nine workshops and their attendees

organizations attended and the more opportunities for establishing informal contacts they had.

A survey was subsequently conducted to find out whether attendees of a higher number of workshops were more likely to establish collaborations. Results revealed that satisfaction with both knowledge and partner acquisition were significantly higher when more than four workshops had been attended.

Implications for innovation management may be summarized as follows.

- Organizations that focus on open innovation should motivate and enable their members to attend congresses, trade fairs, and similar external events. People should be trained in and motivated to establish informal contacts with new people who could provide them with knowledge and ideas.
- Event organizers and hosts of events should consider providing sufficient time and space for social activities and informal communication.

## 5 Project networks

In the last example, we evaluated the value of forming networks within and between projects. In the process of open innovation, firms commercialize externally obtained ideas as well as those ideas generated internally. Many firms deploy both external as well as in-house pathways to the market

(Chesbrough 2003). Open innovation projects seek innovative solutions to research and technology problems that impact organizational, scientific, and industrial progress. These projects are conducted by collaboration between internal and external forces, and many of them may involve a number of partners. In this context, collaboration can be defined as "a cooperative, inter-organisational relationship that is negotiated in an on-going communicative process" (Hardy et al. 2005). According to Belderbos et al. (2004), three distinct types of collaboration can be identified: horizontal (with competitors), vertical (with customers and suppliers), and institutional (with universities or research institutes). Hence, it is important that managers of collaborative projects be skilled and experienced in issues specific to these types of relationships.

A process model for collaborative research projects between universities and industry was developed by Philbin (2008). The process-based function of the model was based on a linear sequence of groups of activities, which involved the following stages: (1) terrain mapping, (2) proposition, (3) initiation, (4) delivery, and (5) evaluation. We suggest that network analysis can offer valuable information across all five stages. In addition, analysis of project networks could provide a means for visualization of collaborative ties and offer the opportunity to reveal affiliation of both internal and external project members to more than one project.

To demonstrate the value of network analysis in reviewing collaborative research projects, we applied the analysis to three projects being run within a large enterprise (€3 billion revenue, 13,000 employees). Project partners and the links between them were derived from team lists and organizational charts. The results are depicted in Figure 6.4.

In this figure, each of the three collaborative R&D projects is represented by a central dot, employees of the company are represented by gray dots, and representatives of external organizations like universities and research institutes are represented by black dots. Dots with two or three connections represent members of two or three projects. Figures like this can offer innovation managers answers to some important questions:

- What is the proportion of internal and external members on a project team?
- Are some internal or external people members of more than one project, and could this be problematic in terms of intellectual property protection?
- Do project team members and their connections change over time?
- Where are the team members located geographically?
- Who are the people connecting two or more projects and what is the benefit of this situation?

To conclude, it is notoriously difficult to get results, maintain focus, and deliver profitable or functional products from large collaborative research

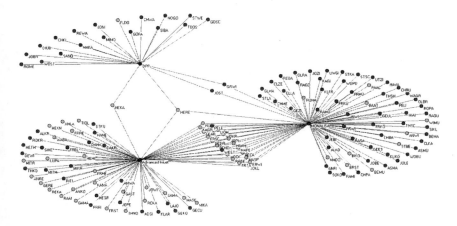

*Figure 6.4* Network of three R&D projects

projects. One of the reasons for these difficulties is the complexity involved in managing these types of projects. Network analysis can provide the innovation manager with an easy tool to interpret by a visual overview the interactions (collaborations) that occur around his/her external collaborative research projects. Although the tool cannot evaluate the quality of the interactions that are occurring per se, it does allow a qualitative overview of the actors that are involved and an overview of the connections that are in place to support the project. This could increase the delivery of value from these complex projects.

## 6  Future research directions

Interest in social network analysis has grown dramatically in recent years and continues to increase. Despite the fact that much research has been conducted into theoretical and practical aspects of networks, the Organization of Economic Co-operation and Development (OECD 2010) stated that "As recent experience by both national and regional authorities has shown, there is scope to develop regional innovation policies that can capture positive local externalities. This may involve improving the efficiency with which partners interact, sharing knowledge and systematizing the relationships between actors."

Most studies on management in the network context have primarily focused on identifying conditions for management in general. From a daily business point of view, complexity, uncertainty, and the importance of innovation management call for future research to reveal more details of both the visible and invisible processes that guide and influence innovation. As one size cannot fit all firms, in-depth studies in different settings will provide innovation managers with insights based on the specific networks that are already available or are still to be established.

In this study we have shown a method that uses network analysis tools to visualize the interactions between people. However, these tools do not reveal whether existing networking models or social capital models accurately describe the social relationships that occur within these networks. In contrast to existing social networking theory that encourages friendship development, technical networking may result in friendships, but is also more significantly linked to expanding one's technical depth. In the literature, it appears that loose networks stimulate innovation, especially in the idea-generation phase, and we would like to further explore the relevance of existing networking theory and social capital development in this context. In addition, it will be of interest to discover how networks could support firms' simultaneous pursuit of exploration and exploitation.

## 7   Conclusions

The main objective of this chapter was to describe some important operational aspects of intraorganizational and interorganizational networks as the basis for the creation of innovative ideas and collaborative activities. Innovation managers should be aware of social network analysis as a tool for visualizing actors and ties within and outside their organizations, and how formal and informal networks shape and influence innovation processes.

Though supported by a vast body of similar literature, this summary of three case studies extracted from industry was not aimed at providing empirical evidence or strict recommendations. Its purpose was simply to demonstrate how social network analysis can help identify organizational conditions and situations that play a role in innovation management.

- Example 1 demonstrated that informal networks between employees could be an efficient path to generating high-quality innovation ideas.
- Example 2 explored how two-mode network analysis can help in understanding how external workshops can support the implementation of collaborative projects.
- Example 3 provided a method for visualizing the composition and members of complex project teams and for identifying the interconnectivity of all the actors. In this study we were able to quantify the interactions between people by using a network analysis tool.

However, these case studies did not reveal whether existing social capital models are relevant in professional contexts, where the main aim is not to build mere relationships, but rather to establish contacts that can meet professional needs in terms of exploration and exploitation. This is an aspect that will be explored in the future.

# References

Abramovsky, L. Kremp, E., Lopez, A., Schmidt, T., & Simpson, H. (2009) Understanding co-operative innovative activity: Evidence from four European countries. *Economics of Innovation and New Technology*, 18(3), April, 243–265.

Adler, P. S., & Kwon, S. W. (2002) Social capital: Prospects for a new concept. *Academy of Management Review*, 27(1), 17–40.

Ahuja, G. (2000) The duality of collaboration: Inducements and opportunities in the formation of interfirm linkages. *Strategic Management Journal*, 21(3), 317–343.

Antikainen, M. Mäkipää, M., & Ahonen, M. (2010) Motivating and supporting collaboration in open innovation. *European Journal of Innovation Management*, 13(1), 100–119.

Axtell, C. M., Holman, D. J., Unsworth, K. L., Wall, T., & Waterson, P. (2000) Shopfloor innovation: Facilitating the suggestion and implementation of ideas. *Journal of Occupational and Organizational Psychology*, 73(3), 265–286.

Belderbos, R., Carree, M., Diederen, B., Lokshin, B., & Veugelers, R. (2004) Heterogeneity in R&D cooperation strategies. *International Journal of Industrial Organization*, 22(8–9), 1237–1263.

Bjerregaard, T. (2009) University-industry collaboration strategies: A micro-level perspective. *European Journal of Innovation Management*, 12(9), 161–176.

Björk, J., & Magnusson, M. (2009) Where do good innovation ideas come from? Exploring the influence of network connectivity on innovation idea quality. *Journal of Product Innovation Management*, 26(6), 662–67.

Borgatti, S. P., & Foster, P. C. (2003) The network paradigm in organizational research: A review and typology. *Journal of Management*, 29(6), 991–1013.

Brass, D. J., Glaskiewicz, J., Greve, J., & Tsai, W. (2004) Taking stock of networks and organizations: A multi-level perspective. *Academy of Management Journal*, 47, 795–817.

Caloghirou, Y., Vonortas, N. S., & Ioannides, S. (2002) Science and technology policies towards research joint ventures. *Science and Public Policy*, 29(2), 82–94.

Caloghirou, Y., Vonortas, N. S., & Ioannides, S. (2003) Research joint ventures. *Journal of Economic Surveys*, 17(4), 541–570.

Chesbrough, H. (2003) *Open Innovation: The New Imperative for Creating and profiting from Technology*. Boston: Harvard Business School Press.

Cohen, W. M., & Levinthal, D. A. (1990) Absorptive capacity: A new perspective on learning and innovation. *Administrative Science Quarterly*, 35(1), 128–152.

Coleman, J. S. (1988) Social capital in the creation of human capital. *American Journal of Sociology*, 94(Supplement), S95–S120.

Drexler, G. (2010) *Interne Netzwerke als Ideenpromotoren*. Master thesis, Danube University, Krems.

Dundon, E. (2002) *The Seeds of Innovation: Cultivating the Synergy that Fosters New Ideas*. New York: American Management Association.

Elmquist, M., Fredberg, T., & Ollila, S. (2009) Exploring the field of open innovation. *European Journal of Innovation Management*, 12(3), 326–345.

Faems, D., VanLoog, B., & Debackere, K. (2005) Interorganizational collaboration and innovation: Toward a portfolio approach. *The Journal of Product Innovation Management*, 22(3), 238–250.

Fetterhoff, T. J., & Voelkel, D. (2006) Managing open innovation in biotechnology. *Research Technology Management*, 49(3), 14–18.

Granovetter, M. S. (1973) The strength of weak ties. *The American Journal of Sociology*, 78(6), 1360–1380.

Hagedoorn, J. (2002) Inter-firm R&D partnerships: An overview of major trends and patterns since 1960. *Research Policy*, 31(4), 477–492.

Håkansson, H. & Ford, I. D. (2002) How should companies interact in business networks? *Journal of Business Research*, 55(2), 133–139.

Hardy, C., Lawrence, T. B., & Grant, D. (2005) Discourse and collaboration: The role of conversations and collective identity. *Academy of Management Review*, 30(1), 58–77.

Hargadon, A. B., & Bechky, B. A. (2006) When collections of creatives become creative collectives: A Fiels study of problem solving at work. *Organization Science*, 17(4), 484–500.

Heikkinen, M. T., Mainela, T., Still, J., & Tähtinen, J. (2007) Roles for managing in mobile service development nets. *Industrial Marketing Management*, 36(7), 909–925.

Horn, P. M. (2005) The changing nature of innovation. *Research Technology Management*, 48(6), 28–33.

Huisman, M., & vanDuijn, M. A. J. (2005) Software for social network analysis. In P. J. Carrington, J. Scott, & S. Wasserman (Eds), *Models and Methods in Social Network Analysis* (pp. 270–316). New York: Cambridge University Press.

Kaufmann, A., & Tödtling, F. (2001) Science-industry interaction in the process of innovation. *Research Policy*, 30(5), 791–804.

Kelley, T. (2001) *The Art of Innovation: Lessons in Creativity from IDEO, America's Leading Design Firm*. New York: Random House.

Koen, P., Ajamian, G., Burkart, R., Clamen, A., Davidson, J., D'Amore, R., Elkins, C., Herald, K., Incorvia, M., Johnson, A., Karol, R., Seibert, R., Slavejkov, A., & Wagner, K. (2001) Providing Clarity and a Common Language to the "Fuzzy Front End." *Research Technology Management*, 44(2), 46–55.

Kutschker, M. & Schmid, F. (2008) *Internationales Management*. München: Oldenburg Wissenschaftsverlag GmbH.

Lee, R. (2009) Social capital and business management: Setting a research agenda. *International Journal of Management Reviews*, 11(3), 247–273.

Madjar, N., Oldham, G. R., & Pratt, M. G. (2002) There's no place like home? The contributions of work and non-work creativity support to employee's creative performance. *Academy of Management Journal*, 45(4), 757–767.

March, J. G. (1991) Exploration and exploitation in organisational learning. *Organization Science*, 2, 71–80.

Marsden, P. V. (2005) Recent developments in network management. In P. J. Carrington, J. Scott, & S. Wasserman (Eds), *Models and Methods in Social Network Analysis* (pp. 8–30). New York: Cambridge University Press.

Obstfeld, D. (2005) Social networks, the tertius lungens orientation, and involvement in innovation. *Administrative Science Quarterly*, 50(1), 100–130.

OECD (2010) *The OECD Innovation Strategy: Key Findings*. Paris: OECD.

Ozman, M. (2009) Inter-firm networks and innovation: A survey of literature. *Economic of Innovation and New Technology*, 1(1), 39–67.

Perry-Smith, J. E. (2006) Social yet creative: The role of social relationships in facilitating individual creativity. *Academy of Management Journal*, 49(1), 85–101.

Perry-Smith, J. E., & Shalley, C. E. (2003) The social side of creativity: A static and dynamic social network perspective. *Academy of Management Review*, 28(1), 89–106.

Philbin, S. (2008) Process model for university-industry research collaboration. *European Journal of Innovation Management*, 11(4), 488–521.

Putnam, R. D. (2000) *Bowling Alone*. New York: Simon & Schuster.

Pyka, A., & Küppers, G. (2002) *Innovation Networks: Theory and Practice*. Cheltenham: Edward Elgar Publishing.

Reinholt, M., Pedersen, T., & Foss, N. J. (2011) Why a central network position isn't enough: The role of motivation and ability for knowledge sharing in employee networks. *Academy of Management Journal*, 54(6), 1277–1297.

Roberts, B. E. (2007) Managing invention and innovation. *Research Technology Management*, 50(1), 35–54.

Rothwell, R. (1992) Developments towards the fifth generation model of innovation. *Technology Analysis and Strategic Management*, 4(1), 73–75.

Simard, C., & West, J. (2006) Knowledge networks and the geographic locus of innovation. In H. Chesbrough, W. Vanhaverbeke, & J. West (Eds), *Open Innovation: Researching a New Paradigm*. Oxford: Oxford University Press.

Sternberg, R. J., & Lubart, T. I. (1996) Investing in creativity. *American Psychologist*, 51(7), 677–688.

Suseno, Y., & Ratten, V. (2007) A theoretical framework of alliance performance: The role of trust, social capital and knowledge development. *Journal of Management & Organization*, 13(1), 4–23.

Teasley, S. D., & Roschelle, J. (1993) Constructing a joint problem space. In S. P. Lajoie, and S. J. Derry (Eds), *The Computer as A Cognitive Tool*. Hillsdale: Lawrence Erlbaum.

Tidd, J., Bessant, J., & Pavitt, K. (2005) *Managing Innovation: Integrating Technological, Market and Organizational Change* (3rd edn). Chichester: Wiley & Sons Ltd.

Tsai, W. (2001) Knowledge transfer in intraorganizational networks: Effects of network position and absorptive capacity on business unit innovation and performance. *Academy of Management Journal*, 44(5), 996–1004.

Uzzi, B. (1997) Social structure and competition in interfirm networks: The paradox of embeddedness. *Administrative Science Quarterly*, 42, 35–67.

Verburg, R. M., & Andriessen, E. J. H. (2011) A typology of knowledge sharing networks in practice. *Knowledge and Process Management*, 18(1), 34–44.

Wasserman, S., & Faust, K. (1994) *Social Network Analysis*. New York, Cambridge University Press.

Wenger, E., & Snyder, W. (2000) Communities of practice: The organizational frontier. *Harvard Business Review*, 78(1), 139–145.

Zhou, S., Siu, F. & Wang, M. (2010) Effects of social tie content on knowledge transfer. *Journal of Knowledge Management*, 14(3), 449–463.

# 7

## The Evolution of Mobile Social Networks through Technological Innovation

*Vanessa Ratten*

## 1  Introduction

Communication technologies have encouraged the acceleration of information flows as people share information and interact across temporal and geographic boundaries with their mobile phones (Humphreys 2010). People use mobile technology to connect with people both locally and globally through their social networks. Despite the increase in mobile communications, which enable social interaction to take place in any geographic and time location, many mobile phone users are socially selective in the people with whom they communicate (Matsuda 2005). This selectivity has enabled mobile social networks to be used to build social ties by enabling members to access their network of friends, which rearranges social practices (Humphreys 2008). This change in social practices has influenced people to capitalize on their tendency to reinforce bonds among existing social ties (Sutko & Silva 2010). Social ties are increased by mobile social networks, which have the advantage of offering potential value creation to businesses that seek like-minded individuals that are in the current Internet-led communication revolution (Varnali et al. 2010). In addition, there is a social and technological shift in the ability of mobile communication devices to help build and connect new relationship ties (Sutko & Silva 2010). Humphreys (2008, p. 342) states that "new services for mobile phones have been developed that purport to create, develop and strengthen social ties" as a result of technological advancement. Therefore, recent mobile phone software has been designed to help people socially network in different geographic locations by encouraging people to build valuable contacts to source and share information or resources (Boyd 2004).

The use of new media such as mobile communications can shed light on how the social environment consumes technology (Humphreys 2007). This is because new communication technology reflects the cultural climate existing in the environment (Williams 1975). Communities often use new technologies to solve time and space issues by reinvesting concepts of social trust (MacKenzie & Wajcman 2002). The role of trust is influenced by the early adopters of a technology in a community, who are social leaders that shape future use of the technology (Bijker et al. 1987). An understanding of how a technology can influence others is needed to help encourage other adopters (Rogers 1995). However, the adoption of technological innovations is a complex social process that can differ among people, and is based on the product life cycle (Preece 2004).

The mobile phone is a technological innovation that has transformed modern cities into real-time systems that constantly monitor environmental conditions important to the operation of the structure (Townsend 2000). Mobile technology allows micro systems to exist, as not everyone with a mobile phone wants to be accessible to other people at all times. Moreover, although there has been research on mobile phone usage to conduct a variety of transactions from banking to paying parking fines, only recently has research begun to focus on mobile social communications (Lenhart 2009).

Mobile social networking acts as a form of micro-coordination to inform multiple participants about a particular action or plan of attack (Ling & Yttri 2002). Mobile social networks were designed to facilitate sociality in public spaces by focusing on the spatial practices of the technology. Certeau (1984) refers to spatial practices as the everyday lived experience of and movement through social and physical space. By looking at the common practices associated with mobile social network usage, the perceived effects of the technological adoption can be better understood. There is a gap in the literature on how technological innovations are adopted by mobile social networks, which this chapter aims to fill. As little is known in the technology innovation literature about the use of mobile social networks to adopt new interactive technology such as Twitter and Facebook, this chapter aims to review the literature on mobile social networks and technology innovation behavior in order to propose a set of research propositions to understand how mobile social networks are adopted by individuals. Therefore, the research question that this chapter seeks to address is:

How are mobile social networks adopted by users of mobile social communications?

This chapter begins by discussing the literature on mobile communications and technology innovation. I examine the research on mobile social networks and I look at the interdependency of social networks and mobile technology. I discuss the use of mobile social networks as a way to facilitate technology marketing and to capture the communication that exists in mobile technology. I conclude by highlighting the practical implications for technology marketers and I discuss how future research might further address new technological innovations in mobile communication.

## 2 Literature review

Technological innovations involve a set of processes undertaken to generate a new idea, product, or service (Cantisani 2006). Some technological innovations involve an alternative use of an idea or object that can change existing services (Rycroft 2006). These ideas are created by knowledge changes that make innovations come into the market faster (Griffin 1997). As the time taken to develop technological innovations has become shorter, more organizations are now competing, based on their social network capabilities. Some organizations are making substantial changes in their structures and design processes to encourage better implementation of technology in the marketplace (Ehrnberg 1995). Some companies take the approach of facilitating scientific developments by encouraging collaboration with individuals to lessen expenditures of time and monetary resources (Ehrnberg 1995). As the product life cycle for technologies has become shorter, companies have focused on constant change as a way to stay innovative (Lundvall & Boras 1999). This renewed interest in the capabilities of technology innovation has lead to consumers developing an interest in the commercialization of products as a strategic tool used by companies to facilitate economic progress (Scholnikoff 2001).

The process of adopting a technology involves purchasing a product or service and then using it (Revels et al. 2010). This process has emotional outcomes, as it affects a person's involvement with and satisfaction with the technology (Schouten & McAlexander 1995). The emotional connections people have to objects are linked to their behaviors and intentions to adopt new products or services (Thomson et al. 2005). In addition, emotional connections are important indicators of how people adopt technology as the connections involve their commitment to learning about the technology (Schouten & McAlexander 1995). People are emotional because of the attachment they have with regard to other people, products, and services, which influences the rate at which they socially network with other individuals about mobile communication technologies. These emotions involve people experiencing a feeling about something that guides their thoughts and actions (White et al. 2005). Therefore, these emotions can relate to past or future outcomes that influence an individual's behavior (Bagozzi & Dholakia 1999).

Past research by Ha et al. (2007) and Gruen (2009) indicate that emotional attachment to technology is an important component of understanding the adoption process. The perception of how useful a technology is to an individual is linked to the person's evaluation of the benefits of the technological innovation (Davis et al. 1989). These benefits include the willingness of a person to use a technology, as that will be partly determined by the sense of emotional attachment to the innovation (Thomson et al. 2005). The relationship between a person's emotions and behavior was highlighted by Schouten and McAlexander's (1995) research on Harley Davidson motorbike sellers, who found it hard to sell their motorbikes because of their emotional connection with both the product and the brand name.

Technology marketing is a feature of social networks as it encourages people to share their experiences about a technological innovation (Snowden et al. 2006). As technology changes at a rapid rate, the more marketing a person receives about a technology, the more inclined the person is to adopt it (Lin et al. 2007). Technology marketing is part of the theories existing in the literature that explain the technology adoption process. The main technological innovation theories are the stages model, the technology acceptance model, the theory of planned behavior, the theory of reasoned action, and social cognitive theory, which I will now discuss.

*The stages model* was one of the first theories to explain the adoption process of a technology; it focuses on the different steps a person goes through in receiving information and then using a product or service (Rogers & Beal 1958). The stages of the adoption process are: awareness, information, application, trial, and adoption (Beal et al. 1957). More recent applications of the stages model have included emotional and contextual factors to highlight how the technological adoption process is developmental and complex. The stages model assumes that there are certain components of the adoption process that involve a person's increasing his or her understanding of the benefits of a technology.

*The technology acceptance model (TAM)* was developed after the stages model to counteract criticism of the stages model for being inflexible and too deterministic in its approach (Ajzen & Fishbein 1980). In addition, the stages model was developed before the invention of the Internet. The TAM originated at the same time the personal computer was gaining in popularity and was first used in the context of technology, unlike the stages model, which was first applied to study the health sector (Davis 1986). The TAM incorporates three variables (perceived usefulness, perceived ease of use, and attitude towards using), which affect a person's behavioral intention. The TAM was developed before the popularity of mobile social networking, in order to understand the variety in adoption behavior caused by human behavior differences (Yousafzai et al. 2010). The TAM is a simplistic model that focuses on the usefulness of a technology (Venkatesh & Davis 2000). The TAM has been used to explain the adoption behavior of technology innovations, including Internet banking (e.g., Yousafzai et al. 2010), mobile services (e.g., Revels et al. 2010), and consumer technology products (e.g., Kulviwat et al. 2007). The weakness of the TAM is that it does not incorporate cognitive aspects of a person's behavior nor the antecedent factors, such as age and experience, that influence the adoption process (Ratten

2009). Although the TAM is a widely validated model in the technology innovation literature when used to explain and predict a person's behavior, it still does not take into account social processes that are beyond a person's control (Venkatesh & Davis 2000).

The theory of reasoned action (TRA) extends the TAM by including the variable subjective norm to show how a person's behavior is influenced by outside attitudes. The TRA includes five variables (*beliefs and evaluation, attitude toward the behavior, subjective norm, normative beliefs,* and *motivation to comply*) that affect a person's behavior and decision to perform an action (Fishbein & Ajzen 1975). The TRA is used to explain human behavior by understanding how a person's perception and intention will lead to a change in the person's behavior (Bagozzi 1992). The TRA assumes that people are systematic in their behavior and decisions about whether to adopt a technological innovation. Recent research by Lippert and Forman (2005) adopted the TRA to help explain the adoption behavior of supply chain members in the United States automobile industry. They found that prior experience with a similar technology, prior technical knowledge, opportunity to experiment, and training effectiveness affected the supply chain members' perceived ease of use and perceived usefulness of the technology.

The theory of planned behavior (TPB) is an adaptation of the TRA, but argues that a person's behavior is preplanned and strategically executed. The TPB includes an additional variable not included in the TRA, which is perceived behavioral control (Ajzen 1991). Perceived behavioral control is an indicator of the resources a person has that influence the person's behavior. A person's behavior is related to his or her perceived behavioral control, which influences how they behave and act in different circumstances and is a result of reflex action. Research by Morris and Venkatesh (2000) used the TPB to examine how older workers are influenced by perceived behavioral control in their decision to adopt a technological innovation in their workforce. In a similar vein, Karahanna et al. (1999) found that preadoption attitude towards Windows technology was a result of a consumer's perceptions of usefulness, ease of use, result demonstrability, visibility, and trialability.

Social cognitive theory incorporates the TAM, TRA, and TPB by providing a comprehensive understanding of how a person adopts technology. Social cognitive theory focuses on the reciprocal relationship between a person's environment and the person's behavior (Bandura 1986). It focuses on both current and future behavior of a person, which is the result of the learning process. Social cognitive theory acknowledges that behavior is sometimes not under a person's control, but rather is a result of events or occurrences that influence a person (Compeau et al. 1999). A person's behavior is affected by the social environment, which determines the person's actions. With technological innovations, the important influencers of this behavior are a person's skills, confidence, and ability to adopt a technology (Compeau et al. 1999). Social cognitive theory is adopted in this chapter as the theoretical framework, as it incorporates factors affecting both a person's internal and external environment, which the other technological innovation theories do not include. In addition, as this chapter focuses on the intention of a person to adopt mobile social networks as a result of the person's mobile communications, mobile networks, and mobile social communities, social cognitive theory is useful as it is a learning theory that describes the process and networks a person goes through in learning about his or her social environment. The next section will discuss in more detail the role mobile technology has in creating social networks for the adoption of technological innovations like mobile social networks.

## 3  Mobile technology

Mobile communication technologies connect people to spaces and to other individuals (Sutko & Silva 2010). This connection is enabled by mobile technologies having synchronous mediated communication while the person is in transit, which other technologies could not easily accomplish (Humphreys 2010). Mobile phone users are free to communicate private information as they are not bounded by a specific location (Tomita 2005). Mobile phone users have been found to be comfortable sharing intimate communication details in public spaces because the other people there are anonymous (Fortunati 2002). In addition, mobile technologies enable people to coordinate congregating in a social space by a process of redirection. "Redirection" refers to a person's ability to act on information obtained through the mobile phone to change the progress of his or her current trip (Ling & Yttri 2002).

Sutko and Silva (2010) argue that mobile phones act as a technology filter by helping users manage their interactions in social locations. A person's geographic location in a social network can be expanded through mobile communications. Silva (2009) found that people seek more opportunities to play computer games when they travel to unfamiliar locations to find other players. In another study, Licoppe and Inada (2006) argue that computer game players make excursions with other players in order to be brought physically together to support their online socialization efforts. Ling (2008) also highlighted in his research that some mobile phone users reinforce their social networks by primarily communicating with friends. This form of bounded solidarity to connect with family and friends can be to the detriment of unknown people (Sutko & Silva 2010).

Mobile technology changes public spaces by accelerating the exchange of information in a real-time format (Townsend 2000). Mobile technology can be used in a number of social spaces, the three main types being public, private, and parochial (Humphreys 2010). Public spaces are open to anyone and are usually characterized by a group of unknown people using the same geographic area. Private spaces are areas that are for a specific group of people, which includes intimate acquaintances and personal networks (Humphreys 2010). Parochial spaces include neighborhoods that operate in a geographic area that has a sense of commonality (Lofland 1998). Humphreys (2010, p. 768) defines this process as parochialization as it is "the process of creating, sharing and exchanging information, social and locational, to contribute to a sense of commonality among a group of people in public space." Social spaces involve different types of people in interpersonal networks that operate in the same realm and are highly contextual, depending on the usage of the space.

A public space is a nondomestic physical site that can be accessed by anyone (McCarthy 2001). In a public space, people experience their space based on interactions with others, which can influence information and communication shared (Humphreys 2008). Public spaces are sometimes referred to as "third places," as they are places beyond the home or workplace in which people congregate (Oldenburg 1991). Part of urban socializing is conducted in public spaces, as it allows for informal and voluntary communication. In public spaces, mobile communication can occur through media that act as a connection between technology and space (Couldry and McCarthy 2004). Mobile social networks are important builders of social spaces as they enable the sharing of information that contributes to a sense of familiarity among users (Humphreys 2008). In addition, mobile social networking is conducted through communication technology in public spaces by encouraging virtual pathways of contact (Humphreys 2007).

Mobile social communication is helpful to redirect a person's path, depending on the social and locational information available. The way mobile social communication is transmitted depends on travel distance, time, and spatial orientation. The travel distance indicates whether it would be feasible to move to a different locations based on geographic area. Timing is important as it determines when a person receives a message and whether the person can immediately act on its contents. Spatial orientation refers to how close the person who sent the message is to the recipient and whether the sender has any first-person direct knowledge of the message, or if it has been received through a third party.

Mobile communication is a complex process that can facilitate and promote social practices (Humphreys 2010). Therefore, it is important to explore how mobile communication is interconnected with space and people (Sheller 2004). Mobile communication blurs the boundaries between public, private, and parochial spaces, as it can be used in almost any geographic location (Hoflich 2006). Computer-mediated communication is used to overcome time and space (Castells 2000). In addition, new mobile communication technologies have altered the relationship between communication and interaction by acting as sociospatial transmitters of information (Wilken 2008).

## 4  Mobile communications

Mobile communications are usually with the users, which means the users can stay connected with their direct social acquaintances (Lugano 2008). The location awareness feature or global positioning system on mobile communications provides information that can be shared at the user's discretion with his or her circle of social contacts (Rheingold 2003). As the number of people with mobile phones is greater than the number of people with Internet access, this has important implications for marketing (Shannon 2008). Mobile phones are now socially accepted as everyday necessities, and technological innovations such as the connection of mobile phones to the Internet have increased in popularity (Kleijnen et al. 2009). Mobile communication enables people to be in perpetual contact by promoting social interactions in a person's social environment (Katz & Aakhus 2002).

Most of the existing literature on mobile phones is on how people maintain pre-existing social connections (e.g., Katz & Aakhus 2002; Ito et al. 2005) rather than on promoting new social networks. Some research suggests that mobile phones have encouraged atomization and privatization rather than facilitating interconnectivity among people (Puro 2002). Therefore, this leads to the first proposition:

Proposition 1: The greater a person's knowledge of mobile communications, the more likely is their intention to adopt mobile social networks.

## 5  Mobile social software

Mobile social software can be web-based or client-based. Client-based interfaces require a specific type of software to be installed on a person's mobile phone, whereas web-based interfaces are based on the connection to the Internet via a mobile phone. Mobile social software has increased in recent years; Apple's iPhone introduced a range of applications or "apps" that people can either download for free or purchase to use on their mobile phone. In addition, mobile phones have become more complex, with emergent technologies such as sensor-based systems (e.g., touch) and media-based systems (e.g., video) becoming available. Mobile social software has advanced to include multimedia communication that facilitates interpersonal mobile communication by providing media-sharing services and real-time formats. This form of community interaction enables people to personalize the information they share on social platforms based on their privacy settings and personal circumstances.

Mobile social software has different functionalities and capabilities, which can be adapted to the needs of the person. It can store information such as contact details, update a diary, share geographic locations and status updates, read barcodes to check prices on products on the Internet, and keep track of shops or restaurants in an area. Mobile social software has also become more advanced, with people posting messages online, recording videos and sending them to friends, and receiving directions to a suitable location. This has allowed people to shop more smartly as they have access to pricing data via the Internet, which saves them time and money. In addition, via Facebook profiles, mobile social software has enabled advertisements, such as coupons via Groupon, to be personalized based on a person's profile. Web 2.0 applications have also promoted mobile social software, as they are websites that have participatory information-sharing facilities that enable people to create content and engage with others, as opposed to websites that are created by others that cannot be updated or changed. Therefore, this leads to the next proposition:

> Proposition 2: The more knowledge a person has about the usefulness of mobile social software, the greater the person's intention to adopt mobile social networks.

## 6   Mobile social networks

Mobile social networks enable people to combine their real and digital lives through a mobile communication medium. Mobile social networks are defined in this chapter as the patterns of interconnection among users emerging from the social use of mobile devices (Lugano 2008). Mobile devices enable a person to establish a direct link among communicating parties (Varnali et al. 2010). This communication allows people to communicate in a flexible way that gives them the opportunity to be connected with others as they deem necessary (Balasubramanian et al. 2002).

Mobile phone usage is integrated into everyday life and facilitates communication that is social and unexpected (Lofland 1998). Mobile social networks change the way people communicate as their experience and engagement with the technology is adapted to suit changing environmental factors. Mobile social networks are a realistic technology as they enable public spaces to be more social (Humphreys 2010). Some people suffer from social insularity that can be overcome through mobile technologies. However, mobile social networks can operate to narrow a person's social circle rather than extending it, because people tend to focus on their same demographics (Crawford 2008). A new usage of mobile social networks is to provide social molecularization, which occurs when people move around a city in a collective group (Humphreys 2007).

Mobile social networks have a number of advantages over other technology devices as they enable real-time communication to occur in a direct and accessible format. People can share geographic location information through mobile social networks that enable them to interact in a more open format. In addition, mobile social networks are important mechanisms for a person to share knowledge with others. In a seminal paper, Granovetter (1973) referred to social networks as helping a person establish interpersonal ties based on shared interests, emotions, and reciprocity. The ties can be strong, weak, or absent based on the type of people involved in the social network. In addition, research by Wasserman and Faust (1994) states that a social network comprises a set of users and the relationships that exist among them. Social networks often are formed through the Internet at websites such as Facebook and LinkedIn, which promote social interaction and connectivity. Social network websites operate as a web-based service that enables people to create a profile based on their interests in order to connect with other users in the system (Boyd & Ellison 2008). Mobile social networks go one step further than other types of social networks, as they enable a person to communicate in any location through a bounded system with other people with whom they share a connection. Some websites are now created in a mobile-friendly way, as they allow people to access their Internet-based social network sites onto the mobile medium.

Mobile social software has been created with the specific purpose of supporting social interaction among people connected to the Internet via their mobile phone (Lugano 2007). Mobile social software enables people to transfer information on their mobile phones via text messages, photos (e.g., Flickr), ratings, videos (e.g., Youtube), tags, and e-commerce sites (e.g., eBay). Early types of mobile social networks were supported by e-mail and instant messaging services, which helped people stay in touch.

Mobile social networks enable people to learn from information shared. The information disseminated on a social network will depend on people's level of integration and connectedness. A well-integrated person in a social network is likely to have access to more information from a variety of people in his or her community (Rogers & Kincaid 1981). This integration enables information to be accessible from a person's network in a timely manner, depending on how regularly a person communicates with others (Lievens et al. 1999). The position of people in their social network will be related to the connection they feel in accessing information (Rogers & Kincaid 1981). Moreover, the number of people a person communicates with via his mobile social network will create a source of knowledge that can be applied when needed (Kleijnen et al. 2009). Information dissemination via a person's social network is an important part of creating a reliable mobile community. A person is likely to regard information as higher quality when it is based on user trust and experience of people in the person's mobile social networks (Zhou et al. 2010). Mobile social network systems can determine both information and system quality, which includes the reliability, response rate, and ease of use (Wixom & Todd 2005). The level and quality of information in a person's mobile social network determines her future usage, but also determines the accuracy and comprehensiveness of information provided by a person's mobile social network (Nelson & Todd 2005).

People require information about products and services before they are willing to use those (Moorman et al. 2004). In the context of social networks, the more information people share as members of a mobile social community, the more messages and information they will receive (Counts & Fisher 2008). However, the information received by a person is influenced by a person's anxiety about sharing personal stories because of privacy concerns (Sadeh et al. 2009). This insecurity about information sharing is likely to change as a person finds the technology easier and more useful to use (Stern et al. 2008). This can lead to a person allowing more information to be downloaded and actively engaging in sourcing new mobile social network contacts.

Previous studies on Short Message Service (SMS) usage among people with mobile phones have indicated that personality-based characteristics affect a person's mobile social networking ability. For example, researchers such as Mahatanankoon (2007) found that SMS usage is dependent on a person's perception of playfulness. In addition, Okazaki (2009) found that opinion leadership influences a group's social intention to engage in mobile (electronic) word of mouth. In a similar vein, Zhang and Mao (2008) state that a person will trust SMS advertising more when she perceives it to be useful. Personality characteristics affect a person's mobile social network by encouraging messages to be sent based on different moods and behavior (Hui et al. 2009). The more innovative the mobile service is, the more likely that the perceptions of the altruistic message captured by the social network will be highly regarded (Butt & Phillips 2008). Messages that are based on self-disclosures of a person to his mobile social community will be a factor in facilitating information to be forwarded to relevant parties (Lederer et al. 2003).

Mobile social networks are important information sources of social, performative, and cataloging of behavior (Humphreys 2010). Social usages include sharing, maintaining, and discovering other people who share a common interest (Humphreys 2008). Entertainment may comprise part of social usage when social media helps facilitate face to face or online meetings of a person's social circle. Mobile social community members may be motivated by performative functions that allow users to associate their identity with other users in their network, and cataloging functions enable people to document and see others' social history, which builds up an online portfolio of interests. In cataloging social activities, a person can keep track of others in her social network by sharing their experiences. It also enables a person to increase his effectiveness at being noticed by others, in order to enhance his online reputation, as an online community is a social network of people who interact through online material in order to pursue common interests that transcend geographic locations. This may affect a person's ability to adopt mobile social network technology innovation. Therefore, the next proposition is:

Proposition 3: The more information disseminated through a person's social contacts, the more likely the person will be to adopt mobile social networks.

## 7 Mobile online communities

Mobile online communities are formed by people with a common goal and interest. Members of these communities usually engage in similar activities, which help technology marketers predict common interests. Some marketing firms monitor mobile online social communities as a way of determining user interest and evaluating people's behaviors. Mobile online communities enable firms to measure and keep track of social information that can be strategically used as a source of community-level activity. Communities are formed online as an outlet for a person's social needs and the technological advantages of communicating in a flexible format (Rheingold 2003). Online environments differ from other traditional communication mediums such as face-to-face interactions as they provide a controlled environment in which people can interact (Rheingold 2003).

Mobile social networks are facilitated by multimedia-rich environments in which people organize mobile online communities (Lugano 2008). These communities support social networks by exchanging social contacts with contextual information. Community types existing on mobile social networks are also called online, virtual, and electronic communities. Mobile social networks usually start in young demographics, but are then transferred to older population segments of the market.

The first adopters of mobile social networks are usually technology innovators (Humphreys 2008). These innovators in most cases live in major urban areas and are geographically close to other social network users. People have different expectations and usages of mobile social networks, depending on their demographics and use of innovations (Humphreys, 2008). Some people use mobile social networks as a source of support in facilitating online discussions. These discussions can include promoting a particular product or service that is relevant to the interests and goals of the online community. A mobile online community can act to elevate their members' interests in a visible way that serves as a source of success. By encouraging community participation, the behavioral patterns and norms of the social community can be used in a beneficial manner such as funding a non-profit organization, depending on the source of the community's actions or decisions. This community participation enables decisions to be made quickly that connect users regardless of their location. Therefore, this leads to the next proposition:

> Proposition 4: The greater a person's level of community participation, the more likely they will be to adopt mobile social networks.

## 8   Managerial implications

There is a promising link between mobile social network adoptions and innovation development. The adoption of mobile social networks will increase the use of technological innovations to generate, distribute, and support new ideas. In addition, increasing a person's mobile social networks will enable more companies to promote sales through contact lists that people use on their Twitter and Facebook accounts. This means that there are a number of considerations that managers could take into account when seeking to identify opportunities through mobile social networks. First, they should encourage consumers to participate in the development of mobile applications that can be used to target potential buyers of products and services. Second, by focusing on the commitment and interest of consumers about news and updates relating to their products and services, managers can focus better on achieving source creditability. Third, better functionality of technology marketing materials can be made by stressing the ease of use and usefulness of mobile social networking websites.

For some consumers, the usefulness and emotional attachment to their mobile phone will be an indicator of how quickly they adopt technological innovations. From a management perspective, an understanding of a person's attachment to mobile communications may aid in the development and marketing of new technologies such as mobile social networks. For example, the type of mobile phone people use relates to whether they can adopt mobile social networks to promote social networking. Future advances in mobile communications should endeavor to employ mobile social networks for more usages in order to facilitate higher adoption rates. In the case of mobile social networks, the advantage of this technological innovation might be expressed in terms of convenience and access to information that is only available on mobile phones.

When a people perceive a technology as straightforward to use, they are more likely to anticipate enjoying it (Bruner & Kumar 2005). This chapter highlighted how people's attitude towards technology innovations like mobile social networks is associated with their adoption intentions. Therefore, it is important for managers to focus on both the advantages and disadvantages of mobile social networks as a source of mobile communications, but also to think about how product design and marketing communications can encourage mobile social networks and online communities.

## 9   Future research suggestions

The implications of this chapter are tempered by some limitations. The chapter proposed a set of research propositions in a single service context, that of mobile social networks, which limits their generalizability. This means that future research should replicate the research propositions in this chapter in other technology services to see if they also apply to the development and sustainability of other mobile social networks. Another limitation is that a person's actual adoption behavior was not examined, but rather his or her intentions to adopt mobile social networks. Therefore, the actual use of mobile social networks should be examined to see how regularly a person makes use of the technology.

This chapter suggests several research trajectories on the usage of mobile social networks to promote mobile social networking. Future research could empirically examine the molecularization process of mobile social networks to see if it differs from or is similar to other types of mobile communication technology. It is important to encourage more research on the adoption rates of mobile social networks and to explore how people exchange information. The usage of websites and online community forums to facilitate mobile social networking might be understood better through current mobile technology such as the use of smartphones and tablets, which facilitate easier communication amongst people.

Future research should incorporate situational factors that affect the adoption of mobile social networks. People who use mobile social networks may mostly include social settings such as cafes as they enable mobility of their mobile phone. In addition, different types of people such as innovators or laggards could be compared to see if there is a difference in the inclination to adopt mobile social networks. Given the rapid technological innovations occurring in mobile phones, it would be useful to study the future of mobile phones, due to the expansion of multipurpose devices such as the iPad and Kindle.

Future research questions to be raised include: What are the advantages of mobile social networks in increasing a person's social network? Do people from different cultures and socioeconomic backgrounds utilize mobile social networking in the same way? In addition, future research should take into consideration the different social contexts of mobile communications (e.g., university, work, restaurant) and geographic settings (e.g., city, country). Cross-sectional studies completed over a longer time frame would illustrate how mobile communication technologies affect a person's social network capabilities. Issues concerning privacy should be explored in future studies of mobile social networks, and the development of contacts through mobile social networks is a potentially rich topic for future study. As more uses for mobile social networking develop, the ways in which people adopt technological innovations like mobile social networks will continue to be an important area of research activity.

## 10 Conclusions

The aim of this chapter was to discuss how mobile social networks are adopted to a mobile communication device. The research propositions stated in this chapter highlight several important findings and implications for managers. A person's usage of online communities contributes to his or her intention to use mobile social networks. This chapter has discussed the current academic literature on mobile communications by focusing on mobile social networks. The research on how technological innovations like mobile social networks are adopted in society was discussed. The role of computer-mediated communication was assessed, with a focus on the social implications of technology innovations. As smartphone technologies are becoming more popular, this chapter has highlighted the business potential in providing mobile social networks.

## References

Ajzen, I. (1991) The theory of planned behavior. *Organizational Behavior & Human Decision Processes*, 50(2), 179.

Ajzen, I., & Fishbein, M. (1980) *Understanding Attitudes and Predicting Social Behavior*. Prentice-Hall: Englewood Cliffs, NJ.

Bagozzi, R. P. (1992) The self-regulation of attitudes, intentions, and behavior. *Social Psychology Quarterly*, 55(2), 178–204.

Bagozzi, R. P. and Dholakia, U. (1999) Goal setting and goal striving in consumer behavior. *Journal of Marketing*, 63(4), 19–32.

Balasubramanian, S., Peterson, R. A., & Jarvenpaa, S. L. (2002) Exploring the implications of m-commerce for markets and marketing. *Journal of the Academy of Marketing Science*, 30(4), 348–361.

Bandura, A. (1986) *Social Foundations of Thought and Action: A Social Cognitive Theory*. Prentice-Hall: New Jersey.

Beal, G. M., Rogers, E. M., & Bohlen, J. M. (1957) Validity of the concept of stages in the adoption process. *Rural Sociology*, 22(2), 166–168.

Bijker, W. E., Hughes, T. P., & Pinch, T. J. (Eds) (1987) *The Social Construction of Technological Systems: New Directions in the Sociology and History of Technology*. Cambridge, MA: MIT Press.

Boyd, D. (2004) Friendster and publicly articulated social networking. In *Proceedings of the ACM CHI Conference on Human Factors in Computing Systems* (pp. 1279–1282). Vienna, Austria: ACM Press.

Boyd, D., & Ellison, N. B. (2008) Social network sites: Definition, history, and scholarship. *Journal of Computer-Mediated Communication*, 13, 210–230.

Bruner II, G. C., & Kumar, A. (2005) Explaining consumer acceptance of handheld Internet devices. *Journal of Business Research* 58(5), 553–558.

Butt, S., & Phillips, J. G. (2008) Personality and self reported mobile phone use. *Computers in Human Behavior*, 24, 346–360.

Cantisani, A. (2006) Technological innovation processes revisited. *Technovation*, 26(11), 1294–1301.

Castells, M. (2000) *The Rise of the Network Society* (2nd edn). Malden, MA: Blackwell.

Certeau, M. de (1984) *The Practice of Everyday Life*, trans. S. Rendall. Berkeley: University of California Press.

Compeau, D., Higgins, C., & Huff, S. (1999) Social cognitive theory and individual reactions to computing technology: A longitudinal study. *MIS Quarterly*, 23(2), 145–158.

Couldry, N., & McCarthy, A. (Eds) (2004) *MediaSpace: Place, Scale, and Culture in a Media Age*. London: Routledge.

Counts, S., & Fisher, K. E. (2008) Mobile social networking: An information grounds perspective. In *Proceedings of the 41st Hawaii International Conference on System Sciences*.

Crawford, A. (2008) Taking social software to the streets: Mobile cocooning and the (an)erotic city', *Journal of Urban Technology*, 15(3), 79–97.

Davis, F. D. (1986) *A Technology Acceptance Model for Empirically Testing New End-User Information Systems: Theory and Results*, Unpublished Doctoral dissertation, Massachusetts Institute of Technology.

Davis, F. D., Bagozzi, R. P. and Warshaw, P. R. (1989) User acceptance of computer technology: A comparison of two theoretical models. *Management Science*, 35(8), 982–1003.

Ehrnberg, E. (1995) On the definition and measurement of technological disconti-nuities. *Technovation*, 15(7), 437–452.

Fishbein, M., & Ajzen, I. (1975) *Belief, Attitude, Intention, and Behavior: An Introduction to Theory and Research*. Reading, MA: Addison-Wesley.

Fortunati, L. (2002) Italy: Stereotypes, true and false. In J. E. Katz & M. Aakhus (Eds), *Perpetual Contact: Mobile Communication, Private Talk, Public Performance* (pp. 42–62). New York: Cambridge University Press.

Granovetter, M. (1973) The strength of weak ties. *American Journal of Sociology*, 78(6), 1360–1380.

Griffin, A. (1997) PDMA research on new product development practices: Updating trends and benchmarking best practices. *Journal of Product Innovation Management*, 14(6), 429–458.

Gruen, A. (2009) The implications of denying fear and anxiety and idealizing victim-izers. *Journal of Psychohistory*, 37(1), 59–66.

Ha, I., Yoon, Y. and Choi, M. (2007) Determinants of adoption of mobile games under mobile broadband wireless access environment. *Information & Management*, 44(3), 276–286.

Hoflich, J. (2006) The mobile phone and the dynamic between private and public communication: Results of an international exploratory study. *Knowledge, Technology and Policy*, 19(2), 58–68.

Hui, P., Xu, K., Li, V. O. K., Crowcroft, J., Latora, V., & Lio, P. (2009) Selfishness, altruism and message spreading in mobile social networks. In *Proceedings of NetSciCom'09 First IEEE International Workshop on Network Science For Communication Networks*. Cambridge Press.

Humphreys, L. (2007) Mobile social networks and social practice: A case study of dodgeball. *Journal of Computer-mediated Communication*, 12(1), Url: http://jcmc.indiana.edu/vol13/issue1/humphreys.html

Humphreys, L. (2008) Mobile social networks and social practice: A case study of Dodgeball. *Journal of Computer-Mediated Communication*, 13, 341–360.

Humphreys, L. (2010) Mobile social networks and urban public space. *New Media and Society*, 12(5), 763–778.

Ito, M., Okabe, D., & Matsuda, M. (Eds) (2005) *Personal, Portable, Pedestrian: Mobile Phones in Japanese Life*. Cambridge, MA: MIT Press.

Karahanna, E. D., Straub, W., & Chervany, N. L. (1999) Information technology adoption across time: A cross-sectional comparison of pre-adoption and post-adoption beliefs. *MIS Quarterly*, 23(2), 183–213.

Katz, J. E., & Aakhus. M. (Eds) (2002) *Perpetual Contact: Mobile Communication, Private Talk, Public Performance*. New York: Cambridge University Press.

Kleijnen, M., Lievens, A., Ruyter, K., & Wetzels, M. (2009). Knowledge creation through mobile social networks and its impact on intentions to use innovative services. *Journal of Service Research*, 12(1), 15–35.

Kulviwat, S. Bruner, G. C. Kumar, A., Nasco, S. A., & Clark, T. (2007) Toward a unified theory of consumer acceptance technology. *Psychology & Marketing*, 24(12), 1059–1084.

Lederer, S., Mankoff, J., & Dey, A. K. (2003) Who wants to know what when? Privacy preference determinants in ubiquitous computing. In *Proceedings of extended abstracts of CHI 2003, ACM conference on human factors in computing systems*, Fort Lauderdale, FL, 724–725.

Lenhart, A. (2009) *Teens and Social Media*. Pew/Internet Report.

Licoppe, C., & Inada, Y. (2006) Emergent uses of a multiplayer location-aware mobile game: The interactional consequences of mediated encounters. *Mobilities*, 1(1), 39–61.

Lievens, A., de Ruyter, K., & Lemmink, J. G. A. M. (1999) Learning during new banking service development: A communication approach to marketing departments. *Journal of Service Research*, 2(2), 145–163.

Lin, C. H., Shih, H. Y., & Sher, P. J. (2007) Integrating technology readiness into technology acceptance: The TRAM model. *Psychology & Marketing*, 24(7), 641–657.

Ling, R. (2008) *New Tech, New Ties: How Mobile Communication is Reshaping Social Cohesion*. Cambridge, MA: MIT Press.

Ling, R. and Yttri, B. (2002) Hyper-coordination via mobile phones in Norway. In J. Katz & M. Aakhus (Eds), *Perpetual Contact: Mobile Communication, Private Talk, Public Performance* (pp. 139–69). Cambridge: Cambridge University Press.

Lippert, S. K., & Forman, H. (2005) Utilization of information technology: Examining cognitive and experiential factors of post-adoption behavior. *IEEE Transactions on Engineering Management*, 52(3), 363–381.

Lofland, L. H. (1998) *The Public Realm: Exploring the City's Quintessential Social Territory*. New York: Aldine de Gruyter.

Lugano, G. (2007) Mobile social software: Definition, scope and applications. In *Proceedings of eChallenges Conference*, 1434–1442.

Lugano, G. (2008) Mobile social networking in theory and practice. *First Monday*, 13(11), 3 November.

Lundvall, B. and Borras, S. (1999) *The Globalising Learning Economy: Implications for Innovation Policy*. Office for Official Publications of the European Communities, Luxembourg.

MacKenzie, D., & Wajcman, J. (2002) *The Social Shaping of Technology* (2nd edn). Buckingham, UK: Open University Press.

Mahatanankoon, P. (2007) The effects of personality traits and optimum stimulation level on text-messaging activities and m-commerce intention. *International Journal of Electronic Commerce*, 12, 7–30.

Matsuda, M. (2005) Mobile communication and selective sociality. In M. Ito, D. Okabe & M. Matsuda (Eds), *Personal, Portable, Pedestrian: Mobile Phones in Japanese Life* (pp. 123–142). Cambridge: The MIT Press.

McCarthy, A. (2001) *Ambient Television: Visual Culture and Public Space.* Durham, NC: Duke University Press.

Moorman, C., Diehl, K., Brinberg, D., & Kidwell, B. (2004) Subjective knowledge, search locations, and consumer choice. *Journal of Consumer Research*, 31(3), 673–680.

Morris, M., & Venkatesh, V. (2000) Age differences in technology adoption decisions: Implications for a changing workforce. *Personnel Psychology*, 53, 375–403.

Nelson, R. R., & Todd, P. A. (2005) Antecedents of information and system quality: An empirical examination within the context of data warehousing. *Journal of Management Information Systems*, 21(4), 199–235.

Okazaki, S. (2009) Social influence model and electronic word of mouth PC versus mobile Internet. *International Journal of Advertising*, 28, 439–472.

Oldenburg, R. (1991) *The Great Good Place: Cafes, Coffee Shops, Community Centers, Beauty Parlors, General Stores, Bars, Hangouts, and How They Get You through the Day.* New York: Paragon House.

Preece, J. (2004) Etiquette online: From nice to necessary. *Communications of the ACM*, 47(4), 56–61.

Puro, J. P. (2002) Finland: A mobile culture. In J. E. Katz & M. Aakhus (Eds), *Perpetual Contact: Mobile Communication, Private Talk, Public Performance* (pp. 19–29). Cambridge, UK: Cambridge University Press.

Ratten, V. (2009) Adoption of technological innovations in the m-commerce industry. *International Journal of Technology Marketing*, 4(4), 355–367.

Revels, J., Tojib, D., & Tsarenko, Y. (2010) Understanding consumer intention to use mobile services. *Australasian Marketing Journal*, 18, 74–80.

Rheingold, H. (2003) Mobile virtual communities. Published: July 9. Available at: http://www.thefeaturearchives.com/topic/Culture/Mobile_Virtual_Communities.html

Rogers, E. M., & Beal, G. M. (1958) The importance of personal influence in the adoption of technological changes. *Social Forces*, 36(4), 329–335.

Rogers, E. M., & Kincaid, D. L. (1981) *Communication Networks: Toward a New Paradigm for Research.* New York: Free Press.

Rogers, E. M. (1995) *Diffusion of Innovations* (4th edn). New York: The Free Press.

Rycroft, R.W. (2006) Time and technological innovation: implications for public policy. *Technology in Society*, 28(3), 281–301.

Sadeh, N., Hong, J., Cranor, L., Fette, I., Kelley, P. Prabaker, M., & Rao, J. (2009) Understanding and capturing people's privacy policies in a mobile social networking application. *Personal Ubiquitous Computing*, 13, 401–412.

Scholnikoff, E. B. (2001) International governance in a technological age. In De la Mothe, J. (Ed.), *Science, Technology and Governance.* Continuum Press: New York.

Schouten, J. W. and McAlexander, J. H. (1995) Subcultures of consumption: An ethnography of the new bikers. *Journal of Consumer Research*, 22(1), 43–61.

Shannon, V. (March 6, 2008). Social networking moves to the cell phone. *The New York Times*.

Sheller, M. (2004) Mobile publics: Beyond the network perspective. *Environment and Planning D: Society and Space*, 22(1), 39–52.

de Souza e Silva, A. (2009) Hybrid reality and location-based gaming: Redefining mobility and game spaces in urban environments. *Simulation & Gaming*, 40(3), 404–424. DOI: 101177/1046878108314643.

Snowden, S., Spafford, J., Michaelides, R., & Hopkins, J. (2006) Technology acceptance and m-commerce in an operational environment. *Journal of Enterprise Information Management*, 19(5), 525–539.

Stern, B. B., Royne, M. B., Stafford, T. F., & Bienstock, C. C. (2008). Consumer acceptance of online auctions: an extension and revision of the TAM. *Psychology and Marketing*, 25(7), 619–636.

Sutko, D. M., & de Souza e Silva, A. (2010) Location-aware mobile media and urban sociability. *New Media & Society*, 13(5), 807–823.

Thomson, M., MacInnis, D. J. and Park, C. W. (2005) The ties that bind: Measuring the strength of consumers' emotional attachments to brands. *Journal of Consumer Psychology*, 15(1), 77–91.

Tomita, H. (2005) Keitai and the intimate stranger. In M. Ito, D. Okabe & M. Matsuda (Eds), *Personal, Portable, Pedestrian: Mobile Phones in Japanese Life*. Cambridge, MA: The MIT Press.

Townsend, A. (2000) Life in the real-time city: Mobile telephones and urban metabolism. *Journal of Urban Technology*, 7(2), 85–104.

Varnali, K., Toker, A., & Yilmaz, C. (2010) *Mobile Marketing: Fundamentals and Strategy*. New York: McGraw-Hill.

Venkatesh, V., & Davis, F. D. (2000) A theoretical extension of the technology acceptance model: Four longitudinal field studies. *Management Science*, 46(2), 186.

Wasserman, S., & Faust, K. (1994) *Social Network Analysis: Methods and Applications*. Cambridge: Cambridge University Press.

White, F., Hayes, B. and Livesey, D. (2005) *Developmental Psychology: From Infancy to Adulthood*. Frenchs Forest NSW: Pearson-Prentice Hall.

Wilken, R. (2008) Mobilizing place: Mobile media, peripatetics, and the renegotiation of urban places. *Journal of Urban Technology*, 15(3), 39–55.

Williams, R. (1975) *Television: Technology and Cultural Form*. New York: Schocken Books.

Wixom, B. H., & Todd, P. A. (2005) A theoretical integration of user satisfaction and technology acceptance. *Information Systems Research*, 16(1), 85–102.

Yousafzai, S. Y., Foxall, G. R., & Pallister, J. G. (2010) Explaining internet banking behavior: Theory of reasoned action, theory of planned behavior, or technology acceptance model? *Journal of Applied Social Psychology*, 40(5), 1172–1202.

Zhang, J., & Mao, E. (2008) Understanding the acceptance of mobile SMS advertising among young Chinese consumers. *Psychology and Marketing*, 25, 787–805.

Zhou, T., Li, H., & Liu, Y. (2010) The effect of flow experience on mobile SNS users' loyalty. *Industrial Management and Data Systems*, 110(6), 930–946.

# 8
## Exploring the Role of Early Adopters in the Commercialization of Innovation

*Federico Frattini, Gabriele Colombo, and Claudio Dell'Era*

## 1 Introduction

Effectively commercializing innovation is a fundamental driver of superior performance (Calantone & Di Benedetto 1988; Cooper 1979). In particular, it is crucial in high-technology markets where the ability to effectively commercialize technological innovation is likely to influence the innovation's success and the firm's overall competitive advantage even more directly than it does in traditional markets. In this vein, Nevens et al. (1990) found a strong linkage between a firm's competitiveness and its ability to commercialize new products and services, and showed that, in many high-tech markets, industry leadership plainly depends on higher commercialization skills.

Notwithstanding the influence that commercialization has on a firm's competitive advantage, especially in high-technology industries, many scholars believe that it is often the least well managed phase of the whole innovation process (e.g., Calantone & Montoya-Weiss 1993). On the one hand, it is an intrinsically risky activity, as witnessed by the number of innovations that, although functionally superior to competing ones, turned out to be far less successful, mainly because of a poor launch strategy (Hartley 2005). On the other, it is very often the single costliest stage of the whole innovation process (Cooper & Kleinschmidt 1988). Moreover, effectively commercializing technological innovation appears to be even more demanding in high-technology markets, as witnessed by the remarkably low success rates that new products have been experiencing in these industries (Chiesa and Frattini, 2011). This is likely to be the result of the rapid advance of new technologies (Bayus 1998) and of the shortening of many product life cycles (Nevens et al. 1990), which narrows the window of opportunity into which a new product should be launched and established, as well as the result of the increased interconnection of high-tech markets, which makes it more and more complex to get clients to adopt new products (Chakravorti 2004a).

In spite of these concerns, which make it difficult for high-tech firms to properly commercialize their innovations, academic and practitioner literature so far has not developed a clear and comprehensive understanding of the determinants of successful commercialization in high-technology markets. The technological innovation process can be roughly divided into two macro-phases, development and commercialization. Whereas new product management research has largely investigated the factors underlying superior performance in development activities, commercialization has remained relatively underresearched so far (Schilling 2005). Therefore, an important direction for future research in innovation management is to study the endogenous and exogenous variables affecting success in commercialization of innovation (Frattini 2009). In particular, a promising approach entails the integration of concepts and models applied in distinct disciplines into new product management research (Tidd 2010).

In this regard, we combine insights coming from marketing and the diffusion of innovation to investigate the role of early adopters in explaining the success of innovations launched in high-tech markets. There are indeed a number of studies suggesting that success in commercializing an innovation among its early adopters is critical for improving performance (e.g., Bass 1969; Lesnick 2000). Nevertheless, such contributions do not specifically focus on high-technology markets, are not exhaustive and, above all, have not yet achieved a definite understanding of the commercialization approaches that are the most effective in favoring a positive appraisal by the early client segments.

In this chapter we try to take a step farther in this direction. Specifically, the goals of our study are: (i) to study the role that early adopters have in the commercialization of a technological innovation in high-tech markets and specifically the influence that they exert on the innovation's commercial success; and (ii) to understand the commercialization approaches that are the most effective in high-tech markets for stimulating positive acceptance of the innovation among its early adopters.

The chapter is structured as follows. In the next section we review the relevant literature. Then we develop research questions and describe the methodology used for the empirical analysis. Finally we present and discuss the findings of the empirical investigation and, in the last section we briefly outline the chapter's limitations and discuss avenues for future research.

## 2   Literature background

### Characteristics of high-tech markets

High-technology markets have three main attributes that make the commercialization of technological innovations a very complex activity: (i) turbulence and dynamicity, (ii) interconnectedness and, most of all, (iii) prepurchase uncertainty.

High-tech industries are characterized by a relevant degree of turbulence and dynamicity that permeates several business dimensions. First, the set of players operating in a definite industry changes very rapidly (Mohr et al. 2001); second, technologies are often commercialized by firms coming from outside established market arenas (Christensen 1997). Furthermore, product offerings are subject to steady and profound revolutions, this stemming from a growing proliferation of new technologies and the shrinking life cycle of many products (Bayus 1998; Ofek & Sarvary 2003; Davies & Brush 1997; Nevens et al. 1990). Second, high-tech markets are getting more and more interconnected; because of improved communication technologies and the diffusion of the Internet and because of business's increased reliance on the global market for products, capital, and labor, more markets have taken on the characteristics of networks (Chakravorti 2004b). In an interconnected market, the success of a new technology depends on the decisions of a set of players that constitute its "adoption network." Basically, this network comprises: (i) companies that supply products or services complementary to the innovation; (ii) companies that compete with the firm that is actually commercializing the innovation; (iii) companies involved in distributing the innovation or information about it. Once the technical and functional content of the innovation is established, the choices made by these three classes of players are crucial to the innovation's success.[1] Finally, high-tech markets are characterized by a significant degree of prepurchase uncertainty, which strictly descends from the previous two features. Moore (1991) states that clients in high-tech markets experience the so-called "FUD (Fear, Uncertainty, and Doubt) Syndrome," a soaring level of fear, uncertainty, and doubt about the ability of a new technology (and of the firm commercializing it) to actually satisfy the needs and the problems it promises to address. The state of anxiety that clients experience in front of a new technology, both in consumer and industrial markets (Mohr et al. 2001), depends on several factors: (i) the uncertainty about the support that the players in the adoption network will provide to the new technology; (ii) the possibility that the new technology will be soon replaced by an improved alternative; (iii) the ability of the innovative firm to stay abreast of the technological development and to provide the required support and future upgrades, which are especially critical for industrial innovations; (iv) the complexity of technologies that makes it hard for clients to understand whether an innovation will succeed in delivering the promised value.

**Diffusion of innovation theory**

The prepurchase uncertainty surrounding adoption decisions does not affect the whole set of innovation potential adopters in the same way. Diffusion of Innovation (DoI) Theory (see Rogers 2003 for a review) helps us shed light on this point. Rogers (2003) shows that innovation adopters are characterized by a different level of "innovativeness," defined as the degree to which an individual or firm is relatively earlier in adopting new ideas, products, and services than other members of a system. As an innovation diffuses into the market, it passes through different categories of adopters that are characterized by an analogous innovativeness, i.e., innovators, early adopters, early majority, late majority, and laggards.

This suggests that the adoption decisions of the earliest categories (basically, innovators and early adopters) are only weakly influenced by prepurchase uncertainty, which significantly hinders adoption decisions of less-innovative clients, i.e., early and late majority and laggards. Furthermore, individuals or firms belonging to the same adoption category have similar characteristics (Rogers 2003). In particular, the first individuals or companies adopting an innovation usually serve as opinion leaders, whereby their opinion about the new product or service they have already bought and used has a critical role in affecting subsequent purchases. On the one hand, given that clients purchasing an innovation later than innovators and early adopters perceive a great uncertainty about how the product will perform for them and how easily it can be integrated into existing processes and operations, they try to reduce this uncertainty by asking previous adopters about how they have fared with the product they bought and subsequently used. As a result, the opinion about the innovation of the early adopters, which is disseminated through a word-of-mouth process (Czepiel 1975), affects subsequent purchases. On the other hand, and especially in industrial markets, a bandwagon effect might originate when information about the fact that some companies have already adopted and used the new products or services disseminates in the market (DiMaggio & Powell 1983; Abrahamson & Rosenkopf 1990). When this happens, potential adopters are indeed encouraged to purchase because they perceive the risk of quickly losing competitive advantage relative to early adopters. Accordingly, the role of early adopters is to stimulate imitative and competitive reactions among later adopters (Powell & DiMaggio 1991; Abrahamson & Rosenkopf 1993).

This brief overview suggests that the success an innovation has among the most innovative categories of adopters could be critical in determining its diffusion among less-innovative adopters for the following three major reasons: (i) customer uncertainty has a very limited influence on the adoption decisions of innovators and early adopters, meaning that they can assess the real value of an innovation without suffering from the FUD syndrome; (ii) innovators and early adopters usually hold opinion leadership positions inside the social system in which they live and work; therefore their judgment about an innovation is taken into great consideration by all the other members of the system (i.e., early and late majority, laggards) and potentially influences their adoption decisions; (iii) late adopters are encouraged to adopt a new product or service because they perceive the risk of ending up at a competitive disadvantage in comparison with early adopters. It should be noted that the importance early adopters have in affecting subsequent purchases has been documented not only in consumer markets, which have received more attention in scholarly research so far (see, e.g., Chiesa & Frattini 2011), but also in industrial sectors (Kennedy 1983). This is clearly evinced also if one considers industrial marketing handbooks. They indicate that the use of references documenting early adoptions to encourage subsequent purchases is a very well-established practice in business-to-business markets (Hutt & Speh 1992; Brierty et al. 1998; Kotler 2003).

To give more detail, the positive impact that innovation acceptance by innovators and by early adopters exerts on its overall diffusion is confirmed by many literature contributions. For instance, Rogers (1976) shows that it is critical to make innovators and early adopters accept a new technology, in order to grant its diffusion into the rest of the market. In particular, he suggests that a firm should implement a marketing strategy that is specifically targeted to early adopters in order to leverage their opinion leadership role; such a strategy should be client-oriented, compatible with client's needs, credible in the client's eyes, and should increase client's ability to evaluate the innovation. Lesnick (2000) underlines the importance that a preliminary identification of opinion leaders among the set of potential early adopters had in maximizing the diffusion rate of photovoltaic panels in the Dominican Republic. Additionally, Bass (1969) asserts that imitators' buying attitude is positively influenced by the existence of a consolidated basis of adopters; he says that "imitators learn, in some sense, from those who have already bought." Building on Bass's model, Sultan et al. (1990) undertake a quantitative analysis that is helpful for identifying the factors influencing new-product diffusion. In particular, they show that the diffusion process is affected more by word of mouth or imitation effects than by the innate innovativeness of clients, showing the importance that the innovation's early adoption exerts on its subsequent dissemination.

How important it is that an innovation is positively accepted by innovators and early adopters, in order to gain adequate diffusion among less-innovative adopters, can be better understood by taking into account the usefulness of different communication channels in determining adoption decisions. Different authors (e.g., Rao et al. 1980; Granovetter 1973; Rogers 2000; Abratt 1986) show that the first-adopter categories know about the existence of an innovation mainly through mass channels, and the information that these channels convey is usually effective in stimulating their adoption decisions. Less-innovative adopters get informed about the introduction of an innovation through mass channels too, but their adoption decisions are far more strongly influenced by interpersonal channels, i.e., face-to-face communication and word-of-mouth. The sources of information that these channels convey are basically those individuals or firms that have already adopted the innovation – innovators and early adopters. As a consequence, word-of-mouth communication would be more effective in convincing less-innovative clients to adopt an innovation if word-of-mouth has delivered a positive appraisal among innovators and early adopters. Finally, Mahajan and Muller (1998) advance the idea that the effectiveness of word-of-mouth mechanisms in influencing less-innovative clients' adoption decisions depends on a set of contextual variables, such as the rate at which the word-of-mouth effect declines during the diffusion process, the characteristics of the innovation (whether it is a consumer product or an industrial product), the relative number of innovators and early adopters with respect to early and late majority, and the speed with which the innovation is accepted in the market.

Although established DoI models clearly highlight that new-product acceptance among early adopters affects diffusion into the bulk of the target markets, there are also some controversial views on this issue. In particular, according to Moore (1991), a "chasm" separates most-innovative adopters (i.e., innovators and early adopters) from the early majority; the chasm is caused by the deep differences in the psychographic characteristics of the two. These dissimilarities actually impede an effective information exchange from taking place. However, Rogers (2003) criticizes Moore's position, stating that "although some scholars claim that a discontinuity exists between the innovators and the early adopters versus the early majority, late majority and laggards, past research shows no [empirical] support for this claim of a chasm."

From this brief literature review it clearly surfaces that, although some conflicting points of view have been found, there is empirical and theoretical evidence that the success an innovation among most-innovative adopter categories (i.e., innovators and early adopters) is important in determining its diffusion among early majority, late majority, and laggards in traditional markets, for new consumer products and industrial products and services. However, the role of most-innovative adopters in stimulating the diffusion of innovation in high-tech markets has received little attention. Furthermore, it has been shown that the degree to which the early success of an innovation influences subsequent adoption decisions is dependent upon the characteristics of the innovation, the market in which it is commercialized, and the diffusion process it experiences. Hence, in order to fill this gap, the chapter investigates the role played by early adopters in the commercialization of different technological innovations in high-tech markets.

## High-tech marketing and DoI theory

DoI Theory does not investigate how a firm could actually foster a positive acceptance of the innovation among innovators and early adopters. However, this last issue is at least partially addressed by the literature applying DoI Theory to studying high-technology marketing. Moore (1991, 1998) is one of the most influential scholars to apply DoI Theory to the study of high-technology marketing issues, with particular reference to industrial markets. He suggests that the target market for a high-technology innovation could be split into two groups, which show radically different buying behaviors (Moore 1991): the early market, which includes innovators and early adopters, and the mainstream market, which aggregates early majority, late majority, and laggards. Considering that the early market comprises about 15 to 25 percent of potential innovation-adopting clients (Rogers 2003), it is clear that a significant diffusion in the mainstream is a necessary condition for an innovation to achieve commercial success.[2] As we noted before, Moore does not believe that the success the innovation encounters in the early market is relevant in influencing its commercial success; nevertheless, he provides some interesting insights about the marketing strategies that are the most effective for commercializing the innovation in the early market. First of all, he states that there are three types of competitive advantage that a high-technology firm can pursue: (i) product leadership, (ii) customer intimacy, and (iii) operational excellence. The most important one in the early market is product leadership (Moore 1998), i.e., the ability of a firm to deliver a product that is at the forefront of both technological and functional development. Second, a direct sales force is found to be the best distribution channel in the early market, because of its ability to efficiently and effectively build, rather than satisfy, the demand for a new technology (Moore 1991). Furthermore, in the early market a firm should adopt a value-based pricing strategy, one aimed at optimizing margins, rather than a commodity-based one aimed at optimizing market share (Moore 1998). Finally, in commercializing the innovation in the early market, it seems advisable to execute a strong preannouncement campaign that relies on messages focused on the technical content of the new product and delivered through specialized channels, e.g., professional fairs and the specialized press (Moore 1991).

Other scholars have been studying high-tech marketing from a DoI Theory perspective; they basically highlight: (i) the different buying behaviors between the early and mainstream market and (ii) the importance of planning an adequate marketing strategy, in order to favor the acceptance of the innovation in the early market. For instance, Easingwood and Harrington (2002) state that a technological innovation should be first "launched" in the early market and then "re-launched" in order to foster its diffusion among mainstream clients, thus underlining the necessity to manage the dimensions of the commercialization strategy differently, according to the adopters' characteristics. In particular, the scholars suggest that the "launch" strategy (i.e., the one addressed to early market) should consist of the following steps: (i) market preparation, including the exploitation of alliances and licensing arrangements, the provision of technical information before the launch phase, the delivery of client education; (ii) targeting, which requires focusing on the needs of more innovative adopters and current clients; (iii) positioning, which should emphasize technological superiority; and (iv) execution, which basically consists of cultivating a winner image. Other scholars (e.g., Butje 2005; Viardot 2004; Mohr et al. 2001) provide some general insights about the way in which a firm can maximize the acceptance of the innovation in the early market. They basically agree about the importance of providing a high level of product customization and technical support and using primarily direct distribution channels (i.e., a direct sales force) in order to add value to the delivered product. Moreover, they do not think that branding strategies are effective in spurring an innovation in the early market, because early adopters are capable of critically evaluating the product's technological and functional content and do not care about the associated image or its belonging to a well-known brand or family name.

However, despite the evidence that the marketing choices a firm takes in commercializing a technological innovation are capable of influencing the diffusion of innovation in the early market, there is a lack of contributions to the literature that systematically investigate the marketing variables and the related choices that are the most critical ones for successfully commercializing an innovation in the early market. Hence this chapter helps to fill this gap by trying to understand how marketing choices influence the acceptance that a technological innovation gets in the early market.

## 3   Research questions

As discussed in the previous section, the acceptance that a technological innovation receives in its early market seems to influence its overall commercial success. Although there is theoretical and empirical evidence supporting this assumption, some criticisms have been raised about this conclusion (Moore 1991). Along with the findings presented by Mahajan and Muller (1998), we believe that these contrasting standpoints partially come from the fact that the role played by the early market is contextual and depends upon several variables. As a result, assessing the impact that a positive acceptance in the early market exerts on the innovation commercial success in absolute terms can lead to misleading conclusions.

In this chapter we advance the idea that the characteristics of the innovation, and specifically its degree of radicalness, are fundamental for understanding the role played by the innovation's acceptance in the early market. There are at least two dimensions of radicalness that can potentially influence the uncertainty perceived by mainstream clients and hence the importance that the acceptance of the innovation in the early market exerts on the innovation's commercial success: technology-related radicalness and market-related radicalness (Veryzer 1998). Technology-related radicalness refers to the degree to which the innovation incorporates technological capabilities beyond existing boundaries (Ali 1994; Kleinschmidt & Cooper 1991). Market-related radicalness refers to the degree to which the innovation requires a change in the thinking and behavior of consumers in order to deliver its value (Robertson 1967; Sheth 1981). When an innovation is radical from a technological point of view, mainstream clients' uncertainty is likely to originate from their incapacity to understand the way in which the new product performs its critical functions and hence is capable of satisfying their needs. When an innovation is radical from a market perspective, it is the need to change behaviors and consumption patterns that makes mainstream clients anxious and doubtful about the opportunity to adopt the innovation. This might translate in a postponement of the adoption decision until further evidence about the value of the innovation is available. Therefore we want investigate the following:

Research Question 1: Is early market acceptance equally important for the successful diffusion of incremental innovations and radical innovations?

Then we investigate the impact that marketing strategies have on the success that the innovation encounters in its early market. The major dimensions that comprise a high-tech new-product launch and marketing campaign can be synthesized as shown in Table 8.1 (see, e.g., Viardot 2004; Mohr et al. 2001; Beard & Easingwood 1996; Davidow 1986); these dimensions represent the variables that can potentially influence the acceptance of the innovation in the early market.

In the second section of this chapter, we reviewed theoretical and empirical evidence (e.g., Butje 2005; Viardot 2004; Rogers 2003; Easingwood & Harrington 2002; Mohr et al. 2001) about the following second research question:

Research Question 2: What marketing decisions positively affect the successful diffusion of the innovation in the early market?

## 4  Research methodology

The empirical research employed a multiple case study methodology (Yin 2003), which was chosen since it allows both an in-depth examination of each case and the identification of contingency variables that distinguish each case from the other. Furthermore, multiple case studies are appropriate when attempting to externally validate the findings from a single case study through cross-case comparisons (Eisenhardt 1989). We studied 11 technological innovations commercialized by 9 different companies. We selected firms operating in high-tech industries and selling industrial products.

Specifically, we adopted the definition of high-technology industry provided by Hatzichronoglou (1997), which is based on direct and indirect R&D intensity and is adopted in Organization for Economic Co-operation and Development (OECD) studies. In particular, he identifies four groups

*Table 8.1*   High-tech marketing dimensions

| Variable | Description |
|---|---|
| Timing | – Timing of innovation launch into the market<br>– Timing of innovation preannouncement |
| Targeting and positioning | – Target market for the innovation, i.e., a group of clients who share similar needs and buyer behavior characteristics and who are responsive to the firm's offering<br>– Market position for the innovation, i.e., the image of the innovation in the eyes of the clients, relative to competitors and substitute innovations, on critical attributes of importance |
| Distribution | – Type of distribution channel used (e.g., specialized or generalist)<br>– Critical functions the distribution channel is expected to perform (in particular, client education) |
| Pricing | – Pricing strategy (e.g., skimming vs. penetration)<br>– Pricing of complementary goods and services |
| Communication | – Type of communication channels (e.g., broad/generic vs. narrow/specific)<br>– Type of message conveyed (e.g., focused on innovation's technical features vs. focused on company brand/family product) |
| Whole product configuration[a] | – Complementary products and services included in the innovation's basic offering, commercialized by the firm |
| Partnerships and alliances | – Adoption network players to partner with (i.e., Software developers and complementary goods manufacturers, competitors, distribution channel members)<br>– Type of partnership agreements |

*Notes*: [a] Whole product can be defined as "the minimum set of products and services necessary to ensure that the target customer will achieve his or her compelling reason to buy" (Moore, 1998).

of manufacturing industries, namely (i) high technology, (ii) medium-high technology, (iii) medium-low technology, and (iv) low technology, where "industries classified in a higher category have a higher OECD-average intensity for all indicators than industries in a lower category" (Hatzichronoglou 1997, p. 5). Moreover we decided to exclude consumer products from the sample, because they show specific characteristics, in comparison to industrial products, that make the commercialization of the innovation in the early market a radically different and perhaps less challenging issue, such as the higher number of clients, their lack of technical expertise, and the associated difficulty in defining their needs in technical terms (Millier 1999). Moreover, adoption decisions in industrial markets are done by groups and not by individuals, the influence of opinion leaders is likely to be lower than in consumer markets, and the diffusion process is typically slower (Day & Herbig 1990). For each firm included in the sample, we chose one (or two) technological innovations with the following basic characteristics: (i) they had been already on the market for a time period that allowed a qualitative assessment of their commercial success; (ii) their commercialization and market launch were managed by a person that was still working in the firm, so he or she could be personally interviewed.

We gathered information mainly through direct interviews; specifically, we went after the following steps.

- At the outset of each case, a relationship was established with a senior manager from the selected firm. This person was briefed about the research project through a written project summary and a telephone meeting. During this meeting, we asked the respondent to identify one or two technological innovations that met the above-mentioned selection criteria and to indicate a product manager or marketing manager that was knowledgeable about the commercialization of the innovation and available to be further interviewed.
- We personally and directly interviewed the selected product managers and marketing managers (on average, we undertook two semi-structured interviews for each respondent) in order to gather the information required to pursue the chapter's research objectives;
- Secondary information was collected in the form of company reports, product brochures, business press, and specialized marketing journals. These informed the researchers with background information about the selected firms, the studied innovations, their commercial success, and commercialization approaches. Moreover, these secondary information sources were integrated, in a triangulation process, with data drawn from the direct interviews, in order to enhance construct validity (Yin 2003).
- All interviews were tape-recorded and transcribed; generally, at this stage a telephone follow-up with the respondents was conducted in order to gather some important missing data.

Interviews with product managers and marketing managers followed a semi-structured guide.

- First, respondents were asked to assess the commercial success of the selected innovation(s). Specifically, they were invited to consider the following dimensions:[3]
  - *Sales vs. objectives*: the extent to which the new product's sales exceeded or fell short of sales objectives; a five-point scale was used, where 1 meant "strongly missed sales objectives," 3 meant "met sales objectives" and 5 meant "strongly exceeded sales objectives."

- *Relative sales*: the extent to which the sales of the new product exceeded or fell short of the sales of other recently introduced new products; a five-point scale was used, where 1 meant "strongly fell short of past sales," 3 meant "equaled past sales," and 5 meant "strongly exceeded past sales."
- Second, we needed a method to measure the technology-related and market-related radicalness of the selected innovations. We relied upon the assessments of a panel of experts, gathered using a two-round Delphi Process (Dalkey & Helmer 1963). The panel included 10 experts with different professional experiences: 4 are professors at Politecnico di Milano and have been developing research projects in high-technology industrial markets for many years; 6 are senior managers working for leading companies that operate in high-technology industrial markets. They were asked to classify the 11 innovations according to their technology-related radicalness and market-related radicalness.[4]
- Then, respondents were asked to identify the client segments that were the first in adopting the selected innovation (and that therefore represented its early market) and to assess the diffusion that the innovation had experienced in this early market and the type of consideration it received, i.e., early market acceptance. To this end, a five-point scale was used, where 1 meant "very low early market acceptance," 3 meant "neither high nor low early market acceptance," and 5 meant "very high early market acceptance."
- Finally, for those innovations that experienced a wide diffusion in the early market and raised a really positive appraisal among their early adopters, we asked respondents to identify the commercialization and marketing decisions that were critical in determining the observed early market acceptance. In this phase, we related data to the model reported in Table 8.1.

Data and information gathered through the case studies were manipulated before being analyzed. In particular, we applied the following techniques (Miles & Huberman 1984): (i) data categorization, which requires the decomposition and aggregation of data in order to highlight some characteristics (e.g., acceptance of the innovation in the early market) and to facilitate comparisons; and (ii) data contextualization, which implies the analysis of contextual factors that may reveal unforeseen relationships between events and circumstances. Then, a preliminary within-case analysis was performed; the purpose was to consider each case study as a separate one and to systematically document the variables of interest. For each case study, the manipulated information was aggregated in order to obtain a systematic description of the composition of the early market, the appraisal that the innovation encountered among early adopters, and the marketing approach that was employed for the commercial launch. Then, explanation-building procedures were applied that identified the relationships among the characteristics of the innovation, the marketing decisions, the early market acceptance, and the innovation's commercial success. Finally, a cross-case analysis was undertaken for comparing the patterns that emerged in each case study, in order to reach a general explanation of the observed phenomena. These structured procedures for data collection and analysis help enhance the reliability of the case study research (Yin 2003). Table 8.2 synthesizes some preliminary information about the studied innovations.

## 5  Findings

In this section, the findings from the case studies are used to illustrate: (i) the relevance of the early market acceptance to the commercial success of innovations characterized by different categories of radicalness; (ii) which marketing strategies are the most effective in positively influencing the innovation's early market acceptance.

**Early market acceptance and innovation radicalness**

All the innovations we studied experienced significant commercial success (see Table 8.2); actually, they exceeded or strongly exceeded their sales targets and the average sales level of new products launched by the firm in the last years. This outcome was foreseeable if we consider that respondents were personally interviewed and that they were free to select an innovation for further investigation. Nevertheless, this is not detrimental to our analysis, since having a homogeneous distribution of the sampled innovations on the basis of their degree of commercial success is not fundamental to pursuing our research objectives. The studied innovations were classified by the panel of experts (through a two-round Delphi Process), on the basis of their technology-related and market-related radicalness, as shown in Figure 8.1.

Then, we investigated the acceptance that the innovations experienced as soon as they were launched into the market; in particular, we asked each respondent to assess if the innovation he or she was responsible for initially diffused among a specific group of clients that were particularly satisfied by its characteristics and whether this positive acceptance was important in favoring the diffusion of the new product into the rest of its target market. Comparing the results of this analysis across the four categories identified in Figure 8.1, an interesting pattern emerged (see Table 8.3). The innovations for which respondents acknowledged the existence of a particular group of clients that enthusiastically adopted them after the market launch and whose positive appraisal definitely influenced their diffusion in the rest of the market are those characterized by a high level of market-related radicalness. In the Appendix, we provide a brief description of the early market acceptance for all the innovations considered in the study.

Table 8.2 Preliminary information about the case studies

| Innovation | Brief description | Co. no. | Year oflaunch | Role of person interviewed | Sales vs. objectives | Relative sales |
|---|---|---|---|---|---|---|
| Innovation 1 | Innovation 1 is a technology that multiplexes multiple optical carrier signals on a single fiber by using different wavelengths of laser light (allowing for a multiplication in capacity and bidirectional communication) | Co. 1 | 2003 | Product manager | 4 | 5 |
| Innovation 2 | Innovation 2 are wireless position sensors used for machine control. Characterized by a high reliability and a lack of any physical contact, they inform the controller about the progress of the machine's movement | Co. 2 | 2005 | Product manager | 5 | 5 |
| Innovation 3 | Innovation 3 is a real-time computer control system used for the automation of industrial processes | | 2002[a] | Product Manager | 4 | 4 |
| Innovation 4 | Innovation 4 are systems, designed for industrial applications, that prevent or mitigate the risks of adverse events (e.g., short circuits) | Co. 4 | 2000[b] | Marketing manager | 5 | 4 |
| Innovation 5 | Innovation 5 are measurement systems, based on laser technology, that help professionals determine areas, volumes, offsets, perimeters, distances, and check squares and parallels, at the touch of a button | Co. 5 | 2003[c] | Product manager | 4 | 4 |

| | | | | | | |
|---|---|---|---|---|---|---|
| Innovation 6 | Innovation 6 is a system designed to improve a vehicle's handling and stability, even in the hardest driving conditions | Co. 6 | 1995 | Marketing manager | 5 | 5 |
| Innovation 7 | Innovation 7 are self-contained computer servers, characterized by an advanced high-density architecture | Co. 7 | 2002 | Marketing manager | 5 | 5 |
| Innovation 8 | Innovation 8 is a vertical CNC machine. Innovation 8 represents an improvement in respect to traditional vertical machine centers; it is characterized by higher reliability and precision | Co. 8 | 2001 | Product manager | 5 | 4 |
| Innovation 9 | Innovation 9 is a vertical CNC machine, and represents an innovative platform that also allows the processing of very large parts | | 2003 | Product manager | 4 | 4 |
| Innovation 10 | Innovation 10 is a new line of high-speed digital black-and-white copiers and printers (with the following fundamental performance: 60 ppm scanner, 135 ppm output) | Co. 10 | 1990 | Marketing manager | 5 | 4 |
| Innovation 11 | Innovation 11 is a new range of server computers based on the UltraSPARC 64-bit microprocessor architecture | Co. 11 | 2004 | Marketing manager | 5 | 4 |

*Notes:* [a] Specifically, we studied the second generation of Innovation 3 commercialized by Company 2.
[b] Specifically, we studied the third generation of Innovation 4 commercialized by Company 4.
[c] Specifically, we studied the third generation of Innovation 5 commercialized by Company 5.

| | Technology-related radicalness | |
|---|---|---|
| | **Incremental** | **Radical** |
| **Market-related**<br>**radicalness** — **Radical** | • *Innovation 2*<br>• *Innovation 9* | • *Innovation 6*<br>• *Innovatio 10*<br>• *Innovation 7* |
| **Incremental** | • *Innovation 3*<br>• *Innovation 8*<br>• *Innovation 5* | • *Innovation 4*<br>• *Innovation 1*<br>• *Innovation 11* |

*Figure 8.1* Classification of the innovations studied according to their technology-related and market-related radicalness

*Table 8.3* Relationship between innovation radicalness and early market acceptance

| Innovation | Technology-related radicalness | Market-related radicalness | Early market acceptance |
|---|---|---|---|
| Innovation 1 | Radical | Incremental | 3 |
| Innovation 2 | Incremental | Radical | 5 |
| Innovation 3 | Incremental | Incremental | 3 |
| Innovation 4 | Radical | Incremental | 4 |
| Innovation 5 | Incremental | Incremental | 3 |
| Innovation 6 | Radical | Radical | 5 |
| Innovation 7 | Radical | Radical | 5 |
| Innovation 8 | Incremental | Incremental | 4 |
| Innovation 9 | Incremental | Radical | 4 |
| Innovation 10 | Radical | Radical | 5 |
| Innovation 11 | Radical | Incremental | 4 |

*Notes:* 1 means "very low early market acceptance," 3 means "neither high nor low early market acceptance," 4 means "high early market acceptance" and 5 means "very high early market acceptance."

The first part of our empirical analysis suggests that the influence exerted by early market acceptance on the innovation's commercial success is much stronger for innovations characterized by a high degree of market-related radicalness than it is for innovations showing a soaring level of technology-related radicalness. In other words, the ability to generate a positive appraisal toward the new product among the earliest clients seems to be vital for the commercial success of market-related radical innovations, whereas it does not emerge as a critical success factor for technology-related radical ones. These findings suggest that the uncertainty affecting mainstream market clients' adoption decisions does not depend on their incapacity to understand whether the revolutionary technological solutions embedded in the new product will actually succeed in satisfying their needs or not. Rather, it is the need to change behaviors and consumption patterns that makes mainstream clients anxious and doubtful about the opportunity to adopt the innovation and especially eager to gather further evidence about its real value.

The empirical evidence that was briefly discussed in this section shows that the rise of a positive appraisal in the innovation's early market can be a spontaneous phenomenon that is simply spurred by the new product's characteristics and/or by chance. This is clear for instance in the case of Company 6's Innovation 6, whose adoption on the Mercedes A Class stimulated its diffusion among middle-end car segments. Nevertheless, very often the firm that is commercializing the innovation uses specific marketing approaches for favoring its acceptance among early clients. This is evident in all the other cases of market-related radical innovations that we studied; Company 10, for instance, even defined a procedure called ECE (Early Customer Engagement) through which Innovation 10 was specifically targeted, immediately after launch, to furnishings and household appliance manufacturers. Actually, furnishings and household appliance manufacturers had specific requirements that allowed them to particularly appreciate the characteristics of Innovation 10 and serve as a beachhead into the mainstream market. Therefore, it emerges that a firm can maximize the likelihood of a positive early market acceptance through appropriate marketing strategies. In the following section we will discuss the commercialization and marketing decisions that seem to be most effective in determining this kind of successful early-adopter engagement.

## Early market acceptance and marketing decisions

In the second step of the empirical analysis, the purpose was to investigate the critical decisions that were taken during commercialization with respect to the fundamental high-tech marketing dimensions listed in Table 8.1. In particular, we compared the marketing strategy used by the companies to favor the adoption of their innovation by the early market (see the Appendix for a brief description of the cases). The companies investigated have several similarities as well as a high early market acceptance on average, which allows us to infer that a commercialization strategy that is more effective in achieving an appropriate early market engagement actually exists (see Table 8.4). The dimensions of this commercialization strategy are briefly discussed in the following pages.

### *Targeting*

Proactively targeting the innovation at launch to one or a few well-defined and accurately selected client segments seems to be critical in order to

*Table 8.4* Marketing decisions shown by companies introducing market-related radical innovations

| | Innov. 1 launched by Co. 1 | Innov. 2 launched by Co. 2 | Innov. 3 launched by Co. 2 | Innov. 4 launched by Co. 4 | Innov. 5 launched by Co. 5 | Innov. 6 launched by Co. 6 | Innov. 7 launched by Co. 7 | Innov. 8 launched by Co. 8 | Innov. 9 launched by Co.8 | Innov. 10 launched by Co. 10 | Innov. 11 launched by Co. 11 |
|---|---|---|---|---|---|---|---|---|---|---|---|
| Timing | | | | | | | | | | | |
| Targeting and positioning | X | X | X | X | | | X | X | X | X | X |
| Distribution | X | | X | | X | X | X | | | X | X |
| Pricing | | | | | | | | | | | |
| Communication | | X | | X | X | | | X | | | |
| Whole-product configuration | | | | | | | X | | X | | |
| Partnerships and alliances | | | | | | | | | | | X |

maximize the likelihood of positive early market acceptance. Typically, the segments that are selected at this stage belong to a particular niche of the innovation's target market and show homogeneous needs that can be perfectly satisfied by the innovation's technical content and functionalities. In this way, the chances that the innovation gets a positive appraisal among its early clients are higher than if it is indistinctly launched into the target market. Furthermore, a proactive targeting process allows the firm to devise a communications and distribution campaign and to design an initial configuration of the new product that is the most adequate for satisfying the compelling reason to buy of early clients. Although it may happen that an innovation succeeds in positively engaging its early adopters without a structured targeting approach (this is the case, e.g., of Company 6's Innovation 6), our empirical analysis shows that the structured targeting approach is a critical success factor in most of the cases. This is particularly evident in the commercialization of the Innovation 2 technology by Company 2; the product manager we interviewed explained that at launch, Innovation 2 was deliberately targeted to automotive and cable manufacturers. As far as the automotive industry is concerned, in fact, Innovation 2 was believed to be of particular value since it enabled the measuring of a huge quantity of process parameters on complex assembly plants, without having the impediment of wiring connections. Similarly, cable manufacturers needed a device capable of acquiring critical measures that could be placed on reels without creating any jam between the switch wires and the cables being produced. A similar pattern is clear in the commercialization of Company 10's Innovation 10, which was initially targeted to furnishings and home appliance manufacturers that needed to print, bind, and introduce technical manuals into the packaging of their final products, without interrupting the production flow. In this respect, the flexibility and the low operating costs typical of Innovation 10 were critical buying reasons. Furthermore, targeting the innovation at launch can also be used to support early market acceptance in an international context. This is for instance the case of Innovation 1, where the company clearly targeted firms operating in different countries in order to reach an international audience of clients.

*Distribution*

Empirical analysis revealed that distribution is another marketing dimension that can significantly influence the ability of the firm to raise positive acceptance toward the innovation in the early market. In particular, it emerged that the most effective approach for distributing the new product in the early market consists in exploiting direct distribution channels, and specifically, in involving the firm's own sales force. This seems to come from the fact that earlier clients are familiar with technology and are willing to be supported, during the adoption process, by people who have a deep knowledge of the technical potential and the functionalities of the innovation, but who are capable at the same time of explaining how it can be best exploited from a business and applicative point of view. This client education function can be adequately performed only by a direct sales force, which nevertheless needs to be carefully trained about the potentiality of the new product before it is actually released into the market. With the exception of Company 8, which does not own an internal sales force because of its limited size, all the respondents acknowledged the importance of introducing an innovation into the early market through direct distribution channels. For instance, the Company 7 marketing manager we interviewed believed that sales force education and skill development were fundamental strategic weapons for stimulating the diffusion of Innovation 7 among the earliest clients. This was the reason why the firm heavily invested in delivering continuous training programs to its sales agents. Similarly, Company 10 and Company 6 proactively sent their sales people, together with technicians and engineers, to the first clients that manifested an interest in their new technologies. This was obviously important for understanding the actual needs of the early adopters and for improving the innovative products before they were heavily diffused in the target market, but it also turned out to be critical for stimulating the adoption decisions of the early market through an optimization of the new technologies in light of its specific needs and technical requirements.

*Communications*

Designing and implementing an adequate communication campaign appears to be another critical determinant of a positive early market acceptance. Specifically, the cases that we studied show that the message to be delivered through the communications campaign should focus especially on the business applications the new technology is able to support and, especially, that narrow, specialized, and interactive communication channels are preferred for conveying this message to early clients. Examples of these channels are symposia, seminars, and open-house events that allow the innovating firm to select a priori an adequate panel of attendants, chosen from the most promising early-adopter categories. For instance, for the commercialization of Innovation 9, Company 8 organized an open-house event that, according to the product manager who was interviewed, was very important in stimulating the early adoption of the innovation. Company 8 invited selected clients and key distributors, showed them how the new product could jointly process two pallets, and illustrated what kinds of business applications this functionality was able to unlock. As a result, many purchasing claims were collected during the demonstration. Similarly, for the market launch of Innovation 2 technology, Company 2 equipped one of its own facilities with the new control systems and invited some selected potential clients, who were shown the advantages of the revolutionary technology, especially in terms of the productivity advantages that it allowed.

*Whole-product configuration*

Finally, the analysis put into evidence that early market clients generally achieve their compelling reason to buy (Moore 1998) when the innovation has a simple configuration, characterized by a core product with excellent technical performance, which does not come with too many complementary products or services (e.g., system integration and consultancy services, additional software). The early adopters of a new technology, in fact, typically possess the technical competencies that are necessary to exploit it internally and to adapt the innovation to their specific needs and technical requirements; furthermore and foremost, they are willing to do so, since mastering the technical development of a new technology and integrating it within existing products and systems is perceived as a value-added activity that is worth undertaking for a technologically advanced and leading company. As a result, a firm that is interested in promoting a positive appraisal of an innovation in the early market should not strive to achieve a very complete whole-product configuration before it is launched into the market. It would be much better for development efforts to be directed toward the improvement of the core product's technical performance, for a simple configuration to be supplied initially, and for a complete whole product to be developed as long as the innovation diffuses into the target market. The availability of complementarities and additional services in the innovation's basic offering, in fact, become critical for the diffusion of a new technology among more pragmatic client segments (Moore 1998). An interesting example in this respect is provided by the commercialization of Company 7's Innovation 7; the firm decided to provide its earliest clients with a preliminary version of the machine, called White Box, that was very simple, equipped with only a few plug-ins, and with a software equipage that was much narrower than the servers it would have supplied few months later. The interviewed marketing manager believed this was critical in stimulating a positive appraisal of the innovation in the early market.

The second part of our empirical analysis shows that the way in which an innovation is commercialized is an important determinant of its early market acceptance. Specifically, decisions concerning the targeting, distribution, communications, and whole-product configuration of the innovation are acknowledged to be the most influential things in this respect.

## 6   Conclusions and future research directions

This chapter investigates the role that the early market exerts on the overall commercial success of a technological innovation. In particular, the analysis revealed that positive early market acceptance is critical for the commercial success of innovations characterized by a high market-related radicalness, which refers to the degree to which a new product requires a change in the thinking and behavior of clients in order to deliver its value. Furthermore, this chapter shows that the success of a market-related radical innovation in its early market is influenced by the approaches through which it is commercialized, and especially by decisions regarding its targeting, communication, distribution, and whole-product configuration.

## Implications for managers

We believe the chapter has valuable implications for product and marketing managers who are responsible for the launch of new products and technologies in high-technology industrial markets. Actually, the results we discussed unveil the importance of devising an appropriate early adopters engagement strategy when commercializing a technological innovation. This strategy should be specifically aimed at stimulating a successful diffusion of the innovation into its early market, which generally comprises clients with radically different characteristics from those of the bulk of the established target market. The chapter provides qualitative insights about the decisions that are likely to maximize the effectiveness of this early clients' engagement strategy. Specifically, it emerged that proactively targeting the innovation at launch to one or a few client segments, who show homogeneous requirements that can be perfectly satisfied by the innovation's technical content and functionalities, is a critical success factor. The innovation is more likely to be positively accepted in the early market when it is handed out through direct distribution channels that entail the involvement of the innovating firm's own sales force. The analysis further revealed that the communications campaign should be focused on the business applications that the new technology is able to unlock, and it should intensively rely on narrow and specialized channels such as symposia and open-house events. Finally, it emerged that the best configuration of the innovation at launch should be simple, characterized by a core product with excellent technical performance, which does not come with too many complementary products or services. It is only after diffusing into the early market, in fact, that the completeness of the whole product configuration should be achieved, since it becomes a fundamental driver for mainstream market adoption decisions. Finally, the chapter explains that the need to devise an appropriate early adopter's engagement strategy is contextual upon the characteristics of the new product being commercialized; specifically, the higher the degree of market-related radicalness of the innovation, the greater the need for an effective early adopter's engagement process.

## Implications for future research

We believe that the chapter holds interesting research implications. It is one of the few contributions, to the best knowledge of the authors, that gathers empirical evidence about the role that an innovation's early market acceptance exerts on its overall commercial success. The results discussed in this chapter support Everett Rogers's criticism (Rogers 2003) of Geoffrey Moore's conclusions about the lack of any relationships between the success an innovation encounters in its early market and the diffusion it experiences in the mainstream market, i.e. about the existence of a chasm that hinders the diffusion process for high-technology innovations (Moore 1991). The fact that a positive early market acceptance was found to be strongly associated with the innovation's commercial success, at least for market-related radical technologies, actually stems from a continuity of the diffusion process, where the adoption decisions by the most-innovative adopters spur the adoption by less innovative ones. The chasm that is observed by Moore, then, is probably not dependent on the intrinsic characteristics of the diffusion process for high-technology innovations, but on a poor commercialization strategy, that is incapable of radically evolving along the stages of the innovation diffusion process, as suggested by many scholars (e.g., Easingwood & Harrington 2002). This assumption is supported by Costa et al. (2004), who have studied many failures in the commercialization of biotech innovations and who recognized that these new products actually lingered in some sort of chasm because of poor commercialization choices.

## Limitations

In this section, a brief description of the chapter's limitations is provided, together with a concise outline of some avenues for future research. First, considering the research methodology that was adopted, it is only possible to analytically generalize the chapter's results to the commercialization of innovations with characteristics similar to those included in our research domain. In this regard, an interesting future research step would be to study whether and how these results are valid in consumer high-technology markets, which have different characteristics from industrial markets (Millier 1999; Day & Herbig 1990).

Another limitation is concerned with the decision to measure commercial success by focusing on a single dimension, i.e., on the market acceptance that the innovation experienced. Clearly, this choice was suggested by the fact that high-tech markets are characterized by competitive dynamics, at least during their hyper-growth phase, which ensure that the innovation gaining the largest market share also has a leading profitability edge (Moore 1998). Nevertheless, it would be useful to investigate how much implementing an effective early adopter's engagement strategy costs, which is a prerequisite for understanding the impact that early market acceptance has on the profitability of the innovation.

Moreover, it should be noted that our sample comprised only industrial innovations that had a very positive commercial performance, as is clear from Table 8.2. This prevented us from developing a more fine-grained understanding of the impact of early market acceptance on the subsequent diffusion of new products and services. Future research could use "polar types" samples (Eisenhard & Graebner 2007), including both new products that experienced a very positive and a very negative diffusion in the market, to understand to what extant differences in performance can be attributed to dissimilar degrees of early market acceptance.

Finally, our cases were mostly focused on diffusion processes taking place within a local market, the Italian one. An interesting topic for future research would be to study the mechanisms through which early adopters help the diffusion and commercial acceptance of new products and services on an international scale, outside the geographical market in which they were initially sold. There are reasons to think that the boundary-spanning role played by early adopters can be effective in overcoming geographical barriers toward diffusion. Cross-fertilizing innovation management research with methods and perspectives typical of international business studies is indeed another promising area for future research.

## Appendix

### Innovation 1

Innovation 1, developed by Company 1, is a technology that allows a single optical fiber to carry multiple signals by using different wavelengths of laser light, allowing higher capacity and bidirectional communication. In order to streamline the adoption of the technology on the market, Company 1 carefully designed an ad hoc early market approach. Specifically, it offered the technology to the divisions of service providers (e.g., Telecom Italia, British Telecom) that built and manage communication infrastructure for other firms. Company 1 chose service providers working in different countries in order to favor an early market acceptance in an international setting. In order to reach the targeted early customers, Company 1 heavily employed its direct sales force in order to support the clients during the implementation process. The choice of service providers as early market is due to the ability of these highly skilled clients to deeply appreciate the innovative aspects of Innovation 1, and to favor, through the word-of-mouth, the diffusion of the technology to the mainstream market, i.e., large enterprises that can build and manage a communication infrastructure on their own. According to the marketing manager interviewed, the identification of the right target, as well as the support of the direct sales force, were deemed more important than the adoption of aggressive timing and pricing strategy to favor early market acceptance.

### Innovation 2

The target market for the Wireless Proximity Switch technology comprised manufacturing firms with the need to keep fundamental parameters in industrial processes continuously under control, especially in those conditions where there was little room for the installation and maintenance of control instrumentation. Company 2 believed that this revolutionary technology could hardly diffuse, replacing traditional wired sensors, unless it was enthusiastically adopted by some influential clients. For this reason, it decided to target Innovation 2, immediately after launch, to large and well-known automotive and cable manufacturers, whose assembly and production processes were extremely complex and hence could represent an ideal field for the installation and use of the novel control technology. Company 2 decided to establish a relationship with these potential early clients, mainly through the organization of technical seminars and events; interestingly, the firm also implemented an Innovation 2 solution in one of its plants, to which potential early clients were invited to see a real-world application of the technology.

## Innovation 3

Innovation 3 is an industrial computer developed to manage and control industrial processes. It elaborates the information collected through a series of sensors and actuators and optimizes the industrial process accordingly. In order to establish the product in the market, Company 2 recognized that Innovation 3 needed a significant amount of personalization. Hence, providing complete assistance to the clients was likely to have a positive impact on their acceptance. This in turn required the innovating firm to have access to sales channels that are knowledgeable about the technical characteristics of the product and its functionalities and, above all, willing to promote it through a direct face-to-face activity. In order to achieve these objectives, the company relied heavily on its highly qualified sales force to distribute the products. In particular, Company 2 focused the attention of its sales force on a particular kind of client that was deemed to play an opinion leadership role in the mainstream market, i.e., the producers of advanced manufacturing systems. According to the marketing manager of Company 2, the employment of the company's highly skilled sales force played a crucial role in favoring the adoption of Innovation 3 in the early market.

## Innovation 4

Innovation 4 is an industrial system developed to prevent or mitigate the risks of adverse events within an industrial plant (e.g., short circuits). The marketing manager of Company 4 decided, in the first phase, to launch the product to the producers of distributed control systems (DCS), since this kind of client is always searching products able to improve safety conditions in industrial plants and can greatly support the diffusion of Innovation 4 within the mainstream market. In order to reach these clients, the company leveraged complementary approaches. According the marketing manager interviewed, besides the employment of its own sales force and several face-to-face meetings with the DCS producers, Company 4 started an impressive communications campaign through the organization of events, conferences, and meetings, where it illustrated the importance of safety issues in industrial plants as well as the value that its clients could achieve through the introduction of Innovation 4 in their plants.

## Innovation 5

Innovation 5 is represented by a measurement system that, thanks to the laser technologies, helps professionals to make accurate measurements of areas, volumes, offsets, perimeters, and distances in an easier way than the traditional measurement system. Furthermore, the measurement system developed by Company 5 is smaller than competitors' products. In order to support an adequate diffusion of the product in the mainstream market, Company 5 did not target a specific category of early customers, since it could leverage a large base of "Company 5 fans," i.e., clients that are really devoted to the company and hence buy its products once they are launched on the market. In order to support the diffusion of the product in this early market, Company 5 took particular care with the distribution of Innovation 5, spending considerable effort in the training of its sales force. Moreover, an important communications campaign was launched, with different seminars and presentations during the meetings of engineering associations, since they are considered important opinion leaders. In these ways, the company developed adequate awareness of the benefits related to the adoption of Innovation 5

### Innovation 6

With the Innovation 6 project, Company 6 aimed at developing and commercializing an electronic control device that potentially could be installed on cars belonging to all segments, starting from high-end categories throughout mid- and low-end vehicles. The marketing manager we interviewed recognized that the success of Innovation 6 was favored by an unpredictable event, which is an example of a spontaneous positive early market acceptance. At the time when Innovation 6 was being developed, a new car segment was emerging that comprised high-end minivans (also called multi-purpose vehicles, or MPVs). These vehicles had particular stability problems, due to their high barycenter. In particular, the A Class by Mercedes-Benz did not pass a compulsory test (called the moose test) and the sales of the vehicle were interrupted. Innovation 6 was capable of solving its stability problems, and it was enthusiastically adopted by other car manufacturers (e.g., Audi, BMW, Fiat, and Peugeot), initially for their MPVs and later on for their other models. Although Company 6 did not proactively target a particular market segment at launch where the new technology could be positively accepted, the success that Innovation 6 encountered among the earliest segments in which it diffused turned out to be fundamental for favoring its diffusion. In particular, in order to speed up the diffusion process within the early market, Company 6 sent its sales people and technicians to all the clients potentially interested in the product to discuss and stimulate their willingness to acquire the technology.

### Innovation 7

For the commercialization of its Innovation 7 series, which had a vast and heterogeneous target market comprising medium and large enterprises working in almost all manufacturing and service sectors, Company 7 designed a specific early-adoption approach that, according to the interviewed marketing manager, turned out to be highly effective in stimulating sales growth. Specifically, Company 7 targeted at launch companies that had the need to undertake server consolidation activities; they were informed and educated about the characteristics of the novel Innovation 7 machine, mainly through seminars and similar events, in which Company 7's distributors also took part. These early clients were given a specifically designed version of Innovation 7, called White Box, which was aimed at meeting their compelling reason to buy (i.e., to carry out the server consolidation process efficiently and effectively) and hence helped stimulate their adoption decisions.

## Innovation 8

Innovation 8 is a vertical CNC (Computer Numerical Control) machine that represents an improvement in respect to traditional vertical machine centers, since it is characterized by higher reliability and precision. The higher precision and reliability of the machine in comparison to traditional CNC systems made Innovation 8 particularly suitable to perform very complex production processes such as the ones performed by the producers of molds. Hence, the marketing manager of Company 8 decided to target this particular segment of the market. In particular, the company presented Innovation 8 at two main specialized industrial exhibitions (i.e., BIMU and EMO), pointing out the advantages of the new machine for mold producers. According to the marketing manager interviewed, this communication strategy was highly effective in reaching the early market, which favored an adequate diffusion of Innovation 8 in the early market.

## Innovation 9

Innovation 9 was a novel CNC vertical machine targeted to small and medium enterprises; besides improved cutting performance, reliability and precision, Innovation 9 enabled the processing of parts with very large dimensions. Company 8 believed that a proactive engagement of influential early clients could serve the purpose of stimulating the diffusion of Innovation 9 among more pragmatic adopters; for this reason, it decided to focus on an industrial district in the northeast of Italy and, specifically, on those firms within this district that had the need to process two pallets in parallel, since Innovation 9 was believed to be particularly suited for this kind of application. These firms were invited to an open-house seminar, where the characteristics of the machine and the business applications it was capable of supporting were illustrated. This seminar convinced the chief of a small industrial laboratory to acquire Innovation 9; he was an influential and well-known entrepreneur within the industrial district and, according to the product manager interviewed, he exerted an important opinion leadership role that was very useful for stimulating the diffusion of the new technology among more pragmatic buyers.

## Innovation 10

Innovation 10's target market comprised medium and large firms that had the need to scan, print, fold and bind documents (e.g., instruction handbooks) in very high volumes. Company 10 recognized that the ability to get a positive appraisal among the first clients that adopted the innovation was fundamental for stimulating its diffusion among less innovative and more pragmatic adopters. As a result, it designed a commercialization process for Innovation 10 that included a formally designed stage (named ECE, or Early Customer Engagement), through which the new product was targeted, immediately after launch, to companies that showed a compelling reason to buy it and had high visibility in the market. The purpose was to get their support and to actively exploit the sponsorship they ensured in order to spur the diffusion of the Innovation 10 in the bulk of the target market. In order to reach this objective, Company 10 sent its sales people and technicians to the first clients that manifested an interest towards Innovation 10. The first adopters of the innovation, who exerted a particularly effective opinion leadership function, were furnishings and home appliance manufacturers that needed to print, fold, bind, and introduce technical manuals into the packaging of their final products, without interrupting the production flow.

**Innovation 11**

Innovation 11 is a new range of server computers based on the UltraSPARC 64-bit microprocessor architecture, characterized by higher performances in respect to the competitors' ones. In particular it is up to 28 percent faster in terms of computational speed and it is easily scalable to match different consumers' needs. In order to launch the product on the market, Company 11 developed a detailed marketing strategy based on i) the identification of the right early adopters target, ii) the employment of a highly qualified sales force, and iii) the development of additional tools to customize Innovation 11. In more detail, the company identified large data processing firms as the most suitable early adopters of the new server, since they can easily appreciate the new potentialities offered by the product in terms of computational speed. In order to reach this set of clients, Company 11 heavily employed its highly skilled and qualified workforce to present the product and its benefits to the potential clients. Finally, the company developed also a set of customization tools that enabled its clients to customize the product to their specific needs, fully exploiting the potentialities offered by Innovation 11. The marketing manager of Company 11 judged these initiatives as highly effective in fostering the adoption of the new server in the early market.

## Notes

1. Compare for instance a firm that launched a new technology in the market of traditional photography with one that entered the digital photography industry with a new image format. When launching its innovation, the first firm had to convince only film producers to support its new technology; on the contrary, in the case of digital photography, a wider set of players could affect the innovation success – printer and personal computer manufacturers, companies that produce software for editing and organizing pictures, broadband communication firms, and manufacturers of cellular phones and handsets.
2. Considering the single innovation as unit of analysis (and not the firm's portfolio of innovations, or its innovation program), commercial success can be defined as

a multidimensional concept that basically comprises two dimensions: the degree of customer acceptance and the financial performance achieved by the innovation (Griffin & Page 1993; Hultink & Robben 1995; Cooper & Kleinschmidt 1987). For the purpose of this study, we focused on the first dimension. This choice was suggested by the fact that high-tech markets are characterized by competitive dynamics, at least during their hyper-growth phase, that ensure a leading profitability edge to the innovation gaining the largest market share (Moore 1998).

3. These two items have been extensively used in the literature about New Product Development (NPD) success and failure for operatively assessing the degree of customer acceptance that a new product experienced in the market (see, e.g., Cooper & Kleinschmidt 1987).

4. Heterogeneity in the experts' assessments in the first round of the Delphi Method can be attributed to several factors: a lack of knowledge of the innovations and actual differences in their subjective evaluation. The second round was aimed at eliminating heterogeneity due to the first factor. For each innovation, we provided the experts with the evaluations gathered in the first round. The panelists were asked to review these data and submit a new set of evaluations. In comparison with the first round, the second one determined a convergence in the experts' opinions on all the innovations. A third follow-up round involved two experts with persistently divergent assessments. It showed that they were not willing to change their assessments anymore and therefore further iterations of the process would not increase convergence (Linstone & Turoff 1975).

## References

Abrahamson, E., & Rosenkopf, L. (1990) When do bandwagon diffusion roll? How far do they go? And when do they roll backwards: A computer simulation. *Academy of Management Review*, 15(3), 155–159.

Abrahamson, E., & Rosenkopf, L. (1993) Institutional and competitive bandwagons: Using mathematical modelling as a tool to explore innovation diffusion. *Academy of Management Review*, 18(3), 487–517.

Abratt, R. (1986) Industrial buying in high-tech markets. *Industrial Marketing Management*, 15(4), 293–298.

Ali, A. (1994). Pioneering versus incremental innovation: Review and research propositions. *Journal of Product Innovation Management*, 11(1), 46–61.

Bass, F. M. (1969) A new product growth model for consumer durables. *Management Science*, 13(5), 215–227.

Bayus, B. L. (1998) An analysis of product lifetimes in a technologically dynamic industry. *Management Science*, 44(6), 763–775.

Beard, C., & Easingwood, C. (1996) New product launch: Marketing actions and tactics for high-technology products. *Industrial Marketing Management*, 25(2), 87–103.

Brierty, E., Eckles, R. W., & Reeder, R. R. (1998) *Business Marketing* (3rd edn). New Jersey: Prentice Hall.

Butje, M. (2005) *Product Marketing for Technology Companies*. Elsevier Butterworth-Heinemann: Oxford.

Calantone, R. G., & DiBenedetto, A. (1988) An integrative model of the new product development process: An empirical validation. *Journal of Product Innovation Management*, 5(3), 201–215.

Calantone, R. G., & Montoya-Weiss, M. M. (1993) Product launch and follow on. In W. Souder and D. Sherman (Eds), *Managing New Technology Development*. McGrow-Hill: New York.

Chakravorti, B. (2004a) The new rules for bringing innovations to market. *Harvard Business Review*, 82, 58–67.

Chakravorti, B . (2004b) The role of adoption networks in the success of innovation. *Technology in Society*, 26 (2–3), 469–482.

Chiesa, V. & Frattini, F. (2011) Commercializing technological innovation: Learning from failures in high-tech markets, *Journal of Product Innovation Management*, 28, 437–454, ISSN: 0737–6782.

Christensen, C. M. (1997) *The Innovator's Dilemma: When New Technologies Cause Great Firms to Fail.* Harvard Business School Press: Boston, Massachusetts.

Cooper, R. G. (1979) The dimensions of industrial new products success and failure. *Journal of Marketing*, 43(3), 93–103.

Cooper, R. G., & Kleinschmidt, E. J. (1987) Success factors in product innovation. *Industrial Marketing Management*, 16(3), 215–223.

Cooper, R. G., & Kleinschmidt, E. J. (1988) Resource allocation in the new product process. *Industrial Marketing Management*, 17, 259–262.

Costa, C., Fontes, M., & Heitor, M. V. (2004) A methodological approach to the marketing process in the biotechnology-based companies. *Industrial Marketing Management*, 33(5), 403–418.

Czepiel, J. A. (1975) Patterns of interorganizational communications and the diffusion of a major technological innovation in a competitive industrial community. *Academy of Management Journal*, 18(1), 6–24.

Dalkey, N. and Helmer O. (1963). An experimental application of the Delphi method to the use of experts. *Management Science*, 9(3), 458–467.

Davidow, W. H. (1986) *Marketing High Technology: An Insider's View.* The Free Press: New York.

Davies, W., & Brush, K. E. (1997) High-tech industry marketing: The elements of a sophisticated global strategy. *Industrial Marketing Management*, 26(1), 1–13.

Day, R. L. & Herbig, P. A. (1990) How the diffusion of industrial innovations is different from new retail products. *Industrial Marketing Management*, 19(3), 261–266.

DiMaggio, P., & Powell, W. (1983) The iron cage revisited: Institutional isomorphism and collective rationality in organizational fields. *American Sociological Review*, 48(2), 147–160.

Easingwood, C., & Harrington, S. (2002) Launching and re-launching high technology products. *Technovation*, 22(11), 657–666.

Eisenhardt, K. M. (1989) Building theories from case study research. *Academy of Management Review*, 14(4), 532–550.

Frattini, F. (2009) The future of innovation... managing the fuzzy back-end of the innovation process. In B. von Stamm and A. Trifilova (Eds), *The Future of Innovation* (326–327). Gower publishing: Surrey, England.

Granovetter, M. S. (1973) The strength of weak ties. *American Journal of Sociology*, 78(6), 1360–1380.

Griffin, A., & Page, A. L. (1993) An interim report on measuring product development success and failure. *Journal of Product Innovation Management*, 10(4), 291–308.

Hartley, R. F. (2005) *Marketing Mistakes and Successes.* John Wiley & Sons: Westford.

Hatzichronoglou, T. (1997) Revision of the high-technology sector and product classification. *OECD Directorate for Science, Technology and Industry Working Paper*, 2, OECD, Paris.

Hultink, E. J., &Robben, H. S. J. (1995) Measuring new product success: The difference that time perspective makes. *Journal of Product Innovation Management*, 12(5), 392–405.

Hutt, M. D. & Speh, T. W. (1992) *Business Marketing Management* (4th edn). Fort Worth: Dryen Press.

Kennedy, A. M. (1983) The adoption and diffusion of new industrial products: A literature review. *European Journal of Marketing*, 17(3), 31–88.

Kleinschmidt, E. J., & Cooper, R. G. (1991) The impact of product innovativeness on performance. *Journal of Product Innovation Management*, 8(4), 240–251.

Kotler, P. (2003) *Marketing Management* (11th edn). New Jersey: Prentice Hall.

Lesnick, P. C. (2000) *Technology Transfer in the Dominican Republic: A Case Study of the Diffusion of Photovoltaics*. PhD Dissertation, Union Institute, Cincinnati, Ohio.

Linstone, H. A. & Turoff, M. (1975) Delphi Method: Techniques and Applications. Addison-Wesley Educational Publishers.

Mahajan, V., & Muller, E. (1998) When it is worthwhile targeting the majority instead of the innovators in a new product launch? *Journal of Marketing Research*, 25(4), 488–495.

Miles, M. B., & Huberman, M. (1984) *Qualitative Data Analysis: A Sourcebook of New Methods*. Sage: London.

Millier, P. (1999) *Marketing the Unknown. Developing Market Strategies for Technical Innovations*. John Wiley & Sons: Chirchester.

Mohr, J., Sengupta, S., & Slater, S. (2001) *Marketing of High-technology Products and Innovations*. Person Education: Upper Saddle River, New Jersey.

Moore, G. (1991) *Crossing the Chasm: Marketing and Selling Technology Products to Mainstream Customers*. HarperBusiness: New York.

Moore, G . (1998) *Inside the Tornado. Marketing Strategies from Silicon Valley's Cutting Edge*. Capston: Chichester.

Nevens, M., Summe, G., & Uttal, B. (1990) Commercializing technology: What the best companies do. *Harvard Business Review*, 68, 154–163.

Ofek, E., & Sarvary, M. (2003) R&D, marketing, and the success of next-generation products. *Marketing Science*, 22(3), 355–370.

Powell W. W., & DiMaggio P. J. (Eds) (1991) *The New Institutionalism in Organizational Analysis*. The University of Chicago Press.

Rao, A., Rogers, E. M., & Singh, S. N. (1980) Interpersonal relations in the diffusion of innovation in two Indian villages. *Indian Journal of Extension Education*, 16, 19–24.

Robertson, T. S. (1967) The process of innovation and diffusion of innovation. *Journal of Marketing*, 31(1), 14–19.

Rogers, E. M. (1976) Communication and development: The passing of the dominant paradigm. *Communication Research*, 3(2), 213–240.

Rogers, E. M. (2000) Reflections on news event diffusion research. *Journalism and Mass Communication Quarterly*, 77(3), 561–576.

Rogers, E. M. (2003) *Diffusion of Innovations* (5th edn). Free Press: New York.

Schilling, M. A. (2005) *Strategic Management of Technological Innovation*. McGraw-Hill: New York.

Sheth, J. N. (1981) Psychology of innovation resistance. *Research in Marketing*, 4, 273–282.

Sultan, F., Farley, J. U., & Lehman, D. R. (1990) A meta-analysis of applications of diffusion models. *Journal of Marketing Research*, 27(1), 70–77.

Tidd, J. (2010) *Gaining Momentum: Managing the Diffusion of Innovations*. Imperial College Press: London.

Veryzer, R. W. (1998) Key factors affecting customer evaluation of discontinuous new products. *Journal of Product Innovation Management*, 15(2), 136–150.

Viardot, E. (2004) *Successful Marketing Strategies for High-tech Firms*. Artech House: Norwood.

Yin, R. K. (2003) *Case Study Research: Design and Methods*. Sage: London.

# 9
# Managing Communities of Practice to Support Innovation
*Stefano Borzillo and Renata Kaminska*

## 1 Introduction

In the face of a complex, hypercompetitive environment (D'Aveni 1994), it is particularly important to understand the factors allowing organizations to simultaneously innovate and generate profits (March 1991; Tushman & O'Reilly 1996; Bradach 1997; Sutcliffe et al. 2000; Tushman & Smith 2002; Warglien 2002). Much of the literature on organizational ambidexterity focuses on solving the control/autonomy tension (Lawrence & Lorsch 1967; Burgelman 2003; Doz & Prahalad 1986; Bartlett & Ghoshal 1998) through decentralization of decision-making processes (Brown & Eisenhardt 1997) and redefining control mechanisms (Pascale 1990; Daft & Lewin 1993; Victor & Stephens 1994). However, these solutions, implemented within the scope of project-based structures and other forms of interdisciplinary team-work (Nonaka & Takeuchi 1995; Halal 1999; Miles et al. 1999; Birkinshaw & Gibson 2004), have proven to be of limited efficacy. Why?

One of the reasons is that none of them breaks away from the traditional command-and-control paradigm, considered to be the main source of organizational rigidity. From this perspective, communities of practice (CoPs) are a promising solution. But what are they? Even if the term was coined by Wenger (1998), its origin dates to common research by Lave and Wenger (1991), where the authors studied how new members become part of informal groups and how they learn through sharing of practices and experience. In this perspective, a community of practice can be defined as an informal group of people who share a common interest and want to contribute to and learn from the community. Communities of practice emerge in diverse contexts: online in the form of discussion groups or blogs, as well as in real-life – anywhere in any field setting. In organizations, they are seen as an interstitial space, enabling experiments with new ideas without managers having to fully assume the risks related to their functioning or outcomes (Brown & Duguid 2001; Wenger et al. 2002; Josserand 2004). They therefore transcend the rigidity of organizational boundaries and hierarchies without destabilizing the formal organizational structure.

However, the literature on CoPs seems to have an unresolved tension regarding both the nature of CoPs and the role they play in the organization, focusing mainly on the paradox between autonomy and control from the managerial perspective. While the first studies viewed them as self-emerging processes, spaces of freedom and cooperation in which organizational practices evolve via an unconstrained exchange of knowledge (Lave & Wenger 1991; Brown & Duguid 1991; Orr 1990), more recent research highlights that, in practice, managers not only attempt to exert some control over CoPs (Josserand 2004; Thompson 2005; Probst & Brozillo 2007 2008), but often actively support their formation to enhance the organization's capacity to innovate (Blanchot-Courtois & Ferrary, 2009). However, even if there seems to be a growing consensus that managers can and do play a role in nurturing CoPs, little is known about the CoP dynamics that support innovation.

The objective of this exploratory research is to better understand CoP dynamics and the role managers play in guiding a constellation of communities of practice to enhance innovation. Presuming that there is an interplay between guidance and autonomy, we argue that if managers use the right guidance mode, the desired knowledge processes will occur within and across communities, thus enhancing the organization's capability to innovate.

Our paper is structured as follows: We start off by confronting the conception of CoPs as emergent, unmanageable processes (Lave & Wenger 1991; Brown & Duguid 1991; Orr 1990) with the more recent focus on management's attempt to control their activities (Josserand 2004; Thompson 2005; Probst & Borzillo 2007, 2008). We argue that these two conceptions are, in fact, complementary and both support CoP dynamics. We draw on community of practice theory to identify forces that drive the CoP dynamics. After a description of our research methods, we offer a detailed case study of five CoPs in a multinational firm (Zyc Chemicals) in the chemical industry. We conclude by showing how a deeper understanding of forces driving CoP dynamics may help managers effectively guide CoP activity to support innovation in their organizations.

Our results reveal two distinct interplaying pair of forces that supported innovation processes in all five CoPs at Zyc Chemicals. On the one hand, the "control-driven" pair combines strong sponsorship and leadership that stimulates knowledge exploitation processes within communities. This pair leads to the recombination of knowledge to improve existing product offerings (incremental innovation). On the other hand, the "autonomy-driven" pair combines members' strong desire to enhance cooperation and span the boundaries of their communities. This combination of forces induces knowledge exploration to generate radically new product offerings (radical innovation).

## 2   Communities of practice dynamics

First introduced by Lave and Wenger (1991) in the context of situated learning, CoPs can be defined as tightly knit groups that have been practicing together long enough to have developed into a cohesive community that provides a sense of belonging and commitment (Wenger 1998). Wenger (1998) identified three characteristics associated with CoPs: the members interact with one another, establishing norms and relationships through *mutual engagement;* the members are bound together by an understanding of a sense of *joint enterprise;* and the members produce a *shared repertoire* of communal resources, including language, routines, artifacts, and stories (1998, p. 72).

CoPs are viewed as environments in which knowledge can be created and shared and, most importantly, used to improve effectiveness, efficiency, and innovation (Lesser & Everest 2001; Wenger & Snyder 2000). Initially viewed as eluding all forms of managerial control, they have increasingly been perceived as "productive structures" (Josserand 2004). Notwithstanding this evolution, the causal interactions between CoPs and the organization need to be further elucidated, which we do in the following section.

### Communities of practice as grassroots structures

There is a long tradition of perceiving CoPs as an emergent phenomenon that develops organically through bottom-up processes (i.e., Orr 1996). CoPs were initially presented as grassroots and self-organizing structures that could not be set up nor steered by management (Brown & Duguid 1991; Lave & Wenger 1991; Orr 1990). They were believed to exist naturally throughout the organization, which means that the participants have not been formally brought together in a networked structure (Wenger 1998, McDermott 1999; Hildreth et al. 2000; Wenger et al. 2002). Only their own motivation, stemming from common interests, drives them to remain in a CoP.

CoPs imply sharing experiences and practices among the members. Because membership is based on spontaneous participation rather than on official status, these communities are not restricted by organizational affiliations; they can cross structures and hierarchies. For Liedtka (1999), the community's practice exists and evolves in its social interaction. At the simplest level, a CoP is a group of people who have worked together over a period of time (Wenger & Snyder 2000). It is neither a team, nor a task force (O'Hara et al. 2002). It is not necessarily authorized and sometimes not even identified in the organization (Büchel & Raub 2002). Liedtka (1999) adds that a shared sense of purpose and a strong motivation to know what others know hold the CoP members together.

According to Wenger (1998), in an organizational context, CoP members naturally group around the topics they find the most relevant for the organization and, through their informal interactions, they develop new knowledge. Even if their understanding of the strategic importance of a particular knowledge can be influenced by outside constraints or directives (from management), members develop knowledge and practices that are their own particular responses to these external influences. Wenger adds that even when a community's actions conform to an external mandate, it is the community – not the mandate – that produces the practice.

Thus, CoPs are characterized by their autonomy, and they design their network in the way that suits best their evolution (Wenger et al. 2002). Büchel and Raub (2002) add that because CoPs correspond to informal gatherings of individuals brought together by shared interests, they may appear to be unmanageable endeavors. The self-emerging aspects of CoPs are also pointed out by McDermott (2001), when he states that the specific issues on which CoPs focus change over time, as the needs and interests of their members change. CoPs grow and thrive as their focus and dynamics engage community members (McDermott 2001). This creates a context that enhances knowledge-creation processes. An increasing number of studies, however, have discussed whether managers can play an active role in steering CoPs – initiating a controversy (Thompson 2005; Borzillo et al. 2008).

## Steering communities of practice

As in related literatures on self-managing work teams (e.g., Barker 1993; Ezzamel & Willmott 1998) and open innovation communities (e.g., Dahlander & Wallin 2006; O'Mahony & Ferraro 2007), more recent studies suggest that CoPs can and should be cultivated. Even if there is no consensus on how managers can use CoPs for strategic advantage (Adams & Freeman 2000; Kimble & Hildreth 2005; Loyarte & Rivera 2007; Milne & Callahan 2006), there seems to be an agreement that managerial control has a role to play (e.g., Lesser & Everest 2001; Thompson 2005; Probst & Borzillo 2008). Control refers to "any mechanism that managers use to direct attention, motivate, and encourage organizational members to act in desired ways to meet an organization's objectives" (Long et al. 2002, p. 198).

Wenger and Snyder (2000), for example, believe managers cannot mandate CoPs. Instead, these authors argue that managers should bring the right people together and provide an infrastructure in which communities can thrive. The idea here is to allow these networks to evolve without imposing any particular structure on them (Thompson 2005). The objective is to identify existing networks and encourage them to evolve around topics that are at the heart of the company's business. Wenger et al. (2002) highlight that this requires involving thought leaders as soon as possible to build energy into the community. Along the same lines, Büchel and Raub (2002) stress that CoPs need to be focused on strategic priorities and that network outcomes should be leveraged.

Consequently, it seems that there is an inherent conflict between the various perspectives on CoPs; furthermore, *how* exactly organizations should address the managerial paradox inherent in CoPs remains undecided (Contu & Willmott 2000; Handley et al. 2006; O'Mahony & Ferraro 2007; Thompson 2005). Therefore, some investigations conclude that the concept of a CoP as a self-regulating process is unrealistic and clearly contrary to a normative goal to steer them to increase organizational performance (Fox 2000; Contu & Willmott 2000). In response to this perspective, more recent research, for example, suggests that organizations need to sponsor and manage CoPs so that they can generate usable knowledge in a sustainable way, but that organizations should not try to exert too much control, as this runs the risk of destroying them (Anand et al. 2007; Brown & Duguid 2001; Swan et al. 2002; Borzillo 2009).

Anand and his colleagues (2007) likewise studied the process of community emergence and found both autonomous and control-oriented generative elements in successfully developing communities. Interestingly, Anand et al. further conclude that different combinations of these generative elements may be possible. While there seems to be a growing consensus that managers can have an influence on CoPs development, more fine-grained theorizing is required to understand this process. We draw on CoP theory to understand the interaction of the forces underlying the dynamics driving CoPs' evolution.

### Driving forces from CoP theory

First, researchers claim that senior executives need to provide *sponsorship* to help communities reach their full potential (Büchel & Raub 2002; Lesser & Everest 2001; Wenger & Snyder 2000). For instance, Wenger & Snyder (2000) recommend the support of an "official sponsor" who can, on behalf of the top management, provide the CoP with the necessary time and financial resources. The sponsor must also ensure that the CoP's knowledge activities remain in line with those of the organization (Wenger & Snyder 2000). Moreover, if members of a CoP have clear knowledge objectives, they will participate more actively (McDermott 2003; Wenger & Snyder 2000). Sponsorship should include the necessary resources to maintain the network's continuous operation, i.e., resources to set up a technological communication infrastructure, to organize meetings, to cover members' traveling expenses, and to defray other network activity expenditures (Büchel & Raub 2002).

Second, organizations can designate *leadership* roles to motivate community members and strengthen the group's identity (Lesser & Everest 2001; Thompson 2005; Wenger et al. 2002). Leadership is a critical factor for a CoP's success. In fact, CoP leaders must commit 20 to 50 percent of their working hours to the promotion and supervision of the CoP's activities to ensure that it remains operative (Wenger et al. 2002). The leaders should be accountable for the degree of CoP member participation (Lesser & Everest 2001) and should help members develop their practices by sharing their expert knowledge (McDermott 2001). With that objective in mind, leaders should frequently respond to members' knowledge requests (Wenger & Snyder 2000). Leaders should also create bonds among the members (McDermott 2001) to enable them to exchange knowledge related to the CoP practice (Wenger et al. 2002). Finally, the leader should encourage members to document their experiences and knowledge, using a dedicated CoP database (Wenger et al. 2002).

Third, *members' ongoing cooperation* may help to actively cultivate communities (McDermott 2001, Wenger et al. 2002). More precisely, Wenger et al. (2002) and McDermott (2001) stress the importance of enhanced cooperation, which they define as the regular organization of meetings, teleconferences, web-based activities, and other informal events to keep the CoP active in knowledge and experience sharing. The CoP should, however, be cultivated without killing it. If the quantity of activities becomes too overwhelming, the members will lose their motivation to participate. Likewise, if CoP activities are not presented frequently enough, members will gradually become disinterested (Wenger et al. 2002).

Fourth, organizations can establish linkages beyond the community's boundaries that enable knowledge to be shared throughout the organization (Wenger 2000). This process of *boundary spanning* (Levina & Vaast 2005, 2006; Lindkvist 2005; Holmqvist 2003; Cohen & Levinthal 1990; Rosenkopf & Nerkar 2001), by interacting with actors external to one's network, brings diversity and novelty into the system, allowing it to create new knowledge. In the context of CoPs, members may belong to several CoPs in the same organization; they can interact with CoP members from other organizations; CoPs may also interact with customers, suppliers, and other external actors.

The next stage of our research involved conducting an exploratory study of five CoPs in the chemical division of a multinational organization to test whether the four driving forces that we had extracted from CoP theory supported knowledge creation within them. In other words, the objective was to examine the forces that drive processes of knowledge creation in CoPs and, consequently, to determine how CoPs contribute to innovation. We continue this chapter by discussing the methodology deployed in our empirical investigation. Thereafter, we present and discuss our results.

## 2 Methods

This research took place between 2004 and 2008 within a multinational organization. For confidentiality reasons, we call this organization Zyc Chemicals. Zyc Chemicals is an international industrial group with a workforce of about 41,000, and it is involved in activities throughout the world. The group is active in the chemicals, energy, and real estate sectors. The company is structured around eight business units, which report directly to the group's management board. We conducted our study in the Specialty Chemicals division, which had a leading position in 80 percent of its chemical business (in 2008). Innovation is thus crucial to sustain Zyc Chemicals' competitive advantage.

### Research setting

The initiative to start this research came from Zyc Chemicals' management executives, who wanted to better understand how CoPs could be used to stimulate innovation within the Specialty Chemicals division. They believed that mandating external researchers to conduct this investigation would provide a perspective that was untainted by a corporate perspective.

The management executives at Zyc Chemicals initially provided us with a sample of five CoPs that they believed were highly promising for supporting innovation processes in the field of specialty chemicals for the Specialty Chemicals division. These five CoPs in the Specialty Chemicals Division were: 1) the Industrial Chemicals CoP, 2) the Health & Nutrition CoP, 3) the Polymers CoP, 4) the Inorganic Materials CoP, and 5) the Coating & Additives CoP.

### Data collection and analysis

Data were collected on each of the five CoPs from more than eight informants to ensure that theoretical saturation was reached (Glaser & Strauss 1967). We collected data via three of the six sources of evidence identified by Yin (2003): interviews, documents, and direct observation. Data collection and analysis were done iteratively over the four years of our longitudinal study (Siggelkow 2002). These various sources of information improved the level of completeness and saturation, which are two key internal validity criteria proposed by Mucchielli (1991).

Interviews were also conducted over the four years of the study period. During this timeframe, we conducted six series of interviews, dividing the four years into six periods of observation. During each series of interviews, the informants were asked to provide a detailed account describing the impact that each of the four driving forces had on knowledge creation within the five CoPs. The accounts were aimed at uncovering how the four driving forces stimulated knowledge creation within the CoPs over time. This approach is called a priori theory testing (Yin 2003). This is a process by which a priori constructs – in our study, the driving forces that support CoP dynamics – are tested in a new context (multiple CoPs at different moments in time). The aim is to uncover how these a priori constructs impact, either positively or negatively, a phenomenon under investigation – in our case, how CoPs may help the Specialty Chemicals division create knowledge and support innovation.

On average, each series of interviews (six overall) lasted 23 days. The number of informants varied from 44 to 67 throughout the six periods.

The informants in the interview process included community sponsors[1] (one per community), community leaders (one or two per community), and core members (multiple). These types of informants were presumed to be the most informed about and experienced with CoPs (McDermott 2001; Wenger et al. 2002). Line managers, who occasionally took on external sponsoring, mentoring, or specialist roles, were also interviewed. They participated less frequently in the CoPs and therefore provided us with a different, more external perspective on the dynamics of the CoPs. Selecting multiple informants also helped to mitigate subject biases (Golden 1992) and to provide a broader range of perspectives on the subject of the study (Guba & Lincoln 1989).

Following the transcription of the recorded interviews – which was done after each of the six series of interviews – a detailed written account (Eisenhardt 1989b; Yin 2003) was prepared of each of the periods under observation. The accounts contained a description of the driving forces in action in each of the CoPs as well as selected quotes representing different informants' opinions on how each driving force had impacted the knowledge creation process in each CoP. In most cases, the informants largely agreed on the most relevant issues, such as the nature of the driving forces used and the role they played in dynamizing knowledge creation in their CoP. Identifying and analyzing the interactions between the driving forces was done over the four-year period. This gave our study an iterative approach "feeling" as data were collected and analyzed simultaneously (Siggelkow 2002; Miles & Huberman 1994). Both researchers were involved in analyzing each of the periods under review throughout the study period. They eventually started to recognize patterns in the five CoPs' overall developmental path. Individually, the researchers analyzed the six periods, only comparing their findings after having completed a full assessment of each.

The result was a basic framework revealing two distinct and interplaying pair of driving forces applying to the five CoPs in the course of the six observation periods. Each pair of forces generated a distinct knowledge-creation process and thus supported a specific type of innovation. Using this framework, we reexamined the data and conducted several additional interviews and meetings to challenge our findings. This iterative process helped refine our framework. We next describe our findings.

## 3 Findings

In this section, we present two distinct interplaying pair of forces that supported innovation processes in all five CoPs. On the one hand, the control-driven pair combines strong sponsorship and leadership. This combination engenders knowledge processes that lead to some incremental innovation. On the other hand, the autonomy-driven pair combines CoP members' strong desires to enhance cooperation and participate in boundary-spanning activities. This combination of forces induced knowledge processes that contribute to radical innovation.

We now continue with a detailed presentation of these two pairs and the interacting forces that constitute them.

### Control-driven forces

The control-driven pair of forces is characterized by the management influencing all five CoPs simultaneously. This pair results from an annual governance committee, during which sponsors from the top management and the CoPs' leaders meet. Each CoP has one sponsor and one leader who represent it in the committee. The purpose of the governance committee is to assess whether each of the CoPs' knowledge creation activities have a strategic potential for innovation within the Specialty Chemicals division.

In this respect, management *sponsorship* plays an important role, as the sponsor is involved in setting knowledge objectives for the CoP and approves the quality of its outputs in terms of knowledge creation. He or she provides feedback on the knowledge's usefulness and thus contributes to the knowledge base's quality. The knowledge objectives are jointly determined by the sponsor and the CoP leader. These objectives are directed towards the division's long-term strategy and are kept general, offering the community members considerable autonomy regarding the right approach to realize them is selected. As the Polymers CoP leader recounted, "While our sponsor discusses the overall knowledge objectives with me before we both agree on them, we ensure that we provide the community members with sufficient autonomy to develop innovative ideas and knowledge that can be applied to our products in order to improve them." The Industrial Chemicals CoP leader summarized:

> We couldn't dynamize this community without setting objectives with our sponsor. Community members would simply lose interest. On the other hand, if we set objectives that are too precise, the members act mechanically and experiment less with new ideas. The optimal solution is to set qualitative objectives together with our sponsor and to ensure that tangible results are generated for the company. I should add that some guidance from our sponsor never hurts. Since he belongs to the strategic sphere of our organization, he always has a good idea of what knowledge is relevant to develop and when to start focusing on it.

The ongoing and participative revision of objectives gave the CoPs momentum by continually reactivating interactions among the core members, maintaining a "healthy tension" within the CoPs and allowing each of them to remain at the top of the chemical industry's fast-moving technological evolution. The process supported innovation and creativity through taking on board fresh ideas from members, who often worked in different R&D teams at Zyc Chemicals.

Our data thus shows that managers at the Specialty Chemicals division got involved in the five CoPs by adopting the role of sponsors. The Polymers CoP leader emphasized that sponsors must belong to the strategic sphere of the organization in order to have a good idea of what knowledge is relevant for development and when to start focusing on it.

Sponsors also interact on a continuous basis with clients of Zyc Chemicals and therefore are in the best position to ascertain their precise needs. Sponsors therefore played an essential role in supporting innovation processes within the five CoPs. They defined specific client problems and shared this knowledge with CoP members, who were then able to work on innovative solutions.

Another important role for sponsors is to ensure top management support for the CoP. The motivation is not to control communities, but to motivate and encourage the community members' further activities. As a community member of the Coatings & Additives CoP asserted, "It gives the members a real sense of belonging when the top management recognizes that innovative knowledge has emerged from our communities' activities." A community leader from the Inorganic Materials CoP mentioned, "Our sponsor regularly presents the most promising molecules conceptualized in our community to the top management in order to gain further financial support to concretely develop them within various innovation teams at Zyc Chemicals."

Coupled with sponsorship, *leadership* plays an important role and takes a distributed form represented by a close collaboration between the sponsors and leaders of the community. Sponsors and CoP leaders worked hand in hand. Together they organized group meetings, during which members discussed the imperfections of existing molecules and tried to improve their effectiveness. In the Coating & Additives CoP, for example, some 150 chemists and engineers were guided by a sponsor and were coordinated by a leader in their quest to explore improvements of coating applications for bodywork in the automobile industry. The leader of the Coating & Additives CoP explained that improved solutions for clients were found fastest when sponsors defined specific client problems and members were encouraged by him (the leader) to exchange their ideas and recombine the existing molecules to produce new formulas.

While he represents the sponsor within the community, the CoP leader is not a hierarchical superior who gives orders or sanctions activities. In general, he acts as a coordinator of the community members. The leader of the Health & Nutrition CoP maintained:

As the leader of this community, I act as the community sponsor's direct representative. Yet, I'm not the members' "chief." My role is mainly to coordinate and coach them in their knowledge creation activities.

In the case of the Polymers CoP, for example, the sponsor and the leader jointly organize CoP meetings during which status reports are presented by the community's core members. In between these meetings, however, the members are free to organize their own meetings without interference from the sponsor or the CoP leader. These and various other examples reflect the management's preference for a combination of limited actions to influence the CoP's activities. An explanation for this approach was often that it was based on the fear that more radical measures could undermine the community's creative character. Some leadership, however, was considered essential to direct the community's activities towards desired outcomes. As the community leader of the Inorganic Materials CoP recounted:

> To guide our community of practice, the sponsor and I try to equalize the leadership and self-direction of the community. Too much leadership and pressure to pursue knowledge objectives could destroy the members' spontaneity and creativity.

Furthermore, the leader not only represents the sponsor within the community, he also represents the members' interest and conveys these to the sponsor. The leader of the Coatings & Additives CoP said, "I report outstanding knowledge creation and innovation to our sponsor, who can urge the management to recognize these contributions."

In sum, the control-driven pair, with its interacting forces, is characterized by tight collaboration between sponsors and leaders in the setting of knowledge objectives. Leaders operate as intermediaries between the sponsor and the other community members and attempt to coordinate CoP members to fulfill knowledge objectives proposed by the sponsor. This provides some managerial oversight while preserving sufficient autonomy for creative thinking and experimentation. This pair of forces prevailed simultaneously in the five CoPs and led to the recombination of existing knowledge in order to improve Zyc Chemical's existing products offering.

*The outcome of the control-driven pair of forces: exploitation and incremental innovation*

As indicated above, the control-driven pair of forces was used to improve the existing product offerings. Such incremental innovation required the creation of new and improved knowledge based on existing know-how. The leader of the Coatings & Additives CoP explained, "The knowledge creation and improvement process occurs by helping one another recombine existing solutions to new and better solutions for our clients." The creation of new knowledge through the recombination of bodies of existing knowledge corresponds to the exploitation of knowledge-based processes. Exploitation contributes to incremental innovation, such as the creation of variants of existing products or the launch of existing products in new markets.

The informants described the control-driven pair of forces as appropriate for supporting exploitation processes. A member of the Health & Nutrition CoP, for example, remarked, "This community of practice provides an environment for a continuous and incremental knowledge-building process oriented towards client-focused innovation, as members are very good at focusing on redeploying existing knowledge." The setting of clear objectives by the management (sponsors) and the CoP leaders creates an adaptive tension and a strategy vector that commits all community members to the same goal. Burgelman (1991, 2002) describes this strategy as beneficial for exploiting existing knowledge, but also warns of inertial consequences that may harm innovation. The management was well aware of these risks at Zyc Chemicals. The leader of the Polymers CoP observed, "If I had allowed our sponsor to set strictly quantitative knowledge objectives for our community, it would certainly have destroyed the members' creativity." To counter the risk of management's attempting excessive oversight of communities in terms of knowledge objectives, CoP leaders at Zyc Chemicals ensured that the knowledge objectives set remained generic, adaptable, and of a qualitative nature. The collaborative and arm's-length leadership approach by the CoP leaders and sponsors thus provided sufficient leeway for the development of autonomous knowledge creation processes in addition to the induced knowledge objectives. Burgelman (2002) has described the equilibrium between induced and autonomous knowledge creation processes as critical for an organization's ability to maintain momentum.

### Autonomy-driven forces

The autonomy-driven pair of forces is characterized by its dominant role of enhancing cooperation and boundary spanning.

*Enhancing cooperation* is very important for engaging in exploration activities and mainly consists of regular and direct meetings between community members. A member of the Inorganic Materials CoP recounted:

> Imagine ten people are dropped in the middle of the rainforest. They have to join forces and be creative enough to develop survival practices. In a similar way, members of our community test their creative ideas together, in small groups, to envisage novel molecules that have nothing to do with existing ones. Sometimes, they don't even know what the exact application of these molecules could be for the formal R&D teams to which they belong. This doesn't matter, either. Most important is that we are experimenting in order to think of new and revolutionary molecular structures. This requires ongoing, face-to-face interaction, because building knowledge on something completely new is a very complex task.

Enhanced cooperation between CoP members is thus characterized by a multitude of informal meetings, which are usually small and relatively unstructured. As a member of the Polymers CoP stated, "During our group sessions, we experience unstructured processes of knowledge puzzling, which can be rather time-consuming and tiring." Meetings often took place in remote locations, including outdoor sessions and external workshops. One community member of the Coatings & Additives CoP explained, "The twenty core members of our community meet every two weeks in my house outside town to share experiences and figure out new ways of doing things." So, enhancing cooperation mostly involves ongoing one-to-one interactions and regular roundtable discussions by small groups of CoP members. Some general meetings were held in which the sponsor, leader, and all members participate. However, most interactions occur spontaneously through direct contacts between members. As one member of the Industrial Chemicals CoP mentioned, "Improving our portfolio of molecules is a complex process that requires regular and direct interactions between chemists."

The autonomy left to members to enhance their informal cooperation enabled the CoPs to produce radically new solutions. Members of the Inorganic Materials CoP tested their creative ideas to envisage novel molecules without knowing their exact application for their formal R&D teams. This did not matter because, as they revealed, what was important was to experiment in order to discover revolutionary molecular structures. In the same line, members of the Coatings & Additives CoP told us that conceiving of new coatings required a lengthy learning process, during which it was often necessary to start from scratch and play around with completely new combinations of chemical formulas. Finally, the leader of the Health & Nutrition CoP recounted that the immediate consequence of encouraging members to enhance their cooperation was that members started to experiment extensively with many more "unorthodox" ideas and decided to work on chemical catalysts, a different knowledge topic, which was completely new to their community.

More formal general meetings are organized to present new and successful knowledge developed within the community and to discuss ongoing innovation projects. However, such meetings occur much more seldom than the many informal meetings between members.

*Boundary spanning* is also considered highly important for exploring radically new knowledge, as reflected by the extensive interrelations between these communities and people external to the community. The intensive exchange with outsiders is perceived as a valuable source of fresh ideas and new insights. A community member from the Health & Nutrition CoP said, "I am always inspired by the new perspectives from the external experts whom we invite each week, which keeps me from being overwhelmed by the rigorous mental schemes that are usually characteristic of the R&D team to which I belong in the organization." Inviting customers to participate in CoPs enables a better grasp of their needs and expectations in terms of improving the product offerings. As the leader of the Coatings & Additives CoP recounted:

> It is almost impossible to come up with new coating applications without regular feedback from our customers. I have therefore extended our community of practice's borders to include our most loyal customers. This enabled us to explore customer needs in an introspective way and to generate new ideas that perfectly match their needs. I can then take these ideas back to my formal R&D teams and put them to work.

In the Polymers CoP, for example, members stress the importance of sharing insights on polymers' possible applications with experts from organizations in various industries (e.g., medical, automotive, optics, textile, sports, and apparatus). These experts are often Zyc Chemicals' clients, and they intervene frequently in the CoPs' knowledge creation activities. A community member recounted:

> The strength of this community is undoubtedly its permeability to knowledge coming from outside the organization's boundaries. Our community of practice invites our most collaborative clients to participate in our knowledge sessions. This way, we can discuss their product development projects with them. This is a good way for us to discover what their needs are in terms of synthetic materials, such as plastics, for instance. It's only through repetitive face-to-face meetings with these external experts that we can anticipate their needs and think of completely new ideas for polymers that suit their new product development projects.

There are also numerous exchanges of knowledge between members of the different CoPs from the Specialty Chemicals division. Members move across different communities to experiment with a broader range of topics. People start experimenting with a broader range of knowledge topics, and this moving increases the knowledge diversity in the communities. As a result, the creation of radically new knowledge is stimulated. Chemical catalysts, for instance, are important to the health and nutrition industry, but also have applications in the industrial chemicals and polymer markets. Therefore, not surprisingly, people from the Industrial Chemicals and Polymers CoP often participate in the Health & Nutrition CoP meetings to share insights on the catalysts. The leader of the Health & Nutrition CoP said that this "creative chaos" was positive because people did not feel bad about shifting from one community to another to "knowledge-shop." Furthermore, we encountered a member of the Industrial Chemicals CoP who recounted:

> The "Industrial Chemicals" community explores new molecules to create additives for high quality synthetics and coatings. I happen to know a bunch of folks who participate in the "Coatings & Additives"

community. This means that I sometimes shift myself to that community to pick up complementary knowledge, and bring it back to my "Industrial Chemicals" community. This is how inter-community collaboration emerges within our Specialty Chemicals division.

In sum, the autonomy-driven pair is characterized by the interacting forces of enhanced cooperation among CoP members and boundary-spanning. Members cooperate extensively within and across the boundaries of their CoP in order to generate radically new knowledge. This pair of forces prevailed simultaneously in the five CoPs. This creates a space for free and unstructured floating of ideas, where members are able to explore radically new ways of recombining knowledge related to Zyc Chemical's products offering.

*The outcome of the autonomy-driven pair of forces:*
*exploration and radical innovation*

As indicated above, through the autonomy-driven pair of forces, communities explore radically new knowledge. The community leader of the Coatings & Additives CoP explained, "Conceiving new coatings requires a lengthy learning process, during which we start from scratch and play around with new knowledge that comes from our members and external sources." Exploration may contribute to achieving radical innovations and can also help the company see its business processes in a new light. In some ways, exploration activities reduce the risk of knowledge deterioration and contribute to the firm's long-term performance and survival (Hamel 1994; March 1991).

The informants described the autonomy-driven pair of forces as an appropriate combination to support exploration. A member of the Industrial Chemicals CoP stated,

Of course, in terms of knowledge objectives, the sponsor doesn't exert a full control. Everybody agrees on that. As soon as the sponsor stops making suggestions on what knowledge to focus on, this community becomes a perfect space to think outside corporate norms and create unconventional knowledge.

The strong use of face-to-face meetings, often outside the normal work environment, was found to be particularly important for creating and testing the viability of new ideas. Prior studies have indicated that strong ties created by frequent and personal meetings support the exchange of tacit or complex knowledge (Hansen 1999; Reagans & McEvily 2003).

An important role was also attributed to boundary spanning as it multiplies external contacts. Being permeable to knowledge from outside the group has been found to elicit new ideas and insights from new sources of knowledge (Ancona & Caldwell 1992; Hansen 1999). The same may be true of the interrelations between communities and external sources, and among communities, as they foster strong internal diversity. As the leader of the Coatings & Additives CoP remarked, "Thanks to the many different perspectives that we import from other communities into our own community, we are able to come up with new and better approaches." Structural diversity in work groups, defined as the differences in the members' organizational affiliations, roles, or positions, was found to positively affect the communities' ability to exchange and absorb new knowledge from external sources (Cummings 2004). Finally, the degree of autonomy left to the CoPs to evolve spontaneously around different knowledge topics may further contribute to knowledge probing. A member of the Inorganic Materials CoP explained, "We try to explore new paths that don't exactly fit in our top management's current process framework." Prior studies have shown that the exploration of new knowledge is best supported by a high degree of managerial latitude (McGrath 2001) and occurs in an undirected and chaotic process (Daft & Weick 1984; Lumpkin & Dess 1996).

Figure 9.1 illustrates the two pair of forces that drive knowledge creation in the five CoPs at Zyc Chemicals.

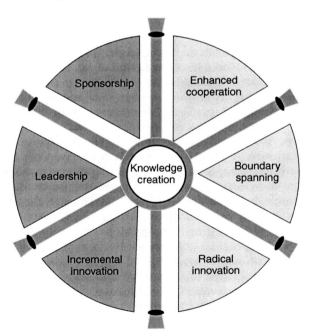

*Figure 9.1*   **The "steering wheel" to drive knowledge creation in communities of practice (CoPs) shows the two pair of forces that drive knowledge creation in the CoPs**

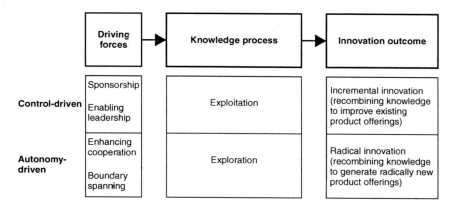

*Figure 9.2*   The relationship between driving forces and innovation outcomes

Figure 9.2 summarizes the relationship between sets of driving forces (control-driven versus autonomy-driven) with their knowledge processes and their related innovation outcomes in Zyc Chemicals' Specialty Chemicals division.

### Innovating through the interplay between the control-driven and autonomy-driven pair of forces in CoPs

We found that the five CoPs all managed the interplay between these two pair of driving forces throughout the period of our study (four years), thus allowing the Specialty Chemicals division to stay at the edge of chaos by continually redefining its knowledge objectives and its product offerings within the constraints of managerial structures (Ouchi 1980)

The results of our interviews revealed that, regarding the control-driven pair of forces, the governance committee (via the sponsors' influence) suggests the knowledge objectives that each of the five CoPs has to pursue, while the sponsors convey these. This creates adaptive tension and pushes the CoPs to create knowledge coherent with the existing innovation path in the Specialty Chemicals division at Zyc Chemicals. Through the control-driven pair of forces, namely sponsorship and leadership, the five CoPs focus on creating knowledge in existing domains by recombining existing skills. As previously described, this set of forces reinforces exploitation.

Linked to the autonomy-driven pair of forces, a nonspecific knowledge search prevails in the CoPs, and inter-CoP boundary-spanning is the rule. This mode supports exploration, which focuses the CoPs on creating radically new knowledge through the search of new domains.

## 4   Discussion

In this study, we analyzed the dynamics of CoPs at Zyc Chemicals. The aim was to deepen our understanding of CoP dynamics and its link to innovation. The results show the existence of four forces driving CoP dynamics: management sponsorship, leadership, enhancing cooperation, and boundary-spanning. We found that the five CoPs at Zyc Chemicals encompass two distinct interplaying pairs of driving forces: control-driven and autonomy-driven. Each pair corresponds to a specific knowledge strategy. The control-driven pair combines management sponsorship and leadership and is focused on refining knowledge to improve existing products (incremental innovation). This pair is characterized by some support from the organization (management). The autonomy-driven pair relies on enhancing cooperation and boundary spanning and is focused on generating new ideas and radically new knowledge to create innovative products (radical innovation). We also found that at Zyc Chemicals, CoPs need to rely on these two pairs of forces to help the division combine two complementary types of innovation – incremental and radical.

These findings are in line with and contribute to the literature in various significant ways. First, the five CoPs we studied in the Specialty Chemicals division at Zyc Chemicals is a real illustration of Brown and Duguid's (1991) perspective of the firm – according to which an organization behaves as a community of communities of practice. Indeed, the co-evolution and interplay of the five CoPs showed how they supported the organization's Specialty Chemicals division in its innovation.

Second, the findings will help researchers look beyond the fruitless debate between the two opposing concepts of CoPs, as either self-emerging/unmanageable structures or controllable managerial tools. Our view is that through a better understanding of the forces driving the dynamics of CoPs, it is possible to align their activities with the organization's innovation needs. This idea is in line with previous findings on the importance of clear knowledge objectives (Josserand 2004; Wenger et al. 2002; Probst & Borzillo 2007, 2008). However, it provides a more realistic perspective in that CoPs should not be steered by top–down control but by creating adaptive tension. In this view, enabling leadership consists of creating the right amount of adaptive tension between managers and CoP leaders for the CoPs to continue their knowledge-creating activities, instead of imposing a top–down control. Enabling leadership that is properly shared between sponsors and CoP leaders creates a context from which distributed intelligence or social capital emerges.

Third, the example of the Specialty Chemicals division may suggest that in reality the concept of communities of practice may be drifting away from Lave and Wenger's (1991) initial definition, where these were described as fully spontaneous and autonomous grassroots structures. The literature on communities of practice has to date remained curiously silent on the role of leadership and power relationships (Contu & Willmott 2003; Fox 2000). There is a long tradition of perceiving communities as emergent phenomena that develop organically through bottom-up processes (i.e., Orr 1996). Much as in the related literature on self-managing work teams (e.g.,

Barker 1993; Ezzamel & Willmott 1998) and open innovation communities (e.g., Dahlander & Wallin 2006; O'Mahony & Ferraro 2007), the insight that power and control have their role is also slowly emerging in the discussion on communities of practice (e.g., Lesser & Everest 2001; Thompson 2005). We contribute to this perspective by showing that some control is exerted in communities, while simultaneously pointing to the need for autonomy to preserve the communities' emergent character. Our longitudinal study conducted in the corporate world suggests that a constant pressure to innovate may push management to try and steer the "most promising" CoPs that it identifies within the organization, in order to achieve strategic advantage. Corporate reality may hence dictate the necessity to broaden the concept of community of practice – where focus is put on individual learning (Lave & Wenger 1991; Brown & Duguid 1991) – to the notion of *strategic community of practice,* where focus is no longer only on individual learning, but also on creating value for the organization.

Fourth, we contribute to the body of literature on balancing control and autonomy within organizations. A lively debate emerged on the conceptualization of autonomy and control as coexisting governance processes that have to be balanced in order to promote an organization's long-term success and survival (Cardinal 2001; Cardinal, Sitkin, & Long 2004; O'Mahony & Ferraro 2007; Sutcliffe, Sitkin, & Browning 2000). The need for balance between autonomy and control has been described for different organizational levels, including the corporate level (e.g., Pearce & Zahra 1991; Sundaramurthy & Lewis 2003), the business unit (e.g., Gibson & Birkinshaw 2004), and the group or initiative level (e.g., Feldman 1989; Puranam et al. 2006; Trevelyan 2001). We contribute to this body of work by revealing that similar balancing processes occur within communities of practice.

Fifth, our work also relates to questions surrounding the fundamental nature of knowledge processes in organizations. The literature claims that organizations utilize multiple learning processes when knowledge is created and transferred (e.g., Argote et al. 2003). More exactly, March (1991) argues that exploitation and exploration are two fundamentally different knowledge activities, between which firms divide their attention and resources. Whereas exploitation is associated with activities such as "refinement, efficiency, selection, and implementation," exploration refers to notions like "search, variation, experimentation, and discovery" (March 1991, p. 102). March concludes that organizations need to engage in both exploitation and exploration to ensure long-term survival and success. A fundamental debate evolved around whether exploitation and exploration can be pursued simultaneously within an organization. Whereas March considers the two types of knowledge processes as fundamentally incompatible, subsequent studies often conceptualize exploitation and exploration as orthogonal variables that can be achieved simultaneously (Baum et al. 2000; Gupta et al. 2006; Katila & Ahuja 2002).

Interestingly, we found instances of both simultaneous and separated pursuit of exploitative and explorative knowledge processes in our analysis of communities of practice. The control-driven pair of forces, for example, is clearly focused on the exploitation of existing knowledge with an orientation towards product refinement. Conversely, the autonomy-driven pair pursues the exploration of new knowledge for variation and discovery of innovative products. While these two pairs of forces are strongly oriented towards either exploitation or exploration, both pairs interplay with one another, enabling CoPs to achieve both. This is, for example, reflected in these communities' dual focus on internal (existing) and external (new) knowledge. Our findings may thus suggest that the communities of practice we investigated are *ambidextrous* – aligned and efficient in managing today's business demands, while simultaneously being adaptive to changes in the environment (e.g., Duncan 1976). This is consistent with earlier papers that explore organizational ambidexterity at the corporate level (e.g., Lubatkin et al. 2006; Smith & Tushman 2005; Tushman & O'Reilly 1996). Our chapter extends this perspective by describing the dual knowledge processes within less formal communities of practice at the subsystem level.

## 5 Managerial implications

The results of our research have several practical implications for managers. We clearly show that managers can and should use CoPs to enhance knowledge creation and thus overcome the exploitation/exploration dilemma. Also, though the right dosage of control and autonomy-driven forces, managers can stimulate innovation dynamics and align it with strategic objectives of their organization.

In order to do this, it is essential to identify the right sponsors and leaders. Their close collaboration is particularly crucial. Sponsors need to create a link with the outside environment and inform CoP leaders about the most relevant research topics. Leaders must create the innovation momentum in CoPs and orient their activities in the direction of company's strategy.

In order to sustain employee engagement in CoPs, managers should preserve the informal aspect of these communities. To preserve the creativity of CoP members, they also need to make sure not to exert too much pressure on CoPs in terms of timely attainment of innovation objectives, which would run the risk of transforming them into more formal project teams. This is particularly important during periods of exploration, when the organization is searching for radically new knowledge. This is probably the most delicate aspect of dealing with CoPs, as managers usually feel more comfortable in traditional command-and-control contexts. As innovation is a complex and uncertain process, managers must provide CoPs with both time and resources for the creative outcomes to happen.

## 6 Limitations and avenues for future research

We conclude this chapter with some comments on the study's limitations. The generalizability of the findings is limited by the size and scope of the sample. We focused exclusively on one large, multinational organization. Larger-scale empirical efforts are necessary to statistically assess the relationships presented here and to help define the boundary conditions according to which these relationships vary. Given the dynamism of today's market environments, future research should continue to investigate different driving forces that may stimulate new knowledge to emerge in communities of practice and support innovation processes within organizations.

Finally, our study is also limited in terms of the performance effects that different pairs of driving forces may exert. We asked informants to describe each community's contribution to knowledge creation, but we did not investigate how the latter affect performance. While earlier studies found a positive relationship between knowledge creation and performance, future research should establish and formally test the direct or indirect relationships among driving forces in communities of practice, knowledge processes, and firm performance. Given the limitations of an inductive study with a limited number of cases, we hope that this study will stimulate further research and empirical tests, thus contributing to the emergence of a more comprehensive and accurate model for steering communities of practice to support innovation in organizations.

### Note

1. Sponsors are senior management members who have a stake in communities of practice. They play an intermediary role between the CoPs and the organization's corporate sphere. They assess whether CoPs have been built and focus on knowledge topics that are relevant to corporate strategy. Accordingly, CoPs may receive funding to support their knowledge activities.

### References

Adams, E., & Freeman, C. (2000) Communities of practice: Bridging technology and knowledge assessment. *Journal of Knowledge Management*, 4(1), 38–44.

Anand, N., Gardner, H., & Morris, T. (2007) Knowledge-based innovation: Emergence and embedding of new practice areas in management consulting firms. *Academy of Management Journal*, 50(2), 406–428.

Ancona, D. G., Caldwell, D. F. (1992) Bridging the boundary: External activity and performance in organizational teams. *Administrative Science Quarterly*, 37(4), 634–661.

Argote, L., McEvily, B., & Reagans, R. (2003) Managing knowledge in organizations: An integrative framework and review of emerging themes. *Management Science*, 49(4), 571–582.

Barker, J. R. (1993) Tightening the iron cage: Concertive control in self-managing teams. *Administrative Science Quarterly*, 38(3), 408–437.

Bartlett, C. A., & Ghoshal, S. (1998) *Managing across Borders* (2nd edn). Harvard Business School Press: Boston, MA.

Baum, J. A. C., Li, S. X., & Usher, J. M. (2000) Making the next move: How experiential and vicarious learning shape the locations of chains' acquisitions. *Administrative Science Quarterly*, 45(4), 766–801.

Birkinshaw, J., & Gibson, C. (2004) Building ambidexterity into an organization. *Sloan Management Review*, 45(4), 47–55.

Blanchot-Courtois, V., & Ferrary, M. (2009) Valoriser la R&D par des communautés de pratique d'intrapreneurs. *Revue française de gestion*, 195–93–110.

Borzillo, S., Raisch, S., Probst, G. J. B. (2008) The governance paradox: Balancing autonomy and control in managing communities of practice. OMT Division of the 2008 Academy of Management Meeting in Anaheim. Best Paper Proceedings of the 2008 Academy of Management Meetings.

Borzillo, S. (2009) Top management sponsorship to guide communities of practice. *Journal of Knowledge Management*, 13(3), 60–72.

Bradach, J. L. (1997) Using the plural form in the management of restaurant chains. *Administrative Science Quarterly*, 42(2), 276–303.

Brown, J. S., & Duguid, P. (1991) Organizational learning and communities of practice: Toward a unified view of working, learning, and innovation. *Organization Science*, 2(1), 40–57.

Brown, J. S, & Duguid, P. (2001) Structure and spontaneity: Knowledge and organization. In I. Nonaka & D. J. Teece (Eds), *Managing Industrial Knowledge: Creation, Transfer and Utilization* (pp. 44–67). Sage: London, UK.

Brown, S. L., & Eisenhardt, K. M. (1997) The art of continuous change: Linking complexity theory and time-paced evolution in relentlessly shifting organizations. *Administrative Science Quarterly*, 42, 1–34.

Büchel, B., & Raub, S. (2002) Building knowledge-creating value networks. *European Management Journal*, 20(6), 587–596.

Burgelman, R. A. (1991) Intraorganizational ecology of strategy making and organizational adaptation: Theory and field research. *Organization Science*, 2(3), 239–262.

Burgelman, R. A. (2002) Strategy as vector and the inertia of coevolutionary lock-in. *Administrative Science Quarterly*, 47(2), 325–357.

Burgelman, R. A. (2003) Strategy making and evolutionary organization theory: Insights from longitudinal process research. *Working Paper*, Stanford University, Palo Alto, CA.

Cardinal, L. B. (2001) Technological innovation in the pharmaceutical industry: The use of organizational control in research and development. *Organization Science*, 12(1), 19–36.

Cardinal, L. B., Sitkin, S. B., & Long, C. P. (2004) Balancing and rebalancing in the creation and evolution of organizational control. *Organization Science*, 15(4), 411–431.

Cohen, W. M. and Levinthal, D. A. (1990) Absorptive capacity: A new perspective on learning and innovation. *Administrative Science Quarterly*, 35(1), 128–152.

Contu, A., & Willmott, H. (2000) Comment on Wenger and Yankow. Knowing in practice: A delicate flower in the organizational learning field. *Organization*, 7(2), 269–276.

Contu A., & Willmott, H. (2003) Re-embedding situatedness: The importance of power relations in learning theory. *Organization Science*, 14(3), 283–296.

Cummings, J. N. (2004) Work groups, structural diversity, and knowledge sharing in a global organization. *Management Science*, 50(3), 352–364.

Daft, R. L., & Lewin, A. Y. (1993) Where are the theories of the "new" organizational forms? An editorial essay. *Organization Science*, 4(4), i–vi.

Daft, R. L., & Weick, K. E. (1984) Toward a model of organizations as interpretation systems. *Academy of Management Review*, 9(2), 284–295.

Dahlander, L., & Wallin, M. W. (2006) A man on the inside: Unlocking communities as complementary assets. *Research Policy*, 35(8), 1243–1259.

D'Aveni, R. A. (1994) *Hypercompetition: Managing the Dynamics of Strategic Maneuvering*. New York: Free Press.

Doz, Y., & Prahalad, C. K. (1986) Controlled variety: A challenge for human resource management in the MNC. *Human Resource Management*, 25(1), 55–71.

Duncan, R. (1976) The ambidextrous organization: Designing dual structures for innovation. In R. H. Killman, L. R. Pondy, & D. Sleven (Eds), *The Management of Organization* (pp. 167–188). New York: North Holland.

Eisenhardt, K. M. (1989a) Building theories from case study research. *Academy of Management Review*, 14(4), 532–550.

Eisenhardt, K. M. (1989b) Making fast strategic decisions in high-velocity environment. *Academy of Management Journal*, 32(3), 543–576.

Ezzamel, M., & Willmott, H. C. (1998) Accounting for teamwork: A critical study of group-based systems of organizational control. *Administrative Science Quarterly*, 43(2), 358–398.

Feldman, S. P. (1989) The broken wheel: The inseparability of autonomy and control in innovation within organizations. *Journal of Management Studies*, 26(2), 83–102.

Fox, S. (2000) Communities of practice, Foucault and actor-network theory. *Journal of Management Studies*, 37(6), 853–867.

Gibson, C. B., & Birkinshaw, J. (2004) The antecedents, consequences and mediating role of organizational ambidexterity. *Academy of Management Journal*, 47(2), 209–226.

Glaser, B., & Strauss, A. (1967) *The Discovery of Grounded Theory: Strategies of Qualitative Research*. Wiedenfeld and Nicholson: London, UK.

Golden, B. R. (1992) The past is the past – or is it? The use of retrospective accounts as indicators of past strategy. *Academy of Management Journal*, 35(4), 848–860.

Guba, E. G., & Lincoln, Y. S. (1989) *Fourth Generation Evaluation*. Sage: Newbury Park, CA.

Gupta, A. K., Smith, K. G., & Shalley, C. E. (2006) The interplay between exploration and exploitation. *Academy of Management Journal*, 49(4), 693–706.

Halal, W. E. (1999) The infinite resource: Mastering the boundless power of knowledge. In W. E. Halal & K. B. Taylor (Eds), *21st Century Economics: Perspectives of Socioeconomics for a Changing World*. Macmillan: New York.

Hamel, G. (1994) The concept of core competencies. In G. Hamel & A. Heene (Eds), *Competence Based Competition*. New York: Wiley.

Handley, K., Sturdy, A., Fincham, R., & Clark, T. (2006) Within and beyond communities of practice: Making sense of learning through participation, identity, and practice. *Journal of Management Studies*, 43(3), 641–653.

Hansen, M. T. (1999) The search-transfer problem: The role of weak ties in sharing knowledge across organization subunits. *Administrative Science Quarterly*, 44(1), 83–103.

Hildreth, P., Kimble, C., & Wright, P. (2000) Communities of practice in the distributed international environment. *Journal of Knowledge Management*, 4(1), 27–38.

Holmqvist, M. (2003) A dynamic model of intra- and interorganizational learning. *Organization Studies*, 24(1), 95–123.

Josserand, E. (2004) Cooperation within bureaucracies: Are communities of practice an answer? *M@n@gement*, 7(3), 307–339.

Katila, R., & Ahuja, G. (2002) Something old, something new: A longitudinal study of search behavior and new product introduction. *Academy of Management Journal*, 45(6), 1183–1194.

Kimble, C., & Hildreth, P. (2005) Dualities, distributed communities of practice and knowledge management. *Journal of Knowledge Management*, 9(4), 102–113.

Lawrence P. R., & Lorsch J. W. (1967) *Adapter les structures de l'entreprise*. Les Editions d'Organisation: Paris.

Lave, J., & Wenger, E. (1991) *Situated learning. Legitimate peripheral participation*. Cambridge: University Press: Cambridge, UK.

Lesser, E., & Everest, K. (2001) Using communities of practice to manage intellectual capital. *Ivey Business Journal*, 65(4), 37–41.

Levina, N. and Vaast, E. (2005) The emergence of boundary spanning competence in practice: Implications for implementation and use of information systems. *MIS Quarterly*, 29(2), 335–363.

Levina, N. and Vaast, E. (2006) Turning a community into a market: A practice perspective on information technology use in boundary spanning. *Journal of Management Information Systems*, 22(4), 13–37.

Liedtka, J. (1999) Linking competitive advantage with communities of practice. *Journal of Management Inquiry*, 8(1), 5–16.

Lindkvist, L. (2005) Knowledge communities and knowledge collectivities: A typology of knowledge work in groups. *Journal of Management Studies*, 42(6), 1189–1210.

Long, C. P., Burton, R. M., and Cardinal, L. B. (2002) Three controls are better than one: A computational model of complex control systems. *Computational & Mathematical Organization Theory*, 8(3), 197–220.

Loyarte, E., & Rivera, O. (2007) Communities of practice: A model for their cultivation. *Journal of Knowledge Management*, 11(3), 67–77.

Lubatkin, M. H., Simsek, Z., Ling, Y., and Veiga, J. F. (2006) Ambidexterity and performance in small- to medium-sized firms: The pivotal role of top management team behavioral integration. *Journal of Management*, 32(5), 646–672.

Lumpkin, G. T., & Dess, G. G. (1996) Clarifying the entrepreneurial orientation construct and linking it to performance. *Academy of Management Review*, 21(1), 135–172.

March J. G. (1991) Exploration and exploitation in organizational learning. *Organization Science*, 2(1), 71–87.

McDermott, R. (1999) Learning across teams: How to build communities of practice in team organizations. *Knowledge Management Review*, 8(May/June), 32–36.

McDermott, R. (2001) How to design live community events. *Knowledge Management Review*, 4(4), 5–6.

McDermott, R. (2003) Building spontaneity into strategic communities: Eight tips to put excitement into management-created COPs. *Knowledge Management Review*, 5(6), 28–31.

McGrath, R. (2001) Exploratory learning, innovative capacity, and managerial oversight. *Academy of Management Journal*, 44(1), 118–131.

Miles, M. B., & Huberman, A. M. (1994) *Qualitative Data Analysis*. Sage: Thousand Oaks, CA.

206    *Stefano Borzillo and Renata Kaminska*

Miles, R., Snow, C., Matthews, J. A., & Miles, G. (1999) Cellular-network organizations. In W. E. Halal & K. B. Taylor (Eds), *21st Economics Century Economics: Perspectives of Socioeconomics for a Changing World*. Macmillan, New York.

Milne, P., & Callahan, S. (2006) ActKM: The story of a community. *Journal of Knowledge Management*, 10(1), 108–118.

Mucchielli, A. (1991) *Les méthodes qualitatives*. Paris : PUF (QSJ).

Nonaka, I., & Takeuchi, H. (1995) *The Knowledge-Creating Company*. Oxford University Press: Oxford, UK.

O' Hara, K., Alani, H., & Shadbolt, N. (2002) Identifying communities of practice: Analysing ontologies as networks to support community recognition. *Proceedings IFIP World Computer Congress. Information Systems: The E-Business Challenge*, Montreal.

O' Mahony, S., & Ferraro, F. (2007) The emergence of governance in an open source community. *Academy of Management Journal*, 50(5), 1079–1106.

Orr, J. (1996) *Talking about Machines: The Ethnography of a Modern Job*. IRL Press: Ithaca, NY (originally from Orr, J. (1990), Talking about machines: An ethnography of a modern job. PhD Thesis, Cornell University, Ithaka).

Ouchi, W. G. (1980). Markets, bureaucracies and clans. *Administrative Science Quarterly*, 25(1), 129–141.

Pascale, R.T. (1990) *Managing on the Edge: Companies that Use Conflict to Stay Ahead*. London: Penguin.

Pearce, J., & Zahra, S. (1991) The relative power of CEOs and board of directors: Associations with corporate performance. *Strategic Management Journal*, 12(2), 135–153.

Probst, G. J. B., & Borzillo, S. (2007) Piloter les communautés de pratique avec succès. *Revue française de gestion*, 170, 135–153.

Probst, G. J. B., & Borzillo, S. (2008) Why communities of practice succeed and why they fail. *European Management Journal*, 26(5), 335–347.

Puranam, P., Singh, H., & Zollo, M. (2006) Organizing for innovation: Managing the coordination-autonomy dilemma in technology acquisitions. *Academy of Management Journal*, 49(2), 263–280.

Reagans, R., & McEvily, B. (2003) Network structure and knowledge transfer: The effects of cohesion and range. *Administrative Science Quarterly*, 48(2), 240–267.

Rosenkopf, L. and Nerkar, A. (2001) Beyond local search: Boundary-spanning, exploration and impact in the optical disc industry. *Strategic Management Journal*, 22(4), 287–306.

Siggelkow, N. (2002) Evolution toward Fit. *Administrative Science Quarterly*, 47(1), 125–159.

Smith, W. K., and Tushman, M. L. (2005) Managing strategic contradictions: A top management model for managing innovation streams. *Organization Science*, 16(5) – 522–536.

Sundaramurthy, C., & Lewis, M. (2003) Control and collaboration: Paradoxes of governance. *Academy of Management Review*, 26(3), 397–415.

Sutcliffe, K. M., Sitkin, S. B., & Browning, L. D. (2000) Tailoring process management to situational requirements: Beyond the control and exploration dichotomy. In R. Cole & W. R. Scott (Eds) *The Quality Movement and Organizational Theory* (pp. 315–330). Sage: Thousand Oaks, CA.

Swan, J., Scarbrough, H., & Robertson, M. (2002) The construction of communities of practice in the management of innovation. *Management Learning*, 33(4), 477–496.

Thompson, M. (2005) Structural and epistemic parameters in communities of practice. *Organization Science*, 16(2), 151–164.

Trevelyan, R. (2001) The paradox of autonomy: A case of academic research scientists. *Human Relations*, 54(4), 495–525.

Tushman, M. L., & O' Reilly, C. A. (1996) Ambidextrous organizations: Managing evolutionary and revolutionary change. *California Management Review*, 38(4), 8–30.

Tushman, M. L. & Smith, W. (2002) Organizational technology. In J. A. C. Baum (Ed.), *Companion to Organizations* (pp. 386–414). Blackwell Business: Oxford, UK.

Victor, B., & Stephens, C. (1994) The dark side of the new organizational forms: An editorial essay. *Organization Science*, 5(4), 479–481.

Warglien, M. (2002) Intraorganizational evolution. In J. A. C. Baum (Ed.), *Companion to Organizations* (pp. 98–118). Blackwell Business: Oxford, UK.

Wenger, E. (1998) *Communities of Practice: Learning, Meaning, and Identity.* Cambridge University Press: Cambridge, UK.

Wenger, E., McDermott, R., & Snyder, W. (2002) *Cultivating Communities of Practice: A Guide to Managing Knowledge.* Harvard Business School Press: Boston, MA.

Wenger, E., & Snyder, W. (2000) Communities of practice: The organizational frontier. *Harvard Business Review*, 78(1), 139–145.

Yin, R. K. (2003) *Case Study Research: Design and Methods* (3rd edn). Sage: London, UK.

# 10
## Joining Innovation Efforts Using both Feed-forward and Feedback Learning: The Case of Japanese and Korean Universities
*Ingyu Oh*

## 1 Introduction

In the study of university-firm alliances for organizational learning and new knowledge development, researchers have mainly focused on the issues of the facilitating factors of university-firm alliances (Geisler 1995; Cassiman & Veugelers 2002; Santoro & Chakrabarti 2002; Tether 2002; Fontana et al. 2006). Preceding studies focused on finding structural or firm-level contingencies for preferring university partners over private-firm partners in forming external research and development (R&D) alliances (Teece 1985; Kogut 1988; Rosenberg & Nelson 1994; Berkovitz & Feldman 2005; Cassiman et al. 2005; Oh 2010); the role of the government policies in galvanizing alliance formations between firms and universities (Capron & Cincera 2003; Mohnen & Hoareau 2003; Eom & Lee 2010); and developing the legal and governance framework for such alliances (Cassiman & Veugelers 2002; Cassiman et al. 2005).

However, to many universities in Japan and Korea, a more puzzling question regarding firm-university knowledge sharing is: How do we attract domestic multinational enterprises (MNEs) to our university-firm alliance projects, despite the fact that these MNEs prefer American and European Union (EU) universities for knowledge development? Top-rated universities like Tokyo National University, Kyoto National University, Osaka University, Keio University (Japan), Seoul National University, Korea Advanced Institute of Science and Technology, Yonsei University, and Korea University (Korea) are having a tough time recruiting not only fee-paying foreign students and internationally renowned faculty, but also collaborative research projects funded by leading Japanese and Korean MNEs like Toyota, Sony, Panasonic,

*Table 10.1*  Firm-university income and funding sources for Korean universities (2009)

| | Income | | Funding sources | | | | | |
| | Firm-university Alliances | Private firm | Government | Donation | Overhead | BBF | Others | Total |
|---|---|---|---|---|---|---|---|---|
| Average amount ($US mil.) | 0.76 | 3.37 | 21.82 | 0.20 | 0.14 | 4.34 | 0.69 | 31.32 |
| Percentage | 2.5% | 10.8% | 69.6% | 0.6% | 0.4% | 13.8% | 2.2% | 99.9% |

*Note*: BBF (Balance Brought Forward).
*Source*: MEST (2010).

*Table 10.2*  R&D spending by Japanese firms (in ¥ billions)

| | 2003 | 2004 | 2005 | 2006 | 2007 | 2008 |
|---|---|---|---|---|---|---|
| **Total to Japanese Universities** | 834 | 836 | 900 | 932 | 967 | 948 |
| **Total to foreign R&D Centers** | 1,985 | 2,012 | 2,742 | 2,666 | 3,075 | 3,399 |

*Source*: METI (2010).

NEC, Samsung, Hyundai, LG, Posco, and SK (Son & Lee 2005; Sung 2005; Katô & Enomoto 2006; Itô 2008).

As Tables 10.1 and 10.2 show, the bulk of the research funding awarded to Korean universities in 2009 was from the government (69.6 percent), while that from private firms was only 10.8 percent. Although Japanese firms have increasingly collaborated with Japanese universities for joint R&D in recent years, the total R&D funding paid to Japanese universities by Japanese firms in 2008 was only 28 percent of what they spent with foreign R&D centers, including foreign universities.

Among various theories and models of promoting university-firm alliances in R&D, I focus on the concept of barriers to innovation in this chapter, in order to explain why Japanese and Korean universities are doing miserably in attracting R&D investments from firms. In this chapter I use the concepts of feed-forward (i.e., explorative) and feedback (i.e., exploitive) learning, because it is apparent that Korean and Japanese firms rely on the North American and European Union universities for feed-forward learning, while they might consider working with local universities for feedback learning. On the other hand, Korean and Japanese universities incessantly emphasize feed-forward learning in conjunction with government policies. Therefore, barriers to innovation exist within the interorganizational learning (IOL) model (Lawrence et al. 2005; Schilling & Kluge 2009). Since both Japanese and Korean cases require explanations of IOL failure with domestic and foreign MNEs, especially in the stages of the integration and institutionalization of IOL, I focus on feedback learning more than on the feed-forward learning. First, I briefly discuss how Japanese and Korean universities have successfully overcome the barriers of intuition and interpretation using the feed-forward and feedback model. Second, I discuss how feedback learning can be implemented for the integration and institutionalization of IOL between universities and domestic MNEs. Third, I present two successful cases – one about Waseda University and its leading role in creating university-based venture start-ups in Japan in such a way as to leverage Japanese firm-university research alliances and one case about the Sungkyunkwan University–Samsung collaboration. Finally, I discuss implications of the case studies and present possible propositions.

## 2  Background

To analyze the failure of the university-firm alliance in Korea and Japan, I employ the concept of barriers to (inter-) organizational learning (Coopey 1995; Berthoin-Antal et al. 2003; Lawrence et al. 2005; Schilling & Kluge 2009). The main reason Korean and Japanese universities fail to attract firms' R&D investments is the firms' unwillingness to participate in IOL with Korean or Japanese universities for feed-forward or explorative learning. This means that the barriers to the university-firm alliance for interorganizational learning usually occur at the level of integration, using Crossan et al.'s (1999) 4I model, mainly because MNEs refuse to accept the "interpretation" proposed by the Korean and Japanese universities. The 4I model is the first successful attempt in the innovation literature to combine the concept of barriers to IOL for innovation and the concepts of feed-forward and feedback learning for IOL.

Barriers to integration in interorganizational learning include political obstacles (Coopey 1995; Lawrence et al. 2005) and cognitive biases and mindsets held among actors, groups, and organizations (Berthoin-Antal et al. 2003). Simultaneously, others have also noted structural and environmental factors (March 1991; Van de Ven & Polley 1992; Kim 1993; Nonaka 1994; Zander & Kogut 1995; Edmondson & Moingeon 1996; Inkpen & Crossan 1996; Schilling & Kluge 2009). Barriers therefore occur at the levels of individuals, groups, organizations, and environments. However, the extant studies of barriers to IOL neglected the importance of feed-forward and feedback learning within IOL, which I think are pivotal in conceptualizing the psychological basis motivating each type of learning. For example, studies on the barriers to IOL often emphasize employees' defection (Zell 2001; Szulanski 2003, Beer et al. 2005) from IOL or fear of punishment after IOL (Argyris 1990; Cannon & Edmondson 2001; McCracken 2005; Sun &

Scott 2005). Others highlight lack of organizational support for IOL or lack of organizational authority in IOL (Elliott et al. 2000; Popper & Lipshitz 2000; Starbuck & Hedberg 2003; Lawrence et al. 2005). The so-called not-invented-here syndrome has also been well documented (Zell 2001). However, none of these studies emphasizes the importance of feed-forward and feedback learning in OL or IOL for innovation.

Similarly, although exploration and exploitation have been clearly distinguished by the following studies, they fail to provide the psychological basis of IOL: structural-organizational factors such as low turnover rates and high levels of workforce homogeneity, especially among top management teams (March 1991; Virany et al. 1992); competition with other MNEs that are tied to universities in North America and Western Europe (Sun and Scott 2005); competence traps within the MNEs through long-term success in their cooperation with universities (Levitt and March 1988; Berthoin-Antal et al. 2003); inadequate communication between MNEs and universities (Elliott et al. 2000; Zell 2001); political and power structures (Coopey 1995; Beer et al. 2005; ineffective resource allocation (Beer et al. 2005); lack of learning values (Sun & Scott 2005); and lack of cultural cohesion between innovation needs and organizational culture (Sun & Scott 2005). Likewise, studies on the societal-environmental factors of barriers emphasize industrial structures that do not welcome innovation (Spender 1989) and failure traps, i.e., the time lag between organizational actions and environmental responses (Berthoin-Antal et al. 2003; Hedberg & Wolff 2003), while they have not been able to distinguish between exploration and exploitation.

The 4I model clearly juxtaposes two types of learning in innovation activities, which are divided into exploration and exploitation. Exploration is tied to feed-forward learning, while exploitation requires feedback learning. Crossan et al. (1999) defined intuition as the individual or group level exploration of new ideas and insights for a possible future innovation. In a similar vein, interpretation is the next step of explaining or articulating these new ideas and insights to other individuals and groups within an organization. Finally, integration is the third process of integrating individual and group interests for innovation within an organization. Unlike feed-forward learning (intuition, interpretation, integration), feedback learning includes the institutionalization (e.g., externalizing new knowledge) of new innovations between two organizations and the institutionalization (e.g., standard operation procedures) of new innovations between individuals and organizations.

Although Crossan et al. (1999) put their emphasis mainly on intraorganizational learning, the 4I model can be expanded to IOL, since the same pattern of learning processes occurs between organizations. For this chapter, IOL refers to interorganizational learning between universities and firms, as individuals and groups at a university can explore new ideas and insights and then decide to present them to an MNE for a potential collaboration

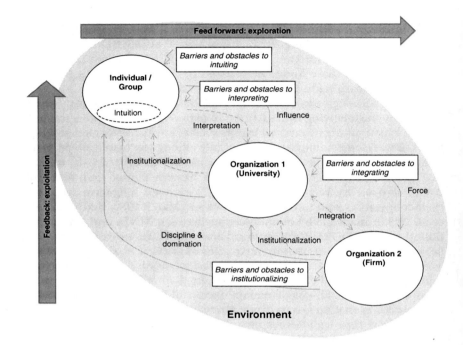

*Figure 10.1* Stages of feed-forward and feedback learning in interorganizational learning (IOL)

*Source*: Schilling and Kluge (2009); modified by the author for the IOL model.

toward technology commercialization. If both parties agree, universities and MNEs can integrate their group interests to implement innovation on an ongoing basis (Fig. 10.1).

## 3   Feed-forward and feedback learning in innovation

Japan and Korea are known for their ability to rapidly narrow the technology gap with the leading Western countries. This process of catching up is often referred to as *kaizen*, although Japan and Korea have also shown tremendous competency in generating new knowledge. As Table 10.3 shows, Japan is still No. 1 in new patent registration in the world, while Korea surpassed Germany and captured the No. 3 place in 2005. Figure 10.2 shows that the impacts of scientific journal articles published by Japanese and Korean scientists are ranked at No. 5 and No. 6 in the world, respectively, followed by the United States, the United Kingdom, Germany, and France. However, this does not mean that Japan and such newly emerging science strongholds

*Table 10.3* International comparison of new patent registration

| Holders' Nationality | 2002 | 2003 | 2004 | 2005 | 2006 | 2007 | 2008 |
|---|---|---|---|---|---|---|---|
| Japan | 174,528 | 182,505 | 187,588 | 186,561 | 217,532 | 232,442 | 239,206 |
| US | 139,059 | 147,728 | 142,979 | 134,482 | 155,978 | 146,046 | 147,154 |
| Korea | 36,917 | 39,301 | 45,365 | 63,865 | 102,668 | 106,599 | 79,567 |
| Germany | 47,216 | 53,021 | 51,448 | 49,600 | 56,920 | 51,916 | 54,032 |
| China | 6,300 | 11,936 | 18,943 | 21,519 | 26,298 | 33,409 | 48,815 |
| France | 21,844 | 24,233 | 23,291 | 21,825 | 26,128 | 24,802 | 25,776 |
| UK | 13,486 | 14,714 | 14,053 | 13,385 | 13,433 | 12,243 | 12,328 |
| India | 949 | 1,133 | 1,497 | 2,124 | 2,789 | 1,025 | 1,282 |

*Source*: METI (2010, p. 42).

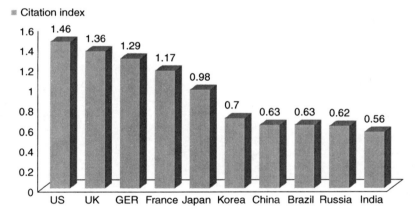

*Figure 10.2* International comparison of citation index (science field only 2008)
*Source*: METI (2010).

as Korea and China are automatically leaders in most science fields. Japanese universities put their names on the Top 5 list only in four natural science fields. China has its name on only three fields, and Korea on none. This implies that Japanese and East Asian universities concentrate their research resources only in scale-oriented industries that require fundamental scientific knowledge in material science, physics, chemistry, and biology.

With the contradictory nature of the Japanese and Korean national innovation system (NIS), highly advanced MNEs in Japan and Korea face a typical leader's dilemma of whether they have to continuously explore new knowledge through feed-forward learning. These MNEs have therefore overcome the barriers to the first three stages of feed-forward learning – namely,

Table 10.4 International citation rankings in natural sciences (1999–2009)

| Ranking | Material Science | Chemistry | Physics | Biology Biochemistry | Pharmacology Toxicology | Immunology | Computer Science |
|---|---|---|---|---|---|---|---|
| 1 | CAS | CAS | MPI | Harvard | UNC | Harvard | MIT |
| 2 | MPI | MPI | Tokyo | MPI | Merck | NIAID | AT&T |
| 3 | Tohoku | UCB | MIT | Tokyo | NCI | UCSF | UCB |
| 4 | AIST | Kyoto | CAS | UCSF | Harvard | Yale | IBM |
| 5 | MIT | Tokyo | RAS | UCSD | Vanderbilt | NCI | Stanford |
| Others | Osaka (9) | Osaka (12) | Tohoku (10) | Kyoto (19) | Tokyo (6) | Osaka (6) | Shuto (11) |

*Note*: East Asian universities and research centers are highlighted. Numbers in parentheses mean rankings.

*Source*: METI (2010, p. 38).

intuition, interpretation, and integration. While Japanese and Korean universities maintain high levels of scale-oriented explorative knowledge through feed-forward learning, their MNEs need more scope-oriented explorative knowledge through feed-forward learning. Furthermore, Japanese and Korean MNEs are in much need of the exploitative knowledge of scale-oriented innovations through feedback learning, whereas their university counterparts are least interested in feedback learning. Last, Japanese and Korean MNEs require exploitative knowledge of scope-oriented innovations through feedback learning, whereas their university counterparts are not interested in such studies at all.

Therefore, the contradictory nature of the Japanese and Korean national innovation system originates from different learning motivations held by universities and firms. To resolve these contradictions, we need to focus on the motivational and opportunity structures of feed-forward and feedback learning during IOL. Since opportunity structures are well documented by the previous studies that I have reviewed above, this chapter focuses on the psychological motivational structures of feed-forward and feedback learning. I am particularly interested in motivating Japanese and Korean universities for feedback learning for exploitation knowledge production in their IOL with MNEs.

## 4  Promoting feedback learning through firm–university collaboration

### Japan

The Japanese path to development in science and technology research during the postwar years has been typified by the ivory tower image of the national universities that boasted their splendid track record of success in fundamental research, along with Nobel Prizes in three science fields. Simultaneously, Japanese MNEs prided themselves in the number of new patent registrations, commercialization of fundamental research, and *kaizen*, resulting in the second highest GDP in the world until the late 2000s. However, to perpetuate the image of the ivory tower, Japanese national universities shied away from cooperating with Japanese MNEs oriented toward feed-forward learning or *kaizen* that involved new product development. This unbalanced arrangement was deemed acceptable by the scientific community, as long as the international competitiveness of Japan hovered over that of other competitor countries. Shockingly, however, Japanese global competitiveness fell to the ninth place in 2011 from first place in the 1980s, mainly because of Japan's devastating twenty-eighth and eleventh ranks in the world, respectively, in the categories of "basic requirements" (infrastructure) and "efficiency enhancers" (social capital) (WEF 2011). The reason Japanese basic requirements are worse than those of Taiwan (15th) or Korea (22nd) derives from the macroeconomic management failure. The snowballing government debt servicing, along with its inability to ease the quantitative shortage of the yen stock, fundamentally harmed the ability of the private sector to make an effective investment and to make other financial decisions. In addition, the reason Japan's ranking was only the eleventh in "efficiency enhancers" mainly derives from its weakness in higher education and training, financial market development, and technological readiness.

The government's response to the overall devastation of the Japanese universities and their incompetence in creating new economic values amid economic recession in the 1990s was to abandon the catching-up model of knowledge development that focused on firm-centered R&D activities. Instead, the government emphasized a leadership model of knowledge management that involves an endless feed-forward and feedback flow among knowledge bases (educational institutions), R&D (firms), and the market where new technologies are disseminated. To ensure this, the government strengthened the institutional infrastructure of intellectual property (IP) protection by strictly enforcing IP law and expanding the range of IP protection. The liberalization and deregulation of the knowledge market entailed the privatization of education, which technically enabled universities and graduate schools to freely open and/or close degree programs based on market demands, actively recruit renowned foreign faculty members, and strictly implement course and teaching evaluations (Daigaku Shingikai 1998).

This new reform blueprint for the Japanese higher learning was put into action in part by the Science and Technology Basic Plan (2001–2006). The Japanese government pledged to provide $240 billion to science and technology research projects over the course of five years, the second-largest sum in the world after the United States' expenditure (Taguchi 2003, p. 156). The program is now in its second phase, continually encouraging the linkage between universities and firms [*sangakurenkei*] toward feed-forward and feedback IOL. Consequently, each university in Japan is now equipped with what many call triple helix organizations, including a TLO (technology licensing office) and campus-based venture or spinoff firms (Katô & Enomoto 2006; Tamura 2006; Nishio 2007).

However, the Japanese triple helix is fraught with structural and institutional problems and/or immaturity. First and foremost, firms and universities maintain no clear agreements as to how they will utilize research results. As mentioned earlier, firms usually do not have a clear idea of commercialization when universities produce research results based on traditional fundamental research. Furthermore, universities still demand that they insert a contractual remedy clause in their contract with private firms, so that they can ask firms to pay settlement fees for the failure of technology/knowledge utilization [*hinagatakeiyaku*]. Second, amidst ambiguity surrounding technology commercialization, firms or universities (or both) abruptly suspend collaborative research projects without any further notice about how they will commercialize new knowledge in the future. Third, university and firm researchers do not always share the same goals, despite the fact that integration in IOL started on the assumption that actors (i.e., universities, government funding agencies, and firms) had agreed upon one common goal. Fourth, given the nature of university-led research and its infrastructure, which are inherently fundamental, few firms can figure out ways of developing tools for commercial applications. Finally, university and firm researchers do not have a commonly agreed upon timeline for completion of projects with a commercialization mandate (Katô & Enomoto 2006; Nishio 2007; Oh 2011a).

### Korea

Albeit similar to Japan in its outcome, the Korean road to successful feed-forward learning was substantially different from that of Japan. First and foremost, the Korean government institutionalized domestic patent laws to protect licensing and innovation rights of leading MNEs. The institutionalization of intellectual property protection leapfrogged MNEs' investments in new knowledge development, as long as (1) the government subsidized corporate R&D costs, and (2) it allowed price increases after patenting a new product in the so-called competitive monopoly market. The creation and heavy subsidization of the public education system by the government also contributed to the corporate innovation system (and later NIS), even as universities in North America, Japan, and Western Europe have continuously provided nearly free advanced education to selected Korean students. Technology licensing with Japanese corporations also opened up a new door for exports of finished goods that utilized Japanese know-how and parts (see *inter alia*, Hong 1993; Lie 2000; Oh et al. 2005). Domestic and foreign capital, education, and intellectual property protection therefore were all helpful for Korean MNEs to overcome hindrances to intuition, interpretation, and integration.

This kind of luxury was not available to Korean universities; a short-lived experimentation with KIST (Korea Institute of Science and Technology), which was modeled after Stanford University under the Park Administration, was the only one that came close to a successful governmental experiment with national education for the planned NIS (Kim & Leslie 1998). Other conventional Korean universities had to provide students with rote-based learning, which more or less aimed at preparing students for the upcoming government exams for entry-level bureaucrats or private sector job interviews.

The abrupt change came when Korean universities began experiencing rapidly declining student enrollments due to the sinking birthrate and the snowballing competition with the universities in North America, Japan,

Australia, and Western Europe over both undergraduate and graduate (especially, PhD) students. By the late 1980s, when international education for Korean youth became fully liberalized, it became obvious that Korean universities could no longer recruit MA or PhD students, as most of them wanted to study in Japan, North America, and Western Europe. Furthermore, second- and third-rate universities, especially in the local areas remote from Seoul or other big urban areas, faced difficulties in recruiting new students, who preferred to go overseas for their undergraduate education instead of wasting tuition money at unrecognized Korean universities. The government therefore used both sticks and carrots to encourage improvements in teaching and research at Korean universities. Professors were denied of tenure or promotion when they failed to meet the required number of publications in international journals. Students not only began filling out teaching evaluation forms about their professors, but also simultaneously sought exchange or study-abroad opportunities at foreign universities at the government's or their home university's expense (Im 2005; Kang 2005).

Particularly noticeable among these radical changes brought about by the government in the realm of Korean higher education was the introduction of the so-called triple helix system, which included alliances between universities and firms. Each university was promised national funding if it established an industry-university alliance center; however, no substantial research or commercialization activities were ever carried out between the two. In fact, the main function of the center still is to carry out an accounting task of calculating and processing bonuses for the professors who published articles in SCI- or SSCI-indexed journals, which the Korean government thinks highly of as world-class achievements by Korean professors (Park & Leydesdorff 2010).

Interestingly, Korea University, Seoul National University, Sungkyunkwan University, Gwangwoon University, and Yonsei University led the race in transferring knowledge to firms, instead of KAIST (Korea Advanced Institute of Science and Technology; formerly, KIST) and Postech, the second most prestigious of all technology universities in Korea, which was built by Posco, the largest steel mill in Korea (MEST 2010, p. 251). However, most of these firms were small and medium-size enterprises (SMEs) that suffered from a chronic shortage of capital with which to run their own R&D team or commission a research project to universities on their own. These SMEs tried to reap the benefit of this new opportunity by aligning with Korean universities and the government that promised high-quality R&D to these SMEs for no substantial fees. This means that participation in the newly established Korean triple helix by big business or MNEs has been trifling, especially since the change of rules in assessing professorial performance, which was solely based on international publications instead of firm-university alliances (Park & Leydesdorff 2010).

In sum, the top–down pressure for publications in international journals and monetary incentives for university-industry alliances easily motivated universities and professors to seek R&D partners with SMEs, if not with leading MNEs. However, since the government-funded R&D ties with SMEs were usually one-shot, the prospect of institutionalizing IOL in Korea still remains improbable in the near future, unless global MNEs enter into and solidify a long-term R&D relationship with universities. The absence of MNEs in the Korean triple helix also signals that the integration of innovative ideas originating from universities has not yet taken place within the MNEs. Therefore, although universities and their professors have initiated the stages of intuition and interpretation, they cannot find big business partners who are interested in the integration or institutionalization of IOL. The Korean situation is not very different from that of Japan. Hesitant Japanese MNEs that would rather rush to foreign R&D centers than haggle with the Japanese national universities for a better R&D contract are a signal that the full integration of innovative ideas originating from universities has not yet taken place within the Japanese MNEs.

## 4 Motivating feedback learning for integration and institutionalization

Feedback learning involves the exploitation or repetitive use of the same knowledge over and over again to master or improve it. Exploitative learners would also try to expand, if possible, the practical and commercial applicability of new knowledge to unexplored areas, mainly through procedural memory that controls tacit knowledge. *Kaizen* could be one example of feedback learning, and it is inherently different from feed-forward learning, which requires explorative knowledge creation through declarative memory that controls explicit knowledge (Cohen 1991; Cohen & Bacdayan 1994; Moorman & Miner 1998; Nooteboom 2000, pp. 273–275). Most studies of organizational learning (OL) emphasize exploration and exploitation without highlighting the feed-forward and feedback nature of learning. This theoretical lacuna results in the neglect of motivational studies of OL, which I think is pivotal in the study of why firms eschew feedback learning.

Motivating feed-forward learning (or intuition and interpretation) is relatively easier for organizations to manage than feedback learning (or integration and institutionalization). As discussed in the Background section, individual and group self-interest and motivation are key barriers to intuition and interpretation. However, motivational problems pose less serious threats to integration and institutionalization than do other forms of barriers (Schilling & Kluge 2009). In other words, intuition and interpretation can be achieved through the manipulation of individual and group motivational structures, whereas that is not the case for integration and institutionalization. It is not only harder to manipulate individual motivation structures for integration and institutionalization, but motivation alone would not remove the barriers to feedback learning. This is why feed-forward learning is individually motivated (i.e., empowering individuals to change organizations) and therefore challenging and fascinating, whereas feedback learning is organizationally motivated (i.e., de-powering individuals to accept new organizational norms and procedures) and therefore top–down in nature, monotonous, and repetitive.

Also, a huge difference exists in the resulting monetary rewards for each type of learning. A typical Toyota or Samsung worker who is a leader in in-house or interorganizational *kaizen* in his/her factory would be paid far less than what Steve Jobs and his employees make for feed-forward learning to develop iPhone apps and/or operating systems. As long as explorative knowledge brings about a bigger monetary return through intellectual property rights than exploitative knowledge does, agents of OL and IOL will be easily motivated toward feed-forward learning. Concurrently, it is more difficult to motivate individuals toward feedback learning for the same reason.

Against this backdrop, we can easily discern why Japanese and Korean universities were interested in feed-forward learning. Therefore, a possible answer to the problem of IOL in Japan and Korea is: (1) to separate feed-forward and feedback learning in IOL, and (2) to allow East Asian universities to participate in the feedback learning part of IOL, while MNEs continue to pursue feed-forward learning within their own R&D facilities or with universities in North America and the European Union. How can we motivate Japanese and Korean universities toward feedback learning or institutionalization in alliance with Japanese MNEs?

According to some of the findings in psychology, melancholia is a stronger motivator of feedback learning than simple nostalgia (Berlant 1988; Greenfeld 1990; Streip 1994; Eng 2000; Žižek 2000; Diaz 2006). Nostalgia occurs simply with aging, a natural psychological tendency of people to idealize things of the past or things that occurred in their lives during their youth. Melancholia, on the other hand, can occur in all ages, whether people are in their twenties or sixties. Melancholia is a psychological disorder diagnosed among the patients who are not allowed to express mourning, even at the time of utmost sadness (Freud [1917] 1963; Kristeva 1989). For example, even though her mother is executed by the government for treason, the daughter cannot openly express the sadness of losing her mother for fear of political repression.

What is interesting about melancholia is that many melancholic patients pursue learning while they suffer from the psychological disorder (Oh 2011b). A famous example is Bill Murray, the depressed (and possibly melancholic) hero in the movie, *Groundhog Day* (1993); while locked up in his fate of waking up again on the same Groundhog Day endlessly, he finally discovers that his only joy is learning to play the piano and reading French poetry every day. Similar examples are many. McDonald's employees, who find it impossible to express to anyone their tragic lifestyle of flipping burgers every day from early morning to late at night, decide to seek exhilaration through learning how to flip burgers while dancing to the music they listen to, using headphones (Garson 1988). In these two cases, learning involves repetitive and monotonous memorization of already existing routines, although slight variation is allowed.

Researchers found that three types of melancholia exist, namely, gendered, racial, and postcolonial, although all forms of melancholia originate from the gendered (Butler 1988; Kristeva 1989; Gilroy 2006; Oh 2011b). Gendered melancholia is the psychological state of people who are not able to openly express their sadness, despite the fact that they were forced to choose their designated gender roles. Although men's choice of their male gender identity is socially rewarded with women as their sexual partners, women's choice of their female gender identity produces no other reward than sexually and socially subjugating themselves to men (Butler 1988). Gendered melancholia therefore occurs when women cannot openly mourn about the loss of their preferred gender identity after reluctantly choosing their socially approved gender role as women. Racial melancholia is similar to the gendered one, because the individual choice of socially approved racial roles unfairly discriminates against the minority, a fact that minorities cannot openly mourn about. Finally, postcolonial melancholia occurs to many former imperialist rulers who have to give up their colonies against their wish but cannot openly mourn about it.

Postcolonial melancholia leads to the study of former colonies, even though the former imperialists lost their political control over the territories (Gilroy 2006). Racial melancholia motivates many minority youths to succeed in white majority societies through education, sports, and entertainment. Typical examples are the Caribbean students who excel in schools in the United Kingdom, Asian minority students who fill up the school honor rolls in the United States, and the African American athletes, who dominate American professional sports leagues such as NBA. Gendered melancholia also motivates selected women to pursue and achieve high levels of "exotic" or extraordinary learning, as championed by Jane Goodall, Birutė Galdikas, and Diane Fossey, all of whom devoted their life to the study of primates. Finally, experiencing *jouissance* or flow through feedback learning that is motivated by melancholia is the final reward that learners can claim after a long period of painstakingly repeating the routines that were permeated by their roles choreographed from above. This is a clinically interesting process of transforming *ressentiment* or deep-rooted psychological anger against males, the majority population, and those that took their colonies away into *jouissance* or flow (Lacan [1966] 2001; Csikszentmihalyi 1996; Žižek 2000).

Japanese and Korean universities can also motivate themselves toward feedback learning in order to transmute their *ressentiment* against world-class universities in the West into *jouissance*. Catching up automatically means regurgitation, repetition, and fine-tuning of the knowledge that has been developed by the West. Like melancholia, *han* or *urami* (Korean and Japanese terms of *ressentiment*) can easily be escalated into a psychological disorder when victims are not allowed to redress their wounds by punishing the aggressors. Like melancholia, *han* or *urami* therefore works as a strong catalyst of learning (Lie 2000; Oh 2010).

## The case of Waseda University Ventures

The above discussion about melancholia as a new facilitator of feedback learning can be applied to the Japanese triple helix. The Japanese triple helix has so far had a hard time attracting Japanese and foreign MNEs into the firm-university alliance oriented toward feed-forward learning due to the relative limitations of the Japanese universities in carrying out explorative research. However, if Japanese universities change their strategy and instead focus on follow-on or feedback research about the knowledge that has already been developed by Japanese MNEs and/or Western universities, Japanese MNEs would find it easier to participate in the firm-university alliance than under the current conditions. In fact, some of the leading Japanese universities have already started university-based ventures [*daigakuhatsubenchâ*] and TLO-approved projects that heavily utilize feedback vis-à-vis feed-forward learning, such as creating technologies for developing economies, routine IT software development, and the refinement of the existing technologies in manufacturing. Among these, Waseda University deserves a close look.

As Table 10.5 shows, Waseda is the only private university in Japan that is ranked second in Japan in terms of the total number of university-based ventures it has created as of 2010. Considering that national universities receive most of the governmental funding and firm-level support, the second place in 2007 and 2010 that Waseda garnered is phenomenal. If I calculate the percentage based on the information on Table 10.5, 63.4 percent of all top 10 university-based ventures were created by national universities in 2007, while the number for private universities was only 29.5 percent (Ogura 2008, p. 133). However, since the majority of these start-ups were concentrated in the area of life sciences (35 percent) and IT software (30.2 percent) in 2008, even national universities are now pursuing feedback learning for IOL, as gleaned from the increase in investments in IT software (METI 2010, p. 140). Only 10 percent of these university-based startups carry out the practice of technology commercialization in the area of manufacturing, such as automobiles (Ogura 2008, p. 135). This is clearly a Japanese phenomenon, because the manufacturing sector has traditionally relied on its own R&D centers or connect and develop (C&D) with United States and European Union research centers and/or universities for technology sharing.

*Table 10.5*  University-based venture start-ups

| Ranking | 2007 | 2008 | 2010 |
|---|---|---|---|
| 1 | Tokyo (107) | Tokyo (125) | Tokyo (147) |
| 2 | **Waseda (96)** | Tsukuba (76) | **Waseda (107)** |
| 3 | Osaka (68) | Osaka (75) | Osaka (82) |
| 4 | Tsukuba (61) | **Waseda (74)** | Kyoto (79) |
| 5 | **Keio (52)** | Kyoto (64) | Tsukuba (76) |
| 6 | Kyushu (48) | Tohoku, TIT (57) | Tohoku (66) |
| 7 | Kyoto (45) | N/A | Kyushu (60) |
| 8 | Kobe (42) | Kyushu (55) | TIT (52) |
| 9 | TIT (40) | **Keio (51)** | **Keio (50)** |
| 10 | KIT (39) | KIT (45) | Hokkaido, Kobe (46) |
| Total | 598 | 679 | 811 |

*Source*: Ogura (2008, p. 136); METI (2010, p. 140); JST (2011).
*Note*: Universities in bold are private ones.

The key to Waseda's rise to the second place in Japan derives from the fact that the university focused on feedback learning vis-à-vis feed-forward learning and technology transfers through high-tech human resources vis-à-vis commercialization of patents in creating venture start-ups. Consequently, Waseda's IT software ventures are currently the single most important group of venture firms that the university has created over the years, as they account for 35 percent of total ventures, whereas bio-ventures account for only 14 percent in 2011. Furthermore, technology transfers to private firms through human resource transfers account for 69 percent of all venture types in 2011 (WTLO 2011).

Along with Tsukuba University, Waseda's TLO (WTLO) was created four months after Tokyo University had established its own for the first time in Japan in December 1998. Since the amount of funding from the Japanese MNEs to Waseda was not colossal, the most important financial source for the IT software ventures at Waseda has been the royalty payments from the MNEs after the WTLO was notified about the inventions by the ventures (see Fig. 10.3). What's peculiar about Waseda's TLO is threefold. First, the proportion of IT software ventures to all WTLO-related ventures is the highest among the top ten universities. Second, the ratio of student-created ventures to all WTLO ventures is the highest (44.9 percent) among the top ten universities in Japan (JST 2011). Third, WTLO is the only Japanese university that has received government funding to start WTLO-related international ventures to work with governments, firms, and universities in developing countries for *kaizen* projects of developing and commercializing technologies for their sustainable economic development in the countries. These three facts clearly indicate that Waseda's TLO emphasizes feedback learning more than feed-forward learning for IOL with Japanese MNEs.

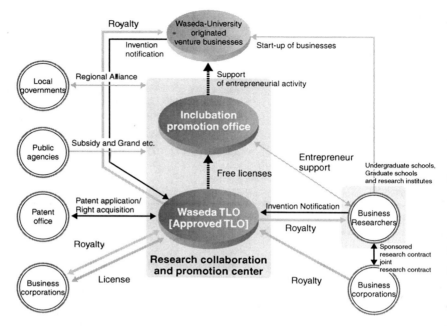

*Figure 10.3*  The structure of the Waseda Technology Licensing Office (TLO) and venture start-ups

*Source*: WTLO (2011).

As Table 10.6 shows, most university-based venture start-ups report huge business losses, especially in the field of biotechnology. However, service-related ventures report lower levels of losses than other sectors, although their profit levels are also low. Among all non-service sectors, information and communication technology (ICT) sectors reported the highest percentage of profitable ventures, although the percentage of companies in the red is similar to that of the biotech service industry. This signals that WTLO's strategic choice of supporting ICT start-ups was the right decision in that it emphasized feedback learning in developing application and solution software for ICT industries. Although the work is tedious in nature, the mobilization of professors and students (mostly, graduate students) could countervail the monotonous tendency of the work and generate income.

The motivation toward feedback learning at Waseda is related melancholia to or *urami*. For many decades, the university and its students have been forced to give up large portions of publicly and privately funded research resources, because it is a private university – resources that in turn were funneled to national universities and to Keio University, a rival private university in Tokyo. By emphasizing the international aspect of higher

*Table 10.6*   Current profit (in yen) of university-based venture firms (2010)

| | In the red (%) | No Profit (%) | Up to ¥10 mil. (%) | Up to ¥30 mil. (%) | Up to ¥100 mil. (%) | Over ¥100 mil. (%) |
|---|---|---|---|---|---|---|
| Total (N = 434) | 53.20% | 6.2 | 32.3 | 5.5 | 2.3 | 0.5 |
| Bioventure | 67.3 | 8.2 | 18.4 | 4.1 | 2 | |
| Manufacturing | 56.4 | 5.4 | 29.5 | 6 | 2 | 0.7 |
| ICT | 46.9 | 6.3 | 40.6 | 6.3 | | |
| Bioservice | 41.9 | 6.5 | 32.3 | 16.1 | 3.2 | |
| Tech service | 51 | 8.2 | 36.7 | 2 | 2 | |
| Other service | 31.8 | 54.5 | 9.1 | 4.5 | | |
| Others | 66.7 | 4.8 | 14.3 | 9.5 | 4.8 | |

*Source*: MEXT (2011).

learning and research, Waseda could secure more foreign students than its rival, Keio, while simultaneously expanding its TLO to Asia, the European Union, and North America. Furthermore, Waseda's venture start-ups are two times those at Keio.

## The case of the Sungkyunkwan University–Samsung collaboration

Unlike the universities in Japan, some of the leading Korean universities are owned and controlled by Korean MNEs. Hanjin et al., for example, purchased Inha University in 1968, Ajou University in 1977, Sungkyunkwan (SKK) University in 1996, and Chung-Ang University in 2008, respectively, while Hyundai and Posco built their own new universities, the University of Ulsan in 1969 and Postech in 1986, respectively. As Table 10.7 shows, among these MNE-owned universities, SKK University is the most successful of the university-industry alliance (with 650 patents) and has been leading the race, among the MNE-owned universities, in the area of the number of technologies transferred to firms. I argue that the main reason for the success of SKK-Samsung alliance vis-à-vis other similar joint ventures is that SKK emphasizes feedback over feed-forward learning by focusing on the specifically designed pre-employment training and job retraining programs for Samsung, among other reasons that I will discuss below in detail. Indeed, as Table 10.8 indicates, the number of firm-university alliances based on feedback learning (e.g., job retraining and pre-employment training programs) has been increasing in Korea, implicitly signaling the increasing recognition of the importance of feedback learning in firm-university alliances.

Samsung's decision to acquire SKK University resolved several problems Samsung had faced in its earlier R&D efforts. First and foremost, Samsung needed a steady supply of workforce for its high-tech industries such as electronics, IT software development, biotech and medical technologies, and other advanced precision technologies. Samsung's sole reliance on the Korean graduates of Japanese, North American, and European Union

*Table 10.7*   University rankings for interorganizational learning (IOL) performance (2009)

| Rank | Total funding science & tech (US$) | No. of patents (Domestic) | No. of transfers |
|------|------|------|------|
| 1 | SNU (358 mil.) | SNU (1,994) | Korea (89) |
| 2 | Yonsei (240 mil.) | KAIST (1,908) | SNU (59) |
| 3 | KAIST (222 mil.) | Hanyang (1,248) | SKK (46) |
| 4 | Hanyang (166 mil.) | POSTECH (1,085) | Gwangwoon (45) |
| 5 | SKK (165 mil.) | Yonsei (1,051) | Yonsei (45) |
| 6 | Korea (142 mil.) | Korea (956) | Gangwon (41) |
| 7 | POSTECH (125 mil.) | Inha (830) | Kyunghee (37) |
| 8 | Jeon Nam (92 mil.) | SKK (650) | POSTECH (35) |

*Source*: MEST (2010).

*Table 10.8*   Increasing feedback learning programs for firm–university alliances in Korea

| Contract type | 2007 | 2008 | 2009 |
|------|------|------|------|
| Re-education training | 38 universities (130 departments) | 45 universities (147 departments) | 70 universities (248 departments) |
| Pre-employment training | 3 universities (3 departments) | 3 universities (4 departments) | 10 universities (15 departments) |

*Source*: MEST (2010).

universities limited the flexibility of Samsung's human resource management (HRM). In order to sustain its leadership position in the global IT and electronic market, Samsung instituted a progressive HRM policy, that intends to educate and employ a large number of technicians with degrees in science and technology for its expanding high-tech business.

Second, by acquiring SKK, Samsung could control the content of education for its future employees and for the current SKK students. Instead of overdosing students with abstract theories, Samsung can now dictate to the SKK professors to teach exactly what Samsung requires from its future employees. Third, by concentrating organizational assets of finance, human resurses, knowledge, and experience in one centralized location, Samsung can control SKK assets in a centralized manner, eliminating threats of asset specificities and information asymmetries. Asset centralization through administrative fiat in fact resolves transaction costs by resolving the corporate governance problems (e.g., agency problems, asset specificities, information asymmetry). By destroying the organizational boundaries between Samsung and SKK, the chances of developing interorganizational assets has also become easier than ever.

To implement the foreseeable benefits of IOL between Samsung and SKK, the two partners established the Department of Cell Phones within SKK's College of IT Engineering. The Department of Cell Phones carries out teaching, research, and human resources (HR) exchanges for Samsung. The entire department's MSc and PhD students receive full scholarships and monthly stipends from Samsung. These students are guaranteed that they will be employed by Samsung upon graduation. The department teaches and researches about the next generation mobile communication technologies, RF (radio frequency) SOC (system-on-chip) technologies, mobile multimedia, and mobile application services (MEST 2010, pp. 320–321).

New degree programs installed at SKK are divided into two types of curricula: job retraining and pre-employment training. The Department of Cell Phones, for example, predominantly carries out the function of pre-employment training. As Table 10.9 shows, SKK and Samsung have established other similar programs with other departments. For example, Samsung Electronics funds pre-employment training for the bachelor's degree students in the Department of Semiconductor System Engineering and funds the master's students in the Department of Embedded Software. Samsung currently is the second-largest manufacturer of cell phones in the world, while competing neck-and-neck with Apple's iPhone in the smartphone industry. It is obvious that the company needs a lot of young human resources for the business. The construction division of Samsung Corp., on the other hand, finances pre-employment training for the master's students

*Table 10.9*   Feedback learning programs at SKK for Samsung (2009)

| Contract Type | Degree | Departments | No. of students |
|---|---|---|---|
| **Retraining** | PhD | DMC Engineering | 0 |
| | MSc | DMC Engineering | 25 |
| | MBA | GSB Executive MBA | 28 |
| | PhD | Semiconductor Display Engineering | 13 |
| | MSc | Semiconductor Display Engineering | 52 |
| | PhD | Industrial Pharmacy | 9 |
| | MSc | Industrial Pharmacy | 5 |
| | PhD | Cell Phones | 2 |
| | MSc | Cell Phones | 3 |
| **Pre-employment** | BSc | Semiconductor System Engineering | 225 |
| | MSc | Embedded Software | 40 |
| | MSc | Mega Buildings & Bridges | 40 |
| | PhD | Cell Phones | 12 |
| | MSc | Cell Phones | 56 |

*Source*: MEST (2010).

in the Department of Mega Buildings and Bridges. Samsung Corp. completed the Burj Khalifa Tower in Dubai in 2010, using its most advanced building technologies, which included ACSF (automatic climbing system form) and lift-up engineering. This is just another reason why Samsung Corp. needs more highly educated engineers in "mega building and bridges," who will perfect the technologies that American architects have developed.

Financing feedback learning projects at SKK in its home country enabled Samsung to begin the integration and institutionalization of the feed-forward learning they achieved through IOL with other researchers in North America, Japan, and the European Union. A long-term alliance between Samsung and SKK might also develop into a new feed-forward learning project after Samsun establishes trust in SKK's capability to develop explorative knowledge. The motivation for the Samsung-SKK alliance derived from melancholia or *han* that Samsung has carried for many decades as a corporation that was forced to keep buying new knowledge from advanced economies. When Samsung first tried to acquire SKK in the late 1970s, SKK students vehemently protested the M&A proposal, resulting in Samsung's reluctant yet pressured withdrawal from the bid. Students' logic back then was more political than economic, because they thought that the capitalist conglomerate would ruin university education by tainting its ivory tower ideal with "obscene" business desires. Samsung's then chairman lamented about the backwardness of students' thinking (Samsung 1998). However, when the MNE tried to acquire the same university one more time in 1996, almost two decades later, most students welcomed the bid. Tenacious bidding to acquire SKK twice in two decades indicates just how serious Samsung was in acquiring a university for their human resource management (HRM) and other R&D purposes. Simultaneously, SKK students also made a priority of improving the school's second-rate brand image in Korea, a form of melancholia or *han* they have been suffering from, believing that Samsung would certainly lift the school's domestic and international brand image, not to mention the fact that Samsung would alleviate SKK's deepening financial woes. As a consequence, Samsung-SKK alliance is now the most successful firm-university venture in Korea that is based on feedback learning.

## 5    Further research directions

Instead of fostering firm-university alliances for an economy that excelled in some industries but lagged behind in the quality of higher education, this chapter has argued that Japanese and Korean universities should envisage an alternative path toward a new East Asian style triple helix that specializes in feedback learning. The Waseda and Samsung-SKK cases suggest that the alternate path to firm-university alliances would result in remarkable success, when the facilitating conditions and other catalysts are present. Therefore,

Proposition 1a. Melancholia or *ressentiment* motivates individual, group, or (inter)organizational feedback learning, despite the illiberal nature of the top–down learning arrangement.

Proposition 1b. Postcolonial melancholia facilitates a catch-up mode of innovation that is an example of feedback learning, as long as it produces *jouissance* for the learners.

Furthermore, the firm-university alliance we examined using the case of Waseda reveals the fact that Japanese MNEs would not purchase university-based technology development, unless universities market-test its commercial value on an ongoing basis (i.e., follow-on research). Therefore,

Proposition 2a. TLO projects and venture start-ups at universities increase the chance of institutionalizing firm-university collaboration in feed-forward learning, if the main barriers to the institutionalization of feed-forward learning are structural rather than technological.

Proposition 2b. University ventures are more conducive than TLO incubation projects to firm-university R&D collaboration in feed-forward learning, as long as it is the venture start-ups that are directly exposed to fluctuating market demands.

However, the case of Samsung-SKK reveals the fact that acquisition was necessary to foster such a joint venture, which was initially protested by stakeholders of the wider community, including university students. The current acquisition of Chung-Ang University by Doosan was also met with strong oppositions from the students and professors. Therefore,

Proposition 3a. The acquisition of university partners by firms is a precondition to IOL that is based on feedback learning, because the two organizations suffer from high levels of asset specificity, a lack of institutional complementarity, informational asymmetry, and have lenient intellectual property protection.

Proposition 3b. The acquisition of university partners by firms allows the latter to control the contents of feedback learning, which can be tailored to the needs of the firm, such as alleviating the firm's HRM shortage.

Proposition 3c. The acquisition of university partners by firms depends on the degree of opposition to the bid from the wider stakeholders of the universities, including students and faculty. Opposition subsides with overall economic hardship and declining university brand image.

Putting feedback learning into action requires manageable programs and learning curricula. From what we see in the Japanese university-based ventures in general and the Waseda case in particular, successful feedback learning involves IT software development and servicing, which have turned out to be more profitable than feed-forward learning projects, such as bioventure and manufacturing. Therefore,

> Proposition 4a. If firm-university feedback learning focuses on the programs that are intended to provide IT software services to the firm, they will be more profitable than feed-forward learning programs (e.g., biotech).

However, the Korean case in general and the Samsung-SKK case in particular revealed that successful feedback learning involves HRM training, either job retraining or pre-employment training. Since these programs are offered at the MSc and PhD levels, feedback learning is not only challenging, but could easily be developed into explorative or feed-forward learning later as well. Therefore,

> Proposition 4b. Firm-university feedback learning usually focuses on the programs that are intended to provide job retraining or pre-employment training services to the firm, because they can be easily implemented using the knowledge and capacity the university has, and their performance can be easily measured by the number of student enrollments.

## 6  Conclusion

This study took up a research question of why firm-university alliances in Japan and Korea were not welcomed by leading MNEs. Based on the literature review of the barriers to the 4I model of OL, I suggested that we modify and apply the 4I model to IOL. I also argued that within the 4I model, integration can be classified as both feed-forward and feedback learning, depending on how one can interpret the learning process that integration requires.

Using the revised 4I model with the distinction between feed-forward and feedback learning, I concluded that the reason East Asian MNEs do not want to participate in firm-university alliances is because universities unanimously pursue feed-forward learning. In order to change this tendency and to galvanize more MNE participation in triple helix projects, I showed that the success of Waseda University in creating and sustaining venture start-ups derived from its emphasis on feedback learning. I also explained that the success of the Samsung-SKK University alliance was possible because of the acquisition of the university by the firm, which forcibly reformed the university curricula and research agendas toward feedback learning. I contended that melancholia or *ressentiment* must be present to motivate universities toward feedback learning, and furthermore, *ressentiment* had to be transformed into *jouissance* among learners at the end of the cycle of feedback learning. Finally, the case studies found that the most successful feedback learning in Japan was either ICT or ICT and other servicing ventures, since they all tend to quickly commercialize new *kaizen* technology using feedback learning in IOL. However, the most successful feedback learning in Korea was either in job retraining or pre-employment training programs at the MS. and PhD levels.

Further empirical and theoretical studies of the same topic can be devised using more generalizable data sets. The limitation of this study is that it did not establish the link between *ressentiment* before feedback learning and the *jouissance* learners have experienced after a cycle of learning. Also, case studies are not meant to generate any generalizable knowledge about IOL motivations. More carefully designed interviews and observations of participation can be planned for future studies, in addition to comparing panel data of successful feedback learning with that of universities and MNEs in other countries.

## Acknowledgment

I thank Jhia Jiun Ho, Ravshan Abdukarimov, and Michael Lamm for editorial assistance with the manuscript. I want to express special gratitude to Alexander Brem and Éric Viardot for encouraging me to write this chapter. This study was partially supported by the Solbridge Travel Fund. All errors are mine.

## References

Argyris, C. (1990) *Overcoming Organizational Defenses*. Boston, MA: Allyn & Bacon.
Beer, M., Voelpel, S. C., Leibold, M., & Tekie, E. B. (2005) Strategic management as organizational learning: Developing fit and alignment through a disciplined process. *Long Range Planning*, 38, 445–465.
Berkovitz, J., & Feldman, M. (2005) Fishing upstream: Firm innovation strategy and university research alliances. Discussion Chapter, DRUID Tenth Anniversary Summer Conference on Dynamics of Industry and Innovation: Organizations, Networks, and Systems.
Berlant, L. (1988) Female complaint. *Social Text*, 19(20), 237–259.
Berthoin-Antal, A., Lenhardt, U., & Rosenbrock, R. (2003) Barriers to organizational learning. In M. Dierkes, A. B. Berthoin-Antal, J. Child & I. Nonaka (Eds), *Handbook of Organizational Learning and Knowledge* (pp. 865–885). Oxford: Oxford University Press.
Butler, J. (1988) Performative acts and gender constitution: An essay in phenomenology and feminist theory. *Theater Journal*, 44(4), 519–531.

Cannon, M. D., & Edmondson, A. (2001) Confronting failure: Antecedents and consequences of shared beliefs about failure in organizational work groups. *Journal of Organizational Behavior*, 22, 161–177.

Capron, H., & Cincera, M. (2003) Industry–university S&T transfers: Belgium evidence on CIS data. *Brussels Economic Review*, 46(3), 58–85.

Cassiman, B., & Veugelers, R. (2002) Complementarity in the innovation strategy: Internal R&D, external technology acquisition and cooperation. CEPR Discussion Chapter, 3284.

Cassiman, B., Di Guardo, C., & Valentini, G. (2005) Organizing for innovation: R&D projects, activities, and partners. *Working paper* No. 597, IESE Business School, Barcelona.

Cohen, M. D. (1991) Individual learning and organizational routine: Emerging connections. *Organization Science*, 2, 135–139.

Cohen, M. D., & Bacdayan, P. (1994) Organizational routines are stored as procedural memory: Evidence from a laboratory study. *Organization Science*, 4, 554–568.

Coopey, J. (1995) The learning organization, power, politics and ideology. *Management Learning*, 26(2), 193–213.

Crossan, M. M., Lane, H. W., & White, R. E. (1999) An organizational learning framework: from intuition to institution. *Academy of Management Review*, 24, 522–537.

Csikszentmihalyi, M. (1996) *Creativity: Flow and the Psychology of Discovery and Invention.* New York: HarperCollins.

Daigaku Shingikai (1998) *21sêki no daigakuzô to kongo no kaikaku hôsaku nit suite.* Tokyo: MEXT.

Diaz, R. (2006) Melancholic maladies: Paranoid ethics, reparative envy, and Asian American critique. *Women and Performance*, 16(2), 201–219.

Edmondson, A, & Moingeon, B. (1996) From organizational learning to the learning organization. *Management Learning*, 29, 5–20.

Elliott, D., Smith, D., & McGuinness, M. (2000) Exploring the failure to learn: Crisis and the barriers to learning. *Review of Business*, 21(3/4), 17–24.

Eng, D. (2000) Melancholia in the late twentieth century. *Signs*, 25(4), 1275–1281.

Eom, B.-Y. and Lee, K. (2010) Determinants of industry–academy linkages and, their impact on firm performance: The case of Korea as a latecomer in knowledge industrialization. *Research Policy*, 39, 625–639.

Fontana, R., Geuna, A., & Matt, M. (2006) Factors affecting university–industry R&D projects: The importance of searching, screening and signaling. *Research Policy*, 35, 309–323.

Freud, S. ([1917]1963) Mourning and melancholia. In P. Rieff (Ed.), *General Psychological Theory* (pp. 164–179). New York: Collier.

Garson, B. (1988) *The Electronic Sweatshop: How Computers Are Transforming the Office of the Future into the Factory of the Past.* New York: Simon and Schuster.

Geisler, E. (1995) Industry–university technology cooperation: A theory of interorganizational relationships. *Technology Analysis and Strategic Management*, 7(2), 217–229.

Gilroy, P. (2006) *Postcolonial Melancholia.* New York: Columbia University Press.

Greenfeld, L. (1990) The formation of the Russian national identity: The role of status insecurity and ressentiment. *Comparative Studies in Society and History*, 32(3), 549–591.

Hedberg, B., & Wolff, R. (2003) Organizing, learning, and strategizing: From construction to discovery. In M. Dierkes, A. B. Berthoin-Antal, J. Child and I. Nonaka (Eds),

*Handbook of Organizational Learning and Knowledge* (pp. 535–556). Oxford: Oxford University Press.

Hong, Y. S. (1993) Leveraging technology for strategic advantage in the global market: Case of the Korean electronic industry. *KIEP Working Paper* No. 93–07. Seoul: KIEP.

Im, Y.-G. (2005) Hanguk daehak gujojojeong jeongchaekeui teukjinggwa jaengjeom. *Gyogyuhaengjeonghak Yeongu*, 23(4), 243–268.

Inkpen, A. C., & Crossan, M. M. (1996) Believing is seeing: Joint ventures and organizational learning. In D. Russ-Eft, H. Preskill & C. Sleezer (Eds), *Human Resource Development Review: Research and Implications* (pp. 299–329). Thousand Oaks, CA: Sage.

Itô, Ken. (2008) *Nihon ni No-berushôga Kuru Riyû*. Tokyo: Asahishinbun Shuppan.

JST (Japan Science and Technology Agency) (2011) Sangakukan kanren dêta. Available online at: http://sangakukan.jp/top/databook_contents/2010. Accessed on Nov. 7, 2011.

Kang, B.-W. (2005) Godeunggyoyuk gyeongjaengryeok ganhwa reul wihan daehak gujogaehyeok banghyang gwa gwaje. *Gyogyuhaengjeonghak Yeongu*, 23(2), 421–446.

Katô, K. and Enomoto, S. (2006) Sangakurenkei kenkyûkaihatsu purojekuto niokeru kadai to sono kaizenjirei. *Purojekuto Manejimento Gakkaishi*, 8(1), 17–22.

Kim, D. H. (1993) The link between individual and organizational learning. *Sloan Management Review* 35(1), 37–50.

Kim, D -W, & Lesline, S. (1998) Winning markets or winning Nobel Prizes?: KAIST and the challenges of late industrialization. *Osiris*, 13, 154–185.

Kogut, B. (1988) Joint ventures: Theoretical and empirical perspectives. *Strategic Management Journal*, 10, 319–332.

Kristeva, J. (1989) *Black Sun: Depression and Melancholia*. New York: Columbia University Press.

Lacan, J. ([1966]2001) *Écrits*. London: Routledge.

Lawrence, T. B., Mauws, M. K., Dyck, B., & Kleysen, R. F. (2005) The politics of organizational learning: integrating power into the 4I framework. *Academy of Management Review*, 30, 180–191.

Levitt, B., & March, J. G. (1988) Organizational learning. *Annual Review of Sociology*, 14, 319–340.

Lie, J. (2000) *Han Unbound: The Political Economy of South Korea*. Stanford: Stanford University Press.

March, J. G. (1991) Exploration and exploitation in organizational learning. *Organization Science*, 2, 71–87.

McCracken, M. (2005) Towards a typology of managerial barriers to learning. *Journal of Management Development*, 24, 559–575.

MEST (Ministry of Education, Science and Technology) (2010) *2009 Daehak Sanhak Hyupryeok Baekseo*. Seoul: MEST.

METI (Ministry of Economy, Trade and Industry) (2010) *Waga Guni no Sangyô Gijutsu ni Kansuru Kenkyû Kaihatsu Katsudô no Dôkô : Shuyô Shihyô to Dêtâ*. Tokyo: METI.

MEXT (Ministry of Education, Culture, Sports, Science, and Technology) (2011) *Heisei 22 Nendo Daigakuni okeru Sangakurenkeitô Jitsijôkyô ni Tsuite*. Tokyo: MEXT.

Mohnen, P. and Hoareau, C. (2003) What type of enterprises forges close links with universities and government labs? Evidence from CIS2. *Managerial and Decision Economics*, 24, 133–145.

Moorman, C., & Miner, A. S. (1998) Organizational improvisation and organizational memory. *Academy of Management Review* 23(4), 698–723.

Nishio, K. (2007) Nihon ni okeru sangakukan no soshikiteki kenkyû kyôryoku ni kansuru kenkyû. *Kenkyû Gijutsu Keikaku*, 22(1), 65–81.

Nonaka, I. (1994) A dynamic theory of organizational knowledge creation. *Organization Science*, 5, 14–37.

Nooteboom, B. (2000) *Learning and Innovation in Organizations and Economies.* Oxford: Oxford University Press.

Ogura, M. (2008) Nihon no daigakutôhatsu bencha- to seifukei kenkyûhatsu bencha. In T. Watanabe (Ed.), *Akademiku Inobe-shon: Sangakurenkei to Suta-to Appusu Sôshutsu* (pp. 129–151). Tokyo: Hakutô Shobô.

Oh, I. (2010) Education and development: Why are Koreans obsessed with Learning? *Comparative Sociology*, 9, 308–327.

Oh, I. (2011a) Not yet triple helix III? Japanese MOT policies and the problem of technology exploitation. In M. Saad and G. Zawdie (Eds), *Theory and Practice of the Triple Helix System in Developing Countries: Issues and Challenges* (pp. 283–304). New York: Routledge.

Oh, I. (2011b) Torn between two lovers: Retrospective learning and melancholia among Japanese women. *Korea Observer*, 42(2), 223–254.

Oh, I., Park, H.-J., Yoneyama, S., & Kim, H.-R. (2005) *Mad Technology: How East Asian Companies are Defending Their Technological Advantages.* New York: Palgrave.

Park, H. W., & Leydesdorff, L. (2010) Longitudinal trends in networks of university-industry-government relations in South Korea: The role of programmatic incentives. *Research Policy*, 39, 640–649.

Popper, M. and Lipshitz, R. (2000) Organizational learning: Mechanisms, culture, and feasibility. *Management Learning*, 31(2), 181–196.

Rosenberg, N. and Nelson, R. (1994) American universities and technical advance in industry. *Research Policy*, 23, 323–348.

Samsung (1998) *Samsung 60 Nyeon Sa.* Seoul: Samsung Hoejang Biseosil.

Santoro, M. D., & Chakrabarti, A. K. (2002) Firm size and technology centrality in industry-university interactions. *Research Policy*, 31, 1163–1180.

Schilling, J., & Kluge, A. (2009) Barriers to organizational learning: An integration of theory and research. *International Journal of Management Review*, 11(3), 337–360.

Son and Lee (2005) Sanhak hyeopryeok eui heo wa sil: Hyeonghwang jindan gwa jeongchaek gwaje. KOTEF Issue Paper 05–08. Seoul: KOTEF.

Spender, J. C. (1989) *Industry Recipes: An Inquiry into the Nature and Sources of Managerial Judgement.* Oxford: Blackwell.

Starbuck, W. H., & Hedberg, B. (2003) How organizations learn from success and failure. In M. Dierkes, A. B. Berthoin-Antal, J. Child and I. Nonaka (Eds), *Handbook of Organizational Learning and Knowledge* (pp. 327–350). Oxford: Oxford University Press.

Streip, K. (1994) 'Just a Cérébrale': Jean Rhys, women's humor, and ressentiment. *Representations*, 45(Winter), 117–144.

Sun, P. Y. -T., & Scott, J. L. (2005) An investigation of barriers to knowledge transfer. *Journal of Knowledge Management*, 9(2), 75–90.

Sung, T.-K. (2005) Firm size, networks and innovative activity: Evidence from the Korean manufacturing firms. *Technological Innovation Studies*, 13(3), 1–20.

Szulanski, G. (2003) *Sticky Knowledge: Barriers to Knowing in the Firm.* Thousand Oaks, CA: Sage.

Taguchi, Toshiyuki (2003) *Sangakukyôdô to Kenkyûkaihatsu Senryaku*. Tokyo: Hakutô Shobô.

Tamura, Yasukazu (2006) MOT to sangakurenkei manejimento. *Ofisu O-tome-shon*, 26(4), 35–44.

Teece, D. (1985) Multinational enterprise, internal governance and industrial organization. *American Economic Review*, 75, 233–238.

Tether, B. S. (2002) Who co-operates for innovation, and why: an empirical analysis. *Research Policy*, 31, 947–967.

Van de Ven, A. H., & Polley, D. (1992) Learning while innovating. *Organization Science*, 3, 92–116.

Virany, B., Tushman, M. L., & Romanelli, E. (1992) Executive succession and organization outcomes in turbulent environments: An organizational learning approach. *Organization Science*, 3, 72–91.

WEF (World Economic Forum) (2011) *The Global Competiveness Report 2011–2012*. Geneva: Switzerland.

WTLO (Waseda Technology Licensing Organization) (2011) Data. Available online: http://tlo.waseda.ac.jp/ABOUT/data.html. Accessed on Nov. 6, 2011.

Zander, U. and Kogut, B. (1995) Knowledge and the speed of the transfer and imitation of organizational capabilities: An empirical test. *Organization Science*, 6, 76–92.

Zell, D. (2001) Overcoming barriers to work innovations: Lessons learned at Hewlett-Packard. *Organizational Dynamics*, 30(1), 77–86.

Žižek, S. (2000) Melancholy and the act. *Critical Inquiry*, 26, 657–681.

# 11
## Innovation Management Reflections: A Brazilian Market Perspective

*Fabian Ariel Salum, Rosana Silveira Reis, and Hugo Ferreira Braga Tadeu*

By reconciling theory and practice, this chapter aims to provide the reader with a view of the innovation process from the perspective of the Brazilian market, as an emerging nation. The case studies presented will illustrate both the theoretical references and the empirical analysis running throughout the sections. This chapter is composed of five sections, and is organized as follows. Section 1, "The context," presents a brief overview of innovation in Brazil and a case study of Azul Brazilian Airlines, named one of the world's 50 most innovative companies by *Fast Company* magazine (2011). The case emphasizes the accomplishments of the Brazilian businessman David Neleman, co-founder of Soutwest, West Canada and Jet Blue. In Section 2, "Process management and its relevance in a strategic context," we detail innovation and process management, as we understand that it is not possible to begin a reflection on innovation management in companies without recognizing that they are managed through systematic processes, in a systematic way. We present the Embraer case study, where we address the importance of customers and the supply chain. In Section 3, "Innovation management," we focus on innovation as a process. Our objective is to analyze the main concepts of innovation and their applicability to emerging countries. We present the model of innovation created by the FDC (Fundação Dom Cabral, a Brazilian business school), which was developed based on research conducted with Brazilian companies. To illustrate this section, we introduce the case study of Fiat Automóveis. In Section 4, "Cost management aligned to innovation in Brazilian companies," we focus on cost reduction and interdepartmental links. The Brazil team is known for its competence in cost reduction and concern for the use of raw materials and their impact on the environment, as noted in the Alstom case study. Finally, in Section 5 we present our conclusions and we outline and synthesize our findings and their managerial implications.

## 1 The context

The international market is now of concern to Brazil's medium-size companies as well as to its big corporations (MDIC 2012). This is a consequence of the economic opening at the beginning of the 1990s by then-president Fernando Affonso Collor de Mello (1990–1992). This opening became the basis for the entry of large international companies into the Brazilian market, which brought technological advances, gains in productivity, and considerable improvement in workforce qualification. This resulted in a significant improvement in the management methods of Brazilian companies, with gains in competitiveness and innovation. For Batista Jr. (1993), all these factors are fundamentally due to globalization, i.e., the characterization of internationally developed cultural interchange, with economic and business models, observed in recent years.

Globalization is a phenomenon of economic integration, a consequence of financial transactions, and is connected with financial system and productivity gains (Stiglitz 2006). This phenomenon has been occurring in Brazil over the past few years. According to Franco (1996), Brazilian prospects could improve after a broader integration with the international market and the entry of competitive organizations, which would bring diversity, capital earnings, and improvements in the regulatory scenario. Following this reasoning, Sera (1992) says that globalization has been providing a greater integration among countries and also a convergence of best practices in search of satisfactory results and improved efficiency.

Since one of the main focuses of Brazilian companies is to seek improved marketing strategies to produce better financial results and efficiency in their processes, it becomes crucial to be active in a globalized, highly competitive market with a focus on perceived value by customers. In this case, perceiving competitors' moves becomes a mandatory task, and organizational benchmarking in search of best practices becomes a market imperative (McNair & Liebfried 1992)

Thus, the use of performance indicators tied to business and to the competitors' behavior became essential. It is known that this task is not easy at all – not only because the competition, as a result of a globalized environment, is becoming more present as time passes, but also because products, services, and processes are getting more like each other and are increasingly perishable. The formulation of organizational strategies is becoming increasingly focused on the generation of value in the production chain. In this case, the term "value" must be related to the preservation of the invested capital; to the rational use of finite resources; to ethical, sustainable and technological practices; and to fulfilling customer demands (Porter 2011).

This whole scenario induces companies to conduct their organizational processes with an international and national market perspective, managing demands with greater agility, and constantly searching for assertiveness and competitive advantages. D'Áveni (1995) argues that it's getting harder to keep competitive static advantages in organizations, because they tend to be rapidly neutralized by the competition, because of constant process and product innovations never seen before. This fact demands that the organizations create new business and innovate in a more nimble way (Porter 1989).

Over the past few decades, the globalization process of the major world economies partially toppled the trade barriers between countries, encouraging the entry of emerging country organizations in the global market. Until then, these organizations were restricted to doing business only in their national territory. In Brazil, this process stimulated the creation of Brazilian companies with medium-sized and large enterprises with a focus and activities in the international market (Silva 2002).

One can say that there were three big "internationalization waves" after the Second World War. The first was dominated by Europe and the United States, until the 1970s; the second was led by Japan and the other countries of the "four Asian Tigers," and lasted until the end of the 1980s; and finally, the last one is represented by the BRICs (Brazil, Russia, India, and China), followed by other emerging countries (Stal & Campanario 2010).

This process of economic opening modified the profile of the 500 biggest organizations in Latin America. Between 1991 and 2001, the number of government-owned corporations dropped from 20 percent to 9 percent. The number of foreign multinational organizations rose from 27 percent to 39 percent (World Development Indicators 2003). In 2011, the rankings by America Economy Intelligence indicated that, of the 500 largest companies from Latin America, 223 are operating in Brazil. Other data, disclosed by the Boston Consulting Group (2009), indicate that between of the list of 100 New Global Challengers Companies, 14 are Brazilian.

According to the National Industry Confederation (2012), Brazil has a modern industrial plant with remarkable performance in sectors such as steel, automotive, oil, and mining, and with great diversity of resources. However, improve process and management strategies in these companies in Brazil were a determinant, observing companies over the world. Data gathered by the Brazilian Central Bank (2012) indicate that both the industry and service sector have been improving, and also highlight investments in new management methods and technologies associated with the financial, educational, and retailer sectors. As a consequence, it is possible to analyze Brazil's innovative advances from an international perspective.

In order to illustrate the concept of innovation as applied to Brazilian companies, we present the case study of Azul Brazilian Airlines, a company which in only three years has opened its doors in a competitive market and established a vision and a brand concept and an operations management with Embraer airplanes never used before with Brazilian airlines companies. In 2012 it operates with 4300 employees, has 40 sites in the country, 40 aircraft, and 16 million satisfied customers. It must be said that innovation of both products and services is one of the key concepts for the Azul brand, which turns this into an internationally important case.

*Box 11.1*: Case study: Azul Linhas Aéreas

To understand the success of a company like Azul Linhas Aéreas, it is necessary to get to know the person behind innovative ideas, such as the one that created this company. First of all, David Neeleman is a Brazilian. Born of American parents who were living in Brazil when he was born, he has had a lifelong appreciation for the country, where he returned to live between age 18 and 20 years. From that period of his missionary life in Rio de Janeiro and in the Northeast to today in 2012, Neeleman has become one of the 100 most influential men in the world, according to *Time* magazine, and has founded no less than three airlines, Morris Air and JetBlue in the United States, WestJet in Canada, and Azul Linhas Aéreas in Brazil. Through these, it was possible to revolutionize the air transportation scene with concepts such as electronic rather than printed tickets, the use of live TV on board, and the advantages and differentials of flying cheaply, but in comfort. In 2007, David decided to invest in the large, still unexplored potential of his birth country and invited a select group of airline executives to lead an innovative project in Brazil's airlines sector. In a short time, the company literally took off. According to the ANAC (Brazilian National Agency of Civil Aviation), between December 2008 and July 2001, the market share of the company rose from 0.31 percent to 9.17 percent. It not only took off, but quickly rose to a high altitude and stayed there, in large part due to the crew – as its employees are called – who form a cohesive, united, proud group that jealously guards its culture. It is not simple to create a strong culture that is well-recognized internally as well as externally in such a short time. But David's team says that this is one of their leader's virtues – a focus on human beings. In addition to its well-executed business rationale, the company esteems and acts according to its values. Together with the team, these values demonstrate the strong concepts of respect for employees and other human beings. To be part of the "blue crew" is valued for several reasons, from good working conditions to small actions that provide incentives related to improved self-image, pride in being part of a winning company, and a sense of belonging. Members are invited to learn and grow along with the company. And everyone wins – the professionals, the company, and the customers, who are better served in their flying experience.

This experience is part of the same set of values that the company brings to its relationship with customers. The company offers its passengers value as expressed by the combination of exceptional quality service and competitive pricing. By "competitive," the company is not necessary saying that it has the lowest price on the market. In some cases this is true. In others, it prefers to compete based on the service it delivers. In every case, it seeks a balance for the customer. The value proposal in this case is located somewhere between what customers pay and what they receive. David's ideal is that customers would feel that their trip with Azul was the best of their lives. This is the perceived value he expects from his company. Thus, people can pay a little more for their tickets, but in return receive a modern new airplane with comfort, space, and excellent service, as well as a television monitor on the seatback in front of them, a snack and coffee on takeoff, a bus to pick them up and transport them from their city to the airport (in this case, Viracopos and Navegantes), on-time flights with 7 options per day for the same route and finally, baggage delivered in less time than other airlines. This set of features that Azul uses to mark its company's differential is what is called the

"Azul experience." This is what Azul considers convenience for customers, and convenience = preference. Satisfaction comes from the overall setup, but convenience is possible to calculate and is easy to use to win over more customers and affect each customer more deeply. That is the reason that from the outset, the company's advertising budget has never been high. It prefers to invest in a differentiated flying experience for passengers, hoping they will always want to come back and will recommend new passengers.

When the Azul proposal was designed, it was clear to the Company's Board that it would not be a company focused on a specific target segment of the public. It wanted to enter the market as a company that everyone would want to fly. The idea was to remain flexible, and in the end everyone would want to fly with a value equation. This was an opportunity in the Brazil air travel sector, with 139 million passengers in 2010, and growing every year. As a marketing strategy, the public was invited to suggest the name of the new company to be launched. It set up a stylized map of Brazil with its corporate logo, symbolizing "the desire to serve, bring together and show Brazilians, nonstop, a new phase in the country's air transportation history." A link with Brazilians was established, a relationship that makes the company proud. This value is perceived by customers, and even more so because the company chose Embraer jets to make up its fleet. This was an act of valuing national products, especially because this firm is a latest generation innovative product, one of the most modern in operation worldwide, capable of broadening the good experience of flying.

Innovation in the product and the service then is one of the fundamental concepts of the Azul brand – in human relations, in communications, and in its manner of serving the market. Everything is subject to the culture of innovation that has been present since its founding. Based on its main service, air transport, the company innovates with everything from passenger comfort (while most airlines are moving in the opposite direction), by removing the middle seat and giving passengers more room, to its concern with sustainability and its attempt to mitigate its own carbon footprint. Azul sponsors a program to replace fossil fuel with renewable fuel and will fly jets with this fuel beginning in 2012. Moreover, it has adopted operating procedures on flights to reduce fuel use by millions of tons per year, and it operates with the highest energy efficiency in Brazil, maintaining cabins with an average 82 percent seat occupancy.

The company believes it is located not in the transportation industry, but in the time-saving industry. What people are looking for when they decide to fly is rapid mobility, to arrive at their destination quickly and to meet their commitments, whether they are professional, family, or leisure. But to "sell time" to their customers, the company needs to invest in other, more tangible variables, such as vehicles and services. Thus, today there is full awareness that the company will be much more efficient and will provide even better service to its customers if it has adequate working conditions. Today Brazil's airport infrastructure is the Achilles heel for the growth of any airline company. This is the big challenge in the near future of Azul: to grow, increase demand, and maintain excellence of service, even when there are outside factors that it can do little to change. However, for everything that depends on the company, it is convinced that it is on the right track.

*Source*: The authors.

It can be considered that Azul Linhas Aéreas was born with an innovative ambience aligned with the cultural traits of its leaders and founders, with a strong cultural influence toward a new business model, considering new airplanes, less oil consumption, concentrated operations in one airport and fitted to the Brazilian context. This sector is highly regulated, with a focus on operations and process management, and commercial aviation companies need supervision to assure compliance (Tadeu & Salum 2012). What makes Azul different is its quest for operational efficacy and efficiency in processes to achieve economic and financial balance. What is the secret of this model? The operation at all organizational levels finds itself aligned to a business strategy that is a new perspective in offering services to an emerging market with an emphasis on the Azul Business Model – which is based in "observe, perceive and serve" customers according to their needs and regionalized demands.

## 2  Process management and its relevance in a strategic context

Process management is a strategic proposition (Paim et al. 2009) and must be understood as such by the organization. Offering products or services to consumers, collaborating with commercial partners, and coordinating the efforts of the employees – in the end, these processes unify the product/service, its brand and value (Malone et al. 1999). According to Gonçalves (2000), process is a concept essential in designing the means by which a company intends to deliver its products and services to customers. Rapid innovation in processes can result in improved organizational capacity (Kanter 1997).

Due to the importance of the subject, there is an increasing demand to modernize management with a consequent use of information systems to support business growth. Therefore there is a special interest in data analysis to achieve process excellence, including performance indicators and the quest for continuous improvement, which is one aspect of methodological innovation (Patterson & Lightman 1993; Ulrich & Eppinger 2004).

In sum, process management and its related set of managerial practices and programs is perhaps the most important managerial innovation of the last 20 years (Cole & Scott 2000; Berner & Tushman 2002). The search for efficient and effective results in fulfilling the wishes of customers and shareholders demands that the strategy, processes, people, and organizational culture be aligned as a whole. The main challenge of a strategy based on consolidated and strong processes is to reach the top of each market scenario, whether national or international. The challenge will, by its own complexity and uniqueness, induce the creation of organizational strategies.

### Strategic alignment

In Brazilian business policy, the concept of strategic alignment is based on the idea that the economic performance of a company is directly related to the ability of management to create a strategic fit, i.e., a company's position in the competitive market supported by an appropriate administrative structure (Tadeu & Salum 2012).

The total amount of business success as a strategy analysis depends on the innovation aspect, in correlation with the business's ability to combine and stimulate ambience and individual behavior (Stabell & Fjeldstad 1998). Some companies have an explicit strategy and knowledge, supported with reports, analyses, and forecast demand. Other companies do not set a conceptual strategy. As a consequence, this practice does not make sense to investors and stakeholders, because it is hard to transmit decision criteria as a strategy for growth or development (Sterman 2000). Consequently, we understand that the organizational focus must be on the continuous improvement of processes, differentiating its profit sources and margins, and continuously redefining the path to be followed.

Organizations create value as contributors transform scarce resources into products and services with a value that is perceived by customers. Standards for interaction, coordination, communication, and decision making, which guide such transformation, are the processes (Christensen & Raynor 2003). According to these authors, processes differentiate themselves not only with respect to their purposes, but also by their visibility. Some are formal and are explicitly documented and systematically monitored. Others are informal, and can be identified as working routines that evolved over time and were adopted because they actually work, or because they constitute the organizational culture. For Bartlett and Ghoshal (2000), the process management model can answer some of the current challenges that are faced by organizations. Usually, organizations have good management when they are inserted in a scenario with various choice dimensions, with choices that are translated into financially viable results.

With so many market factors and processes being analyzed in the current global context, the development and implementation of a strategy depends on strong commitment and leadership by the managers (Porter 1996). The leader must know that choosing what not to do is as important as knowing what to do. The decisions regarding a new positioning of the organization in the market must be oriented by the market's behavior, by the development of sustainable processes, and by the search for competitive advantage.

For example, in 2009, in partnership with the consulting firm Strategos, Fiat Automóveis do Brasil began a study and developed a diagnostic. Involving over 300 people, including factory workers, suppliers, and concessionaires, the survey attempted to understand what innovation is and how it occurs at Fiat. The result was the FIAT's recognition of its creativity capacity but it was not very process-oriented and had a nonsystematized culture of innovation. The largest part of innovation practiced at Fiat was technological and incremental, in a system typical of the automotive industry. With the present panorama of innovation, outlined in 2011, Fiat began to design its new strategic lines and the goals for innovation for the entire organization (see Figure 11.1). Part of the Product Engineering Department, which before had been the "innovation sector," was moved and transformed into a new work area linked to the business development managers. The new positioning of the team that would work directly with innovation was important, because the subject was then treated as strategic and vital to the factory. The Fiat subsidiary in Brazil (FIASA) sought growth whose main pillar of company strategy was innovation. It tried to go beyond its level of technological innovation in products and services and elevate innovation management to new dimensions, to new business models, to the strategic level of the company and one that was cross-cutting, i.e., it meant that innovation must reach all levels of the company.

For Albuquerque and Rocha (2007, p. 36), "fulfilling a request depends on an internal process that involves various areas or departments in the organization. Customers will judge this effort by the final product or service." Also, the quality of this delivery depends on the synchronization among the sectors involved in the process. According to these authors, there are some premises for transforming organizations into integrated systems:

a) Organizational synchronization (synch) implies aligning the strategic indicators of people and processes.
b) To achieve organizational synch it's necessary to work with process management.
c) Every change must be directed by strategic directives that reflect not only financial targets, but also are in accord with the market, internal perspectives, capabilities and learning.
d) External monitoring, especially of customers and competitors, must be continuous, and along with environmental changes must lead to changes in strategic leadership, with consequences for processes and people.
e) All the goals and strategic indicators for processes and people must be quantitative and must have established deadlines.

Given all of the above, it is important correlate business strategies with the company's realistic capacity for implementing them. To do so, it is necessary to understand the context and the importance of managing their processes, along with the ability of the processes to implement strategies since, if this is not feasible, they must be revised and made to work. In the next section we examine process formation and characteristics of the organizational processes.

### The characteristics of process formation
To analyze organizational processes requires a broad understanding of their functions and their market context, since this is a continuous movement of knowledge, which should be shared among teams, customers, suppliers, and investors (Marlyn 2004; Pyke & Whitehead 2006; Jeston & Nelis 2008).

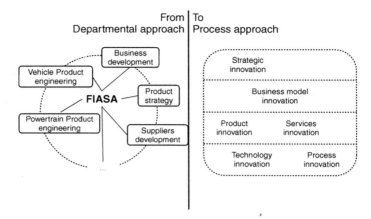

*Figure 11.1*   Changes in innovation management at Fiat Automóveis do Brasil
*Source*: Arruda, Salum and Rennó (2012).

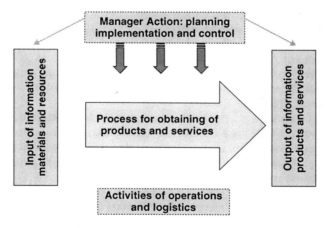

*Figure 11.2*   Generic model process
*Source*: Adapted from Maranhão and Macieira (2004).

Every process has four phases (Lovelock & Wright 2001), with sequential and complementary activities building until the final results are achieved. Figure 11.2 describes the flux of the typical process activities: the inputs correspond to resources and information that will be transformed until the final delivery to customers. Operational and logistics activities aligned to managerial actions add value to customers' demands, thanks to the perception of quality and costs.

It should be emphasized that the concept of added value is not necessarily associated with criteria for financial decision making, but also is associated with intangible aspects (Lovelock & Wright 2001), such as process management, customer service, and sustainability. When the theme is process management, it stresses the need for people management and performance indicators,.

As an example of a company in Brazil concerned with sustainability, we mention Magnesita. Magnesita is the third-largest producer of refractories in the world and is a market leader for integrated solutions for its customers in Brazil (Arruda & Salum 2011). In December 2010, Brazilian Federal Law no. 12.305/2010 went into effect and instituted the National Policy for Solid Residues.

Since 2004, anticipating the new regulations, Magnesita of Brazil has invested in technology and improved its production process to achieve a rational utilization and to follow the order of priorities of nongeneration, reduction, reuse, recycling, treatment, and environmentally adequate final disposal of residues (data from the company).

Increasingly concerned with the final destination of refractory residues, in 2001 the company made an additional investment of over one million *reais* (US $555,555 as 1 US$ = 1.8 reais) to assure an increase of scale and productivity and capture of all the residue generated by its customers at units created and authorized by Magnesita Ecobusiness, thus inaugurating a new phase for industrial refractories.

The company's high investment in research and sustainable technology and development of methodologies for management, logistics, and processes to discover an appropriate destination for refractory materials for over 10 years has made it possible to develop recycling routes for every type of refractory residue with which the company works. The Reverse Logistics Project assures Magnesita customers of an adequate destination and recycling of refractories in a manner that adds value to the supply contract and the application of its materials.

Currently, according to Arruda and Salum (2011), Magnesita is the only refractory company in Brazil that provides the reverse logistics system to its customers, preserving the technology, quality, and safety of the materials in addition to playing an important role in minimizing the environmental impact and increasing the sustainability of the production chain.

As illustrated in Figure 11.3, the 2009 crisis had a strong impact on steel production and consequently on consumption of refractories and their recycling. The goal established for 2011 is 22,000 tons. However, the company thinks it will reach 24,000 tons, almost double that of 2010. The goals for 2012 are still being defined, due to uncertainties in the global macroeconomic scenario that could have a negative impact on the steel market.

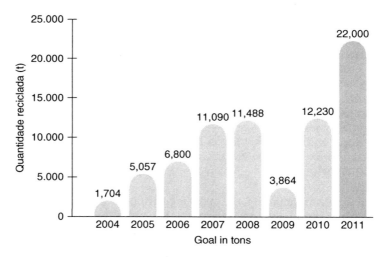

*Figure 11.3*   Amount of steel recycled for the period 2004 to 2011 (in tons)
Source: Arruda, Salum and Rennó (2011).

It is known that many organizations don't fully adopt process management. Many of these organizations insist on keeping power concentrated and verticalized in fiefdoms, and zealously guard their area, personnel, and resources. It would be ideal if organizations reviewed their management practices, focusing on integrated processes, instead of being conducted by approaches based on functions, areas, sectors, or products.

For Maranhão and Macieira (2004), the possible reasons that prevent organizations from adopting management by processes are market complexity and demands for constant changes; difficulties in undoing the hierarchic and functional management paradigm; little or no strategic business vision; lack of knowledge about customers and their needs; and the lack of an organizational culture regarding process management.

The organization that adopts management by processes creates a harmonious internal environment, pleasing its employees and giving them motivation to improve their performance and the way to do their tasks. At the same time, it encourages the development of attributes that are favorable to their competitiveness. It's more likely that this kind of organization will develop positive characteristics in conducting its operational processes and their strategies,.

## Process improvement

Process improvement can be defined as the accomplishment of a set of complementary activities that lead working teams to a better performance.

Despite the differences between organizations that can exist due to market segment, size, and management models, there are some basic procedures for improvement in this environment.

As an example, Whirlpool Corporation is the largest appliance industry worldwide, with 67,000 employees, and it is the holder of recognized brands in the market, such as Whirlpool, Maytag, Kitchen Aid, Jenn-Air, Amana, Bauknecht, Brastemp, and Consul (Arruda & Salum 2012). In Brazil, the company understood that differentiation and an increase in its competitive advantage would occur through developing products with innovative features as well as designs that showed its customers the functional and decorative potential of its appliances. To initiate the process, Whirlpool hired a consulting firm in 2000 to map out and survey the existing processes and competencies in innovation and those necessary to meet this challenge. Their report globally redefined the process of innovation and the organizational culture of the company (data gathered from the company).

Until 2005, there was no department dedicated to innovation, but rather a group of professionals from diverse areas that took concepts in innovation as their work scope. Between 2005 and 2006, the company arrived at the conclusion that it was necessary to create an area dedicated to innovation. The structure implemented had mixed centralization and decentralization, and its proposition was to incorporate the innovation process into the organizational culture, from the start of decision making and formulation of strategy to production, in a way that the challenge to innovation simultaneously would reach the level of leadership as well as the day-to-day production, i.e., the levels where it is decisive, strategic, and where it will really take shape.

According to Arruda and Salum (2012), the organizational and procedural structure is composed in this form by valuing innovative thinking that emerges with the idea of Design Thinking. Design thinking is applicable within the innovation process called "Divergir e Convergir" (diverge and converge). This process seeks the highest possible volume of information, references, and knowledge about the concepts, technologies, and the design that is under construction. The material will converge in some solutions that can be innovations. Based on the strategic guidance of the organization, opportunities to innovate will occur through competencies, methodologies, and tools specific to the process, and this is managed by the marketing area and supported by the innovation area. The solution, then, is made tangible by the design department and by the technology department, which are also responsible for seeking patents.

Before starting the development implementation project, the responsible team must identify and locate the errors, to define control points that permit assessment of their origins, and then look for corrections, which must be redirected to the respective sectors or areas. This information will also be useful for making decisions about investments in tools, training, and changes of procedures.

To conclude this section we present the case study of Embraer, which illustrates the importance of management by processes in all the phases of the crafting of a commercial or military airship. The two main points of this case are: (i) the decision the company made to work with partners from all over the world who were willing to share the risks in identifying opportunities and innovations; and (ii) reducing the number of strategic suppliers of complete subsystem components, which allowed the development and launch of new products. In this case, the main success factor is the management process aligned with business strategy.

---

*Box 11.2*: Embraer

The world's third-largest aerospace company, Embraer achieved its present position in June of 2008, when it passed Canada's Bombardier and moved closer to Airbus and Boeing, the two largest commercial aircraft manufacturers in the world. With 41 years of experience in design, manufacturing, commercialization and post-sales, Embraer has already produced around 5,000 airplanes. In 2012 it operates in 88 countries on five continents. Currently, it has a workforce of over 17,000 employees, with 94.1 percent based in Brazil.

With a global customer base and important world-renowned partners, Embraer has a significant market share and was Brazil's largest exporter in 1999 and 2001, and the second-largest in 2002, 2003, and 2004. Its Brazil headquarters is in São José dos Campos, about 100 km (62 miles) from São Paulo. Overseas it has subsidiaries, offices, technical assistance centers, and parts distribution located in Australia, China, Singapore, the United States, France, and Portugal. Customer satisfaction is the foundation of company action. The company believes that taking off from there, it will begin to develop a successful business with high technology, qualified personnel, global action, capital intensity, and flexibility.

The company is divided into three large business areas: commercial aviation, executive aviation, and the defense and security market. In the commercial aviation market, it works with two groups of aircraft, the ERJ 145 with up to 50 seats, now operating in 17 countries, and the EMBRAER 170/190, with 122 seats, with 39 countries in its customer portfolio. It shows high performance as well in the defense and security market, with many aircraft models for military operations. In the executive aviation market, there are three families of aircraft with up to 22 seats. Of US$5.35 billion in net revenue in 2010, 53 percent came from its commercial division, 13 percent from defense and security, and 13 percent from aeronautical services and others. Embraer has invested US $144 million in research and development, and US $103 million in property, plants, and equipment.

In the global aircraft manufacturing industry, Embraer has large international suppliers such as Honeywell, Liebherr, General Electric, Kawasaki, Parker, Gamesa, and Hamilton Sundstrand. These companies are risk partners who not only supply 85 percent of parts, but help with Embraer's cost control. In this way, the Brazilian company manages to prioritize its dedication to designing aircraft and coordinating product development.

Embraer's experience in working with partners who share risks helps it to pursue opportunities offered by the market, such as the moment it decided to get into the larger jet segment. In 1998, the company perceived a gap in demand for production of jets with from 70 to 110 seats. In 1999 it resolved to invest in a new line, the Embraer 170/190, which would allow it to compete not just with Bombardier (its traditional rival), but also with the line of smaller planes of both Boeing and Airbus. The platform was conceptualized starting from zero, with a custom design that could confer the benefits of airplane weight along with ideal configurations, making the planes significantly more economical than its competitors' aircraft. Executives know that the key factors to success for a new family of planes in the more competitive market are rapidity of development and integration of the most advanced technology into the product.

To develop a new product line, the company expanded its number of strategic partners responsible for delivering complete subsystems instead of components, thus encouraging development of tighter relationships with fewer suppliers. The number of regular suppliers was reduced from 350 to 22, and the number of risk partners was extended to 16. These partners bear 66 percent of the total cost of development and are responsible for designing and supplying structural segments and those for the main aircraft systems. Risk partners receive a previously agreed upon amount per unit sold, updated according to cost indices from the industrial sector, as well as the major share of the spare parts business. However, there is no sharing of profits or revenue with suppliers.

Embraer is responsible for integration of the entire project, as well as the design and development of the aircraft, manufacturing part of the fuselage, and assembling the entire set. It identified and involved partner companies right away to help conceive and develop the other components of the structure, interiors, and systems. The result of all this effort at integration was that fact that the Embraer 170 made its first flight in February of 2002, 30 months after its launch in June of 1999. This was a record for product development in the sector.

Outsourcing allowed Embraer to obtain better quality, less-expensive parts in comparison to its more verticalized and horizontally integrated competitors, such as Bombardier. This strategy also provided greater flexibility in dealing with the inevitable market recessions. With this type of supply chain and assembly process, it becomes simpler to determine the rate of production, and little investment is necessary to increase production speed. The Embraer 170 met with high acceptance on the international market, despite the crisis in the aviation sector that began in 2001.

*Source*: The authors.

## 3 Innovation management

The perception of the importance of innovation and its consequences was felt in delayed fashion in the emerging countries. But there is no turning back on the route that needs to be taken. Innovation is a subject that has imposed an agenda since the beginning of the century in an implacable manner in Brazil as well as in its companies; it brings together all the conditions for ascension to a vanguard position in the process (Salum & Jardim

2010). The need for innovation was seen by consensus among all the actors involved in organizational processes – strategic partners, government, customers, and universities. Innovative behavior become one of the main differentials in the economies, with an impact on their level of development, growth, and dynamism indices, and it stood out as one of the factors responsible for large gains in competitiveness. Innovation came to mean not just technological development, but also development of products, processes, and management and business models.

### Innovation in companies

Innovation is considered a strategic option, useful in creating an environment that is favorable to the development of the maximum number of collaborators and partners, involving the company culture and structure, and also establishing indicators that permit following, encouraging, and developing the company's creativity.

The need for innovation can be determined, among other factors, by the life cycle of products and services, which begins with development, undergoes commercial growth, reaches maturity, and then declines. In a world where the life cycle of a new product or service is getting shorter, it becomes very important to have the capacity to create new product versions, in order to develop differentiated processes that permit the offering of products and services with a perception of value. The lack of time mandates making changes faster, without properly measuring the results.

With this in mind, it can be concluded that management by processes requires constant sequential revisions to identify new opportunities, creating the need for and importance of incremental innovation. Therefore, the act of innovation becomes more difficult and, as a consequence, the focus turns to new opportunities that can mean commercial success for the company.

Another aspect to be considered is determining the influences that market globalization has, leading to direct competition between Brazilian and foreign services, the latter possibly being more competitive regarding price and use of technology. This competitive scenario can be hostile to Brazil's markets, and can result in the withdrawal of domestic companies.

The remaining organizations feel mandatory to keep acting, and some of them aim for innovation management as a continuous improvement. In this sense, new products or services are seen in the market as the top of innovation, the innovative process plays a strategic role that is equally important (Coral et al., 2008). Being capable of doing something new or doing it better than others is a real advantage.

### But what is innovation, after all?

In the 1960s, the approach to innovation management was basically focused upon a structuralist rationale, seeking to identify ideal systems for organizations to deal with environments more propitious to innovation (Burns & Stalker 1968). A more functionalist emphasis was introduced only in the second half of the 1980s, when authors sought to understand less-strategic issues and, therefore, focus on more routine to this process (Rothwell 1992). At that time, innovation management models began to acknowledge the managerial importance not only of choosing the ideal management models and structures, but also of managing internal processes and other tools for the purpose of facilitating organizational innovation, corroborating the view of entrepreneurs being the main process inducers (Schumpeter 1997).

In recent decades, as a result of constant technological advances, of globalized communication, and of intense competition among companies, innovation has acquired even greater importance in organizational studies (Christensen 2001; Motta 2001). Although innovation has been present in routine managerial decisions and major contemporary academic debates, it is neither the subject of consensus nor a systemic theme – as are, for example, studies on the economy, strategy, or finances (Kimberly 1981).

Innovation is strongly linked to business activity and is considered an important factor in leveraging companies, to lead them to outstanding positions in the market, especially the emerging markets (Drucker 2002) However, for this to become a reality in organizations, it is important that companies structure their processes, strategies, and management to favor innovation activities (Arruda and Barcellos 2009). Schumpeter (1961) made a landmark contribution that focuses on the importance of innovation and technological advances in developing businesses and the market, which has served as the basis for numerous innovation theories and models conceived later.

Dissemination of new products and production processes inside and outside companies is clearly one of the fundamental aspects of growth and transformation of the contemporary economy (Silverberg et al. 1988). This context increases the importance of innovation to organizational studies, considering that the innovation process is strongly linked to companies' structures and strategies (Albuquerque 1996; Teece 1986). In this sense, innovation can be understood as the allocation of resources to explore and develop new products and processes aimed at exploring new market opportunities (Silverberg et al. 1988). In the same way, the innovation process is characterized by the innovating company's appropriation of differentiated profits in the market (Teece 1986) and by its creating new products/processes acceptable to the market (Henderson and Clarke 1990; Pavitt 1984; Dosi & Freeman 1992).

Definitions of innovation may vary, but they all include the need for complementary paths of development and exploration new knowledge, not just its creation. Being a good creator does not ensure commercial success, despite the quality of the original idea. The world will only "come to us" if we also take care of project management, commercial development, financial

management, organizational behavior, and other important aspects. Another
common error is confusing innovation and innovative process with continuous improvement and related terms (Arruda and Rossi 2009).

Innovation has to cause a significant impact on revenue growth,; it needs
access to new markets and new customers and must increase profit margins.
Continuous improvement does not usually ensure competitive advantages
in the mid- and long-term, but it can keep a product competitive in terms of
cost. In Table 11.1, a summary of the evolution of the breadth of the concept
of innovation can be seen.

It can not also be said that innovation is just created by a single genius,
and neither does it happen by accident. It is, indeed, a result of a teamwork
that is systematic, organized, tough, shared, and in synch with corporate
strategy.

*Table 11.1*   Evolution of the concept of innovation

| Year | Author | Definition |
| --- | --- | --- |
| 1943 | Schumpeter | Innovation is the introduction of a significantly different product in a market. It demands a new production technique and the opening of a new market |
| 1982 | Chris Freeman | Industrial innovation reunites technical, design, production, management, and commercialization actions that are part of the marketing of a new product, process, or equipment (improvement of an existing one) |
| 1985 | Peter Drucker | Innovation is a businessman's tool, it's the way by which businessmen explore a change and turn it into a business opportunity or a different service. It has the potential to become a discipline that can be learned and practiced. |
| 1990 | Manual de Oslo | Innovation is the implementation of a product (goods or service) that is new or significantly improved, or a process, marketing method, or new organizational business method in a company, or in the relations the company keeps |
| 1990 | Michael Porter | Organizations acquire competitive advantages with innovation acts. They approach the concept in its broad sense, which refers to new technologies and new ways to do something |
| 1992 | Roy Rothwell | An innovation is not always associated with the launch in market of the results of some technological advance (radical innovation), but also involves the use of little modifications in the know-how of technology (improvement or incremental creation). |

*Source*: The authors.

This context was the point of departure for the Center of Innovation Research at the Fundação Dom Cabral (FDC) to develop a strategic vision of the subject, using it as a support methodology to study organizations. The following model (see Figure 11.4) was created based on the observation of the best organizational practices, and also on the relationship between the FDC and innovative companies that are active in Brazil and Latin America with their subsidiaries and research centers around the world. The model shows an integrated process for innovation management, based on strategic decisions made by the companies, incorporating a systematic vision and promoting the integration of the companies' various areas and levels (strategic, tactical, and operational).

According to Arruda and Rossi (2009), professors at the FDC Innovation Center, in their article titled "Creating Conditions for Companies to Innovate,"[1] organizations are dynamic learning systems, which demand innovation management to act in a united, structured way, based on knowledge and commitment to sustainable innovation. To achieve the planned objectives, it is necessary to have a strategic plan with clear goals, plans, objectives, funding, and performance indicators. In a macro vision, innovation is the result of a multifunctional and multidisciplinary work, linking marketing, production, financials, clients, partners, and suppliers. Therefore, it must be aligned with a strategic vision and the market. In this

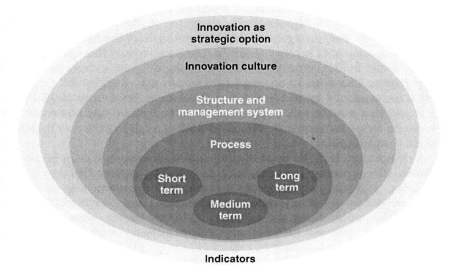

*Figure 11.4*  Fundação Dom Cabral (FDC) innovation model
*Source*: Arruda and Rossi (2009).

process, leadership plays a fundamental role in the creation of a favorable and integrating environment.

It is also up to top management to promote innovation based on disseminating strategic values, sponsoring ideas, and creating and aligning the strategy for innovation with corporative processes. It is also important to involve employees, in a proactive way and across different areas, and to orchestrate them well in their search for signs given by the market, which can help to anticipate problems and find creative solutions.

### Innovation components

Organizational culture seems to be a critical factor in the success of any organization (Martins & Terblance 2003). According to Kenny and Reedy (2006), there is some broad agreement on the type of organizational culture needed to improve creativity and innovation. To stimulate innovation, it's necessary to do more than just define its importance in the organization's strategic positioning or put effort and stock into the production of knowledge. It's fundamental that the organizational culture and environment be favorable to the search for innovation and manifestation of creativity (McLean 2005).

Each organization has its own culture, with peculiar characteristics, and it is different from all the others, because it reflects the employees' attitudes and their values (Schein 1985). To innovate, the collaborators must tend and desire to create and transmit knowledge, as opposite to fearing to take risks. A pro-innovation culture must be fed by strategic objectives that are aligned with the organization's orientation to the future (Crespell & Hansen 2008). The management staff is responsible for the maintenance of an environment that is idea friendly, minimizing aspects that can prevent or inhibit individual creativity, and combining different perspectives and experiences. Therefore, an organizational culture can either inhibit or stimulate the tendency toward innovation (Martins & Terblanche 2003). But it would be a mistake to identify a culture in an isolated way and credit it with the recipe for success or failure, without considering the peculiarities of each organization.

As innovation is at the heart of competitive advantages, it must be associated with opposing elements, such as *promising opportunities* and *risks of failure*, depending on the results it can have or the manner by which it evolved. With efficient innovation management – a multifunctional process that relates opportunities and external and internal needs, ideas, research, development, business models, and implementation – the company can reduce uncertainties, deadlines, and costs that are part of developing new products, services, processes, and businesses.

Most of the companies still act with shortsightedness, focused only on the search for improvement and investments in short-term projects, with low risks and fast results, more control of the variables, and less uncertainty.

This, however, is an obsolete practice and conflicts with integrated innovation management, which, among other things, demands a different posture, with a long-term view and permanent investment. The practice of some of the most innovative Brazilian companies shows that, besides research and continuous investment in development, the big challenge of surviving in the marketplace is the ability to "predict" the future. Using management tools to analyze scenarios, the company can get signs that may ensure its competitive advantage, survival, and exceeding of customer expectations. Those are practices that can make all the difference and can contribute to the success of innovation management and reduction of risks.

According to Rodríguez, Dahlman, and Salmi (2008), the most successful innovation programs are based on a well-articulated vision, common agreement regarding the program, and efforts to address all four pillars of the knowledge economy through a combination of bottom-up initiatives and top–down reforms. Thinking about the future of innovation, it requires the proactive balancing of multiple innovation performance requirements. Striving for sustained innovation and longer-term competitive advantage implies a continuous confrontation between today's work and tomorrow's innovation in the organizing of innovation efforts (Nederhof 2009).

To illustrate this section, we'll present the case study of Fiat Automóveis. After 35 years of operation in Brazil, Fiat Automóveis has its factory in Betim (Minas Gerais – Brazil), with the highest production of automotive vehicles of the group. More than a reproducer, the unit is becoming a developer. This is because Brazil is becoming more important to Fiat's innovation process.

As an assembly plant that uses innovative or post-Fordist techniques, Fiat Automóveis achieved conceptual and technical development in its production chaining, especially with its supplier's park with just-in-time technology for the almost 75,000 cars made monthly at the Brazilian industrial unit.

---

*Box 11.3*: Case study – Fiat Automóveis

In 2010, the Fiat factory in Betim produced the most automobiles in the world, delivering 900,000 units in Brazil, which consumes 90 percent of factory production – and sending them to some Latin American countries, mainly Argentina. Today, FIAT Group considers this factory to be more than an affiliate; it is a developer. The assembler relies on the Brazilian operation to develop differentiated vehicles with the latest features.

The trajectory to innovation for Fiat in Brazil was consolidated with the creation of the Giovanni Agnelli Product Development Center in 2003. Fiat Automóveis is the company's most complete development center outside Italy. About one thousand people work there, distributed among six engineering, departments, and Fiat Automóveis owns all the technology for automobile development, from design through building the prototypes. To accomplish this, these areas were equipped with laboratories with the latest resources, capable of simulations and dynamics testing on a real scale. Although conceiving, designing, and developing a new car model takes months – and even years – from the moment it begins to be sold on the market, it takes about 23 hours to finish its assembly, from flat steel plates to final testing.

The starting point for the 23 hours is the press shed, where the life of the automobile begins. There, sheet steel is transformed into parts for the automobile chassis. They arrive in precut, chemically treated rolls or sheets. Then the presses cut, fold and drill the sheets until arriving at the desired part. Seventeen lines of mid-sized and large presses are functioning, each one with a capacity to make from 15 to 20 different parts. There are 108 presses with impact power between 50 and 1500 tons. Some of these lines are totally automated. From the start of the operation, one can observe interesting points, such as worker safety precautions and the care taken with risky tasks, as well as recycling of 98 percent of all production residues, such as steel shavings left over from the presses that fall on the transport belt, which takes them from the basement to a shed where they are recycled and later re-pressed and used in new metal sheets.

In the chassis production unit, parts that are stamped in the presses are soldered and the chassis begins to take shape. The car floor is the starting point. Then it is welded to the front of the vehicle, and the roof, rear, and sides are added. The chassis is sent to an important factory machine, call the Masherone – in honor of the engineer responsible for its creation – and in the space of a minute it gets its first 48 soldering points. To complete its soldering points, the chassis follows a path along which 3,800 to 4,500 points will be soldered, depending on the vehicle. Finally, the vehicle undergoes a rigorous inspection process and then it is sent for painting. Several UTEs (Elemental Technology Units) are distributed in the chassis section, and throughout the entire factory. In the blue and glass room, human resource, safety, quality and production activities are integrated for the specific task that is being developed at that spot in the factory. The activities of world-class manufacturing (WCM), an international quality program to which Fiat belongs, are also concentrated there.

Workers rotate tasks every two hours to maintain their health. In total they can exercise three different functions, and that is why they need to possess knowledge and achieve autonomy in all processes, as quality control determines for all stages. The task changeovers, breaks, and exchanges among the three factory production shifts cannot interrupt production on the assembly line.

In the paint section, the chassis "travels" by aerial transporters, cabins, and automated applications, very common to the paint process. The heart of durability and beauty for automobile chassis, paint is only applied after treatment to protect the body against corrosion and provide resistance to the elements. Application of noise abatement and waterproofing materials produces comfort and protects the inside of the car against dust, water, noise, and damage from stones. The gases produced during painting, drying, and soldering are collected, filtered and purified to remove polluting substances. Fiat was the first assembler in Brazil to totally cease releasing emissions from solvents into the atmosphere.

The vehicles are finished in the final assembly shed, where they receive their outside and inside parts such as upholstery, seats, windows, panels, lights, motors, suspension, and electrical harnesses. From that moment, the car is produced according to the sale specifications. First the chassis number is engraved. From then on, all the parts, items, and specifics of the car will be applied, according to customer demand. Fiat has adopted post-Fordism and does not work with stock – the parts in stock are determined by their respective chassis. With the just-in-time assembly line, the parts come online along with the chassis for which they are destined. Few errors occur on this line. When a part is missing from the sequence, for example, if it is small, the line continues (it will be placed at the end, in the customer delivery unit) and if the part is large and is a determinant for final assembly, the line will be stopped.

At the end of the line, the automobile goes directly into the experienced hands of the test driver at the customer delivery station. To assure safety and quality, 100 percent of the cars are taken to the test track. It is 3,800 meters (12,350 feet) long and divided into six stretches of higher and lower speed. The vehicles are assessed for noise levels, water tightness, and their transmissions and brakes are checked. On the track, the vehicles undergo tests that submit them to real situations at speeds varying between 70 and 110 km/hour (between 43 and 68 miles/hour) – the maximum speed permitted. The cars run 20 kilometers (12.4 miles), a standard Fiat established. All drivers have at least two years of assembly line experience. They know the production process well, and have advanced knowledge of mechanics. They will discover and repair any possible defects before delivery to customers. Finally, after 23 hours, the vehicle leaves the Betim factory for the dealership and from there it goes to its first owner.

*Source*: The authors.

## 4    Cost management aligned to innovation in Brazilian companies

Brazilian companies commonly reflect on cost quality and the correct implementation of management models to ensure that the management and innovation processes will be respected. In this case, some questions must be examined: What is the expected feedback, considering the competitive national and international contexts? What are the necessary investments to assure meeting customer demands? How is it possible to carry out cost management in the context of Brazilian innovation?

According to Leone (2000), cost management as employed by Brazilian companies must be understood as the rational use of resources, and it depends on the search for differentiation, the correct understanding of the value chain, and the goal of serving customers. The costs of business projects must be justified, as a function of the expectation of return, and also of future savings on operational costs, in accord with the efficiency of the management models. In this case, many segments of the economy display this behavior; strategic sectors such as automotives, aerospace, mining, and petroleum, once very susceptible to the international economy, are now self-sufficient with respect to the dynamics of Brazilian economy.

One of the advances noticed by Tadeu (2011) regarding the opening of market and the increments of innovation is the adoption of total-quality programs, with the remarkable performance of industrial sector in the

1990s, and of the services sector in the following decade. It is demonstrable that quality is related to the economic viability of a business, and it is also quantifiable, measuring the return on investment.

For Reis and Pires (2005), cost measuring done by Brazilian companies was a considerable advance, along with the entry of international companies into the national market and the advance of strategic models. It can be observed that cost measurement helps business people to make high-quality management decisions. In this case, several studies carried out the 1990s by the IPEA, the Brazilian Institute of Applied Economic Research (2012), show that through the use of performance indicators and cost management, Brazilian companies could get to know their financial behavior and properly analyze business results.

The growing of Brazil's economy has been giving to national companies an expectation of feedback, especially in a transition from operation costs models to a strategic paradigm, which allows the capitalization, by open-end funds related to the growing of strategic sectors in the country.

One can imagine that the elaboration of innovation-related projects and their inclusion of strategic and cost planning is something in broad development, intrinsic to the market's expectations. By "inherent" we mean the capacity of Brazilian organizations to maintain short-term and long-term actions for cost and innovation management in the Brazilian context.

A substantial advance, according to Kaplan and Norton (2004), was the implementation of cost management, and also the measurement of the performance of processes, customers, and workers. In the Brazilian context, several companies have been improving cost management and relating it to broader business performance, demanding from the business units an optimization of processes in the attempt to achieve a good return on investment. Several business plans are presented, also in the public sector, as a guarantee of cost management, customer satisfaction, and organizational processes.

Currently, the challenge faced by Brazilian companies is to improve management systems, by improving the quality of services and products delivered to customers. If the Brazilian economy intends to insert its companies into the international market, the advances already achieved can't be lost. In this sense, it's not easy to realize that the use of information systems with costs is relevant to the management models used. It becomes important to develop studies to discover how to develop more efficient cost-management models, with the integration of processes and innovation, a practice in effect in few Brazilian companies.

Knowledge of business costs is essential to efficiently allocate funds, which reduces inefficiencies in the eyes of the customers (Tadeu 2011). Similarly, knowing about the organization itself is an essential element for cost management, highlighting the correct classification of costs and investments and also the correct performance evaluation. This is a recent situation, due to compliance with international tax service standards.

One of the great advances in the Brazilian business scenario comes from the public sector, with innovations related to the development and implementation of the tax regulatory systems, also with the conscious adherence by Brazilian companies and a focus on reducing tax evasion, and increasing financial transparency, as demanded by international legislation. One can conclude that the cost management carried out by Brazilian companies has been improved and, despite not solving all the organization's problems, it will allow a suitable environment for investment, growth, and innovation, associated with the management of information and decision making.

There is an interesting organizational tradeoff between cost and innovation. The solution to problems, investments, growth, and innovation are themes studied by the IPEA (2012), when examining the paradox between innovation and cost reduction. For Viotti and Macedo (2003), Brazilian organizations have evolved significantly in budget management, but need to comprehend as soon as possible the need for more investment in innovation. Data from the Brazilian Ministry of Science and Technology (2012) shows that Brazilian organizations invest low amounts in innovation – about three percent of gross revenue, in part because of past economic practices and the lack of stability during the inflationary period of the 1990s. Cost reduction is also perceived in human capital formation, with low qualification in Brazil (Lastres & Albagli 1999). Brazilian Ministry data also indicate that, along with Brazil's low budget for innovation, with timid advances in sectors such as finance and oil, the national culture is still is wedded to the idea that investments must yield profits in the short term. As a result, in many sectors of the national economy, there has been the presence of international capital of experts from other countries, which has been slowing sustained economic growth.

According to Felippe (2007) argues that investments in innovation and human capital in Brazil are considered expenditures, not costs, with few large size international companies being concerned with developing new products and processes with a view to being competitive on the international and domestic markets, and few with a strategic view for the long term.

To illustrate this section, we present the case study of Group Alstom. How to manage research and development inside industrial units working at 100 percent? The Alstom Competence Centers face the challenge and deliver results, research, and innovation, as well as important reflections about processes, structure, and proximity to the customer. The highpoint of this case study is the relevance of the R&D&I (research, development, and innovation) sector to Alstom's business. For a small team of professionals, the challenges are big and complex, ranging from management by launching new solutions in their quest for efficiency in the management of internal processes of the company to serving customers.

The case further exemplifies the difficulties of doing short-term management with medium- and long-term innovation projects. It also highlights the need to be able to maintain operational efficiency and cost management. It portrays an organization that aims at the future and at differentiation through innovation.

---

*Box 11.4*: Case study – Alstom Group

Grupo Alstom (the Alstom Group) is divided into four companies, located in three types of industrial activity. The organization maintains Alstom Power for thermal generation and Alstom Renewable Power for wind energy. Alstom Transport makes trains, high-speed trains, and subways, while Alstom Grid is responsible for energy transmission equipment and services. Here we will deal particularly with Alstom Grid, with a focus on the Centers of Competence. In Alstom Grid, which is divided by product line, Instrument Transformer is the controller of the Power Compensation subdivision. It has factories in Itajubá, Minas Gerais, and Centers of Competence for reactors and transformers, as well as other lines and factories in Finland, India, and Mexico.

The Center of Competence for Reactors was created in 2006. At the time, the company had no idea what the area would be, but was seeking information and references in the market to set up a robust infrastructure for research and development. By the next year they were developing the third technology for high-power reactors. In 2008, the acquisition of Nokian Capacitors in Finland prompted the integration and transfer of technology that is still ongoing in 2012. With technologies from different parts of the world, the transfer process is delicate and constitutes a big challenge due to the need to process lots of information and technical detail.

The technical challenges in this area are very large, especially in product research and development. The team is attempting is to find solutions for its customers, often the decisive factor for winning a project. However, there is still a level of crucial importance for the company in administrative and financial terms, the limitation of having a small team that can't afford to work by focusing on just one project at a time. The R&D sector is responsible for new products, technology transfer, cost redesign – redesigning equipment to reduce costs – and research projects with technical studies, technical research, and calculation models. This division also deals with improving processes and gives special attention to short-term projects, those for rapid implementation, but that require great care and involve risks to maintain the quality of the business.

Among the major projects undertaken since 2006 was developing a high-power reactor, a new product for the company, but a technology that its competitor already possessed. One of the basic research projects was to develop a program to calculate noise generated by equipment. This was a demand with a lot of specifications and was a joint project with a university. A project to develop a product new to the world market was a response to a special demand from a customer who wanted smaller dry-run equipment – which had only existed previously on oil. This was an important breakthrough for Alstom's Grid R&D sector in Brazil.

Developing the high-power reactor, for example, was an important learning experience for structuring the department and overcoming the difficulty of working without dedicating full time to each project. In this case, it was necessary to create a totally new industrial process, joining equipment from different industries. However, normal production at the factory could not be halted. Day-to-day engineering is always a priority; it has more realistic goals for revenues and results than the new projects. At some point, the lack of a research and development center inside the organization became evident. The Centers of Competence were set up inside a group of specific units that already have a history, priority production lines, logistics, and technology of their own.

The present management model, centralized under the control of Alstom Grid in Europe, is tending to weaken. The organization has changed over the last ten years. The strategy contemplated is to internalizel competencies, especially Research and Development, as close as possible to Communications, and thus to the customer.

In this sense, Alstom Grid is presenting a project that has been under development with the so-called Transmission Users Group. The project is a new experience in Latin America for effective proximity to customers seeking joint technological solutions for its products. The goal is to create a technical community focused on product development and service systems through a direct communications channel among Alstom development teams and users, within the concept of treating the customer as the main focus of all initiatives. Customers are provided with the opportunity to influence future developments in research and development, in addition to interacting among themselves to exchange experiences and knowledge. For Alstom, it is an opportunity to get close to its customers outside the commercial process and discuss everything from the technical parts to the value added to the product.

In practice, the Transmission Users Group was created during an event that took place among Alstom units in India and the Middle East, and again at the beginning of 2011, in Guarujá, São Paulo. The Latin American event brought together 82 customers and 40 companies and united all the elements from the other events: the set of lectures involving technical themes and the objectives of the groups; a plenary session to discuss selected topics of general interest and the needs of the local technical context; work groups divided according to specific technical matters from the technical area of each one; and a Smart Village, an interactive project that brought product models together in a single environment, as well as real products, so that customers and specialists could see and touch them, ask questions, and exchange experiences and ideas.

The Transmission Users Group is scheduled to take place every two years. Alstom considers it a big success, due to the rich exchange of information with customers, its focus on technical solutions and on process and product innovation, and to its proximity to the consumer market. This is one way that the company is challenged to commit with its customers to conceiving of standardized solutions that will deliver the correct balance among technical needs, innovative solutions, and adequate costs.

*Source*: The authors.

## 5  Conclusions

We can conclude that a company known for innovation is capable of mobilizing its collaborators and creating new routines and products that can become differentials and that represent gains for the organization. A structure for innovation includes having this as a strategic option, promoting the culture of innovation, and having management methods that create facilitating processes, accompanied by management indicators.

In the course of the study, we found a tendency for multinational companies to adopt programs that permit technological advances in Brazil also, with the opening of research centers allied to university research programs. Microsoft, which inaugurated its first center for technology in Brazil in 2012, is an example. During the inauguration of the Microsoft Technology Center (MTC), Microsoft signed a statement of intent with the Ministry of Science and Technology and Innovation to create six "incubators" to encourage innovation in the country. According to Paulo Ludicibus, Microsoft's Director for Innovation, the goal is that "each of the six incubators care for the incentives for at least ten start-ups." The sectors that will receive the most attention will be health, education, oil and gas, and games (*O Globo*, January 17, 2012).

In the last 50 years, society and the Brazilian government have made considerable effort to construct a system of Science, Technology and Innovation (CT&I), which currently stands out among developing countries. The world ranking that measures technology worldwide, developed by the Confederation of Industries of India, in partnership with the World Organization for Intellectual Property (OMPI), disseminated in 2011, placed Brazil in 47th place, in front of India, South Africa, and Russia.

Even with full development, many actions and much care must be taken with innovation management in Brazil, because of factors such as: a) few resources in companies that have a focus on innovation, b) linear thinking that investment in innovation must have a quick return on capital (preferably in the same year), and c) lack of knowledge on the part of business people of the lines of tax support, lines of governmental support, and bank financing.

These factors end up generating isolated and distant strategic and budget planning. That is, companies isolate themselves and create innovation budgets founded on their own resources, which are most often very expensive to administer over time. In parallel, whenever there is a market crisis or reduction in product consumption, the Brazilian business class immediately drastically reduces investment in R&D&I.

In fact, Brazilian companies have been giving more importance to cost management, since the economic start of the 1990s and the greater competition observed on the domestic market, with the entry of new competitors into the market (Martins 2001). The expansion of lines of international and domestic financing has required greater operational efficiency from Brazilian companies, resulting in lower fixed costs and the intensive use of information technology. Studies by the Banco Central (2012) indicate that Brazil has received a great amount of financial resources, due to the behavior of the economy and its recent dynamism, but essentially due to the improved condition of management and innovation of the companies, which is a product of the financial results presented over the last few years.

Starting from the premise that innovation is something recent for Brazilian organizations, another important aspect to consider is cultural management. According to Jose Henrique Diniz (Coordinator for Innovation and Creativity of IETEC, BR), in an interview in the newspaper *Estado de Minas* (2012), one must search for a "culture of innovation."

The cultural factor is also evidenced in the Fundação Dom Cabral (FDC) model for innovation (described in Section 3). It states that to encourage innovation in an organization, it is not enough to define its importance in the strategic position and allocate capital and efforts to produce knowledge; it is indispensable for the organizational culture and climate to favor innovation and creative manifestations.

Becoming an innovative company is a strategic option that requires incorporating the concept and culture of pro-innovation into the firm's processes, indicators, planning, and teamwork. But the style of conducting the work teams is influenced by the leadership inside each organization, which can be proven by the management model adopted in Brazil, predominantly by the market leaders in their respective segments. The company management model most often encountered is that which values people who like to work in conventional, predefined systems. Renovating this organizational model requires motivation and a favorable environment (Rossi & Cozzi 2010).

A recent study sponsored by the FDC (Salum et al. 2012) demonstrated that participating mid-size companies did not have a clear direction toward innovation in their business strategies, and the lack of knowledge about the role of innovation could be pointed to as the main reason for this low involvement. An example of this is the contradiction that 71.2 percent of participants believed that innovation is evident in the ideology of their companies, however a large part (48.9 percent) say they do not know of programs that encourage innovation.

Considering the importance of mid-sized companies to Brazil's economy and also to the production chain of large companies, it becomes necessary for the theme of innovation in the country to acquire a highlighted position in the strategy and goals of organizations. It is necessary to examine it in a new light, treating it as a priority and also as an option for growth.

## Note

1. Original in Portuguese, "*Criando condições para a empresa inovar,*" Revista DOM, No. 8, April 2009.

## References

Albuquerque, A., & Rocha, P. (2007) *Sincronismo organizacional: Como alinhar a estratégia, os processos e as pessoas.* São Paulo: Saraiva.

Albuquerque, E. (1996) Sistema nacional de inovação no Brasil: uma análise introdutória a partir de dados disponíveis sobre a ciência e a tecnologia. *Revista de Economia Política,* 16(3), 63–72.

Arruda, C., & Barcellos, E. (2009) Criando Empresas Inovadoras. *Revista DOM,* 1, 43–52.

Arruda, C, & Rossi, A. (2009) Criando as condições para inovar. *Revista DOM,* 8, 37–43.

Arruda, C., Salum, F and Rennó L.(2011) Caso de Inovação: Magnesita Ecobusiness. FDC – Fundação Dom Cabral, view web site: http://www.fdc.org.br/pt/publicacoes/ Paginas/Casos.aspx

Arruda, C., Salum, F. and Rennó, L. (2012). Caso de Inovação: FIAT AUTOMÓVEIS Estratégia de Inovação. view web site http://www.fdc.org.br/pt/publicacoes

Arruda, C. & Salum, F. (2012) Working paper: WHIRLPOOL – Desenvolvimento de Competências para a Inovação. FDC – Fundação Dom Cabral, view web site: http:// www.fdc.org.br/pt/publicacoes/Paginas/Casos.aspx

Banco Central do Brasil, http://www.bc.gov.br. Janeiro/2012.

Bartlett, C. A., & Ghoshal, S. (2000) *A organização individualizada: Talento e atitude como vantagem competitiva.* Rio de Janeiro: Campus.

Batista Jr, P. N. (1993) A armadilha da dolarização. *Estudos Economicos,* 23(3), 37–62.

Berner, M. J., & Tushman, M. (2002) Process management and technological innovation: A longitudinal study of the photography and paint industries. *Administrative Science Quarterly,* 47, 676–706.

Boston Consulting Group (2009). Boston, USA: *The 2009 BCG 100 New Global Challengers.*

Burns, T., & Stalker, G. (1968) *The Management of Innovation.* Tavitock Publications: London.

CAPES (Coordenação de Aperfeiçoamento de Pessoal de Nível Superior), http://www. capes.gov.br. Janeiro/2012.

Castells, M. (2007) Communication, power and counter-power in the network society. *International Journal of Communication,* 1(1), 238–66.

Confederação Nacional das Indústrias, http://www.cni.org.br. Janeiro/2012.

Cole, R. E., & W. R. Scott (Eds) (2000) *The Quality Movement and Organizational Theory.* London: Sage.

Coral, E., Ogliai, A., & Abreu, A. F. (2008) *Gestão integrada da inovação: Estratégia, organização e desenvolvimento de produtos.* São Paulo: Atlas.

Christensen, C. (2001) *O dilema da inovação.* São Paulo: Makron Books.

Christensen, C., & Raynor, M. (2003) The innovator's solution: Creating and sustaining successful growth. *Research-Technology Management,* 46(5), 61–72.

Crespell, P., & Hansen, E. (2008) Managing for innovation: Insights into a successful company. *Forest Products Journal,* 58(9), 6–18.

D'Áveni, R. (1995) *Hipercompetição: Estratégias para dominar a dinâmica dos mercados.* Rio de Janeiro: Campus.

Dosi, G., & Freeman, C. (1992) The diversity of development patterns: On the processes of catching-up, forging ahead, and failing behind. International Economic Association, Varena congress.

Drucker, P. F. (2002) *Inovação e espírito empreendedor: prática e princípios.* São Paulo: Thompson/Pioneira.

Fast Company (2011) The World's 50 most innovative companies, Magazine, March.

Felippe, M. I. (2007). 4 C´s para competir com criatividade e inovação. São Paulo, BR: Qualitymark.

Franco, G. H. B. (1996) *A inserção externa e o desenvolvimento.* Rio de Janeiro: Harbra.

Gonç alves, J. E. L. (2000) As empresas são grandes coleções de processos. *Revista de administração de empresas* (online), 40(1), 6–9, ISSN 0034-7590.

Harrington, H. J. (1998) *O processo do aperfeiçoamento: Como as empresas americanas, líderes de mercado, aperfeiçoam controle de qualidade.* São Paulo: McGraw-Hill.

Henderson, R. M. and Clark, K. B. (1990) Architectural innovation: The reconfiguration of existing product technologies and the failure of established firms. *Administrative Science Quarterly*, 35(1), 9–30.

IPEA 2012 (Instituto de Pesquisas Economicas Aplicadas), http://www.ipea.gov.br. .

Jeston J., & Nelis, J. (2008) *Business Process Management: Practical Guidelines to Successful Implementations.* Butterworth-Heinemann.

Kanter, R. M. (1997) *Frontiers of Management.* Cambridge: Harvard Business School Press.

Kaplan, R. S. & Norton, D. P. (2004) *Strategy Maps: Converting Intangible Assets into Tangible Outcomes.* Boston, USA: Harvard Business School Plushing Corporation.

Kenny, B. & Reedy, E. (2006) The impact of organisational culture factors on innovation levels in SMEs: An empirical investigation. *Irish Journal of Management*, 27(2), 119–143.

Kimberly, J. R. (1981) *Managerial Innovation.* New York: Oxford University Press.

Lastres, H. M. M., & Albagli, S. (1999) *Informação e globalização na era do conhecimento.* Rio de Janeiro: Campus.

Leone, G. S. G. (2000) *Curso de contabilidade de custos* (2nd edn). São Paulo: Atlas.

Lovelock, C., & Wright, L. (2001) *Serviços: Marketing e gestão.* São Paulo: Saraiva.

Malone, T., Crowston, K., Lee, K., Pentland, B., Dellarocas, C., Wyner, G., Quimby, J., Osborn, C. S., Bernstein, A., Herman, G., Klein, M., & O' Donnell, E. (1999) Tools for inventing organizations: Toward a handbook of organizational processes. *Management Science*, 45(3), 425–443.

Maranhão, M., & Macieira, M. E. B. (2004) *O processo nosso de cada dia: Modelagem de processos de trabalho.* Rio de Janeiro: Qualitymark Ed.

Martins, E. (2001) *Avaliação de empresas: Da mensuração contábil à econômica.* Fipecafi, São Paulo, Atlas.

Martins, E. C., & Terblanche, F. (2003) Building organisational culture that stimulates creativity and innovation. *European Journal of Innovation Management*, 6(1), 64–74.

Marlyn, O. (2004) Business process management: A rigorous approach. *The British Computer Society*, January, 360pp.

McLean, L. D. (2005) Organizational culture's influence on creativity and innovation: A review of the literature and implications for human resource development. *Advances in Developing Human Resources*, 7(2), 226–246.

McNair, C. J., & Leibfried, K. H. J. (1992) *Benchmarking: Uma ferramenta para a melhoria contínua.* Rio de Janeiro: Campus.

MDIC (Ministério de Desenvolvimento, Indústria e Comércio Exterior), http://www.mdic.gov.br. Janeiro/2012.

Motta, P. R. (2001) *Transformação organizacional: a teoria e a prática de inovar.* Rio de Janeiro: Qualitymark.

Nederhof, P. W. (2009) The future of innovation is organising for strategic flexibility. In D. B. Von-Stamm & A. Trifilova (Ed.), *The Future of Innovation.* Surrey, England: Cower Publishing Limited.

Paim, R., Cardoso, V., Caulliraux, H., & Clemente, R. (2009) *Gestão de processos: Pensar, agir e aprender.* Porto Alegre: Bookman.

Patterson, M., & Lightman, S. (1993) *Accelerating Innovation: Improving the Process of Product Development.* Van Nostrand Reinhold, New York.

Pavitt, K. (1984) Sectoral patterns of technical change: Towards a taxonomy and theory. Science Policy Unit, University of Sussex, Brighton, UK.

Porter, M. E. (1989) *Vantagem Competitiva.* Rio de Janeiro, BR: Campus.

Porter, M. E. (1990) *The Competitive Advantage of Nations.* Palgrave.

Porter, M. E. (1996) What is strategy? *Harvard Business Review,* 6, 61–78.

Porter, M. E. (2011) Creating shared value. *Harvard Business Review,* 89(½), 62–77.

Pyke, J., & Whitehead, R. (2006) *Mastering Your Organization's Processes: A Plain Guide to BPM.* Cambridge University Press.

Reis, L. G., & Pires, E. A. (2005) *Sistemas de gestão de qualidade: custos inerentes e o problema da descontinuidade.* Anais do IX Congresso Internacional de Custos: Florianópolis.

Rodríguez, A., Dahlman, C. and Salmi, J. (2008) *Knowledge and Innovation for Competitiveness in Brazil.* Washington, DC: World Bank.

Rossi, A., & Cozzi, A. (2010) A cultura empreendedora como aliada da inovação. *Revista DOM,* 11, 25–31.

Rothwell, R. (1992) Development towards the fifth-generation model of innovation. *Technology Analysis & Strategic Management,* 1(4), 73–75.

Salum, F., Arruda, C., Grisolia, L., & Pereira, R. (2012) Caderno de Ideias: Inovação nas médias empresas brasileiras: um desafio para a competitividade. FDC – Fundação Dom Cabral, view web site: http://www.fdc.org.br/pt/publicacoes/Paginas/CadernosDeIdeias.aspx

Salum, F., & Jardim, R. G. A. (2010) *Os Desafios da inovação tecnológica no Brasil.* Website: http://www.fdc.org.br/pt/pesquisa/inovacao/paginas/publicacoes.aspx"www.fdc.org.br/pt/pesquisa/inovacao/paginas/publicacoes.aspx. Accessed on August 16, 2011.

Schein, E. H. (1985) *Organizational Culture and Leadership.* San Francisco, CA: Jossey-Bass.

Schumpeter, J. A. (1961) *Capitalismo, socialismo e democracia.* Rio de Janeiro: Fundo de Cultura.

Schumpeter, J. A. (1997) *Teoria do desenvolvimento econômico.* Editora Nova Cultural: São Paulo.

Sera, K. (1992) Corporate globalization: A new trend. *Academy of Management Executive,* 6(1), 89–96.

Silva, M. L. (2002) *A Internacionalização das grandes empresas brasileiras de capital nacional nos anos 90. (Tese de doutorado).* Universidade Estadual de Campinas.

Silverberg, G., Dosi, G., & Orsenigo, L. (1988) Innovation, diversity and diffusion: A self-organisation model. *The Economic Journal,* 98(393), 1032–1054.

Stabell, C. B., & Fjeldstad, O. D. (1998) Configuring value for competitive advantage: On chains, shops and networks. *Strategic Management Journal*, 19, 413–437.

Stal, E., & Campanario, M. A. (2010). Empresas multinacionais de países emergentes: o crescimento das multilatinas. *Economia Global e Gestão*, 15(1), 55–73.

Sterman, J. D. (2000) *Business Dynamic: Systems Thinking and Modeling for a Complex World*. Boston, McGraw-Hill.

Stiglitz, Joseph, E. (2006) *Making Globalization Work*. New York, London: W. W. Norton.

Tadeu, H. F. B. (2011) Gestão de Estoques - Métodos Quantitativos Aplicados. *Cengage Learning*, 1(1), 1–42: 302 pp.

Tadeu, Hugo F. B., &Salum, Fabian A. (2012) Process management and innovation for the Brazilian airport infrastructure. *African Journal of Business Management*, 6(10), 3745–3755, March 14.

Teece, D. J. (1986) Profiting from technological innovation: Implications for integration, collaboration, licensing and public policy. School of Business Administration, University of California, Berkeley, U.S.A.

Ulrich, K. T., &Eppinger, S. D. (2004) *Product Design and Development*. McGraw-Hill, Boston.

Viotti, E. B., & Macedo, M. M. (2003) *Indicadores de ciência, tecnologia e 30 inovação no Brasil*. Campinas, SP: Editora da Unicamp.

World Development Indicators (2003) Washington DC, USA: World Bank Publications.

# 12
## The Global Importance of Innovation Champions: Insights from China

*Anton Kriz, Courtney Molloy, and Bonnie Denness*

## 1 Introduction

Some call them deviants, others call them heroes, but one commonality for what the literature has deemed "innovation champions" is that these individuals can play a vital role in modern organizations. The role of innovation in creative destruction and national economic growth is well documented. The role of entrepreneurs in this process was identified by Joseph Schumpeter and has since been extensively researched. Schumpeter also realized that it takes a range of individuals to make innovation happen. This chapter is focused on a particular individual in an organization: the innovation champion. The notion of an innovation champion is complex and has been labeled in various ways. Innovation champions (Howell & Higgins 1990a) have many of the characteristics of the independent entrepreneur, but they operate within an existing business, rather than operating a venture of their own. Classifying innovative champions more succinctly and understanding the role they play in global companies and rapidly developing countries like China is the key aim of this chapter. This chapter is conceptual, but refers to considerable investigations and research conducted on management practices and interpersonal and organizational relationships in China. Using China as the unit of analysis and a single case is not without precedent (Styles & Ambler 2003). The lead author has extensive experience in China and has conducted over 200 in-depth interviews and numerous empirical studies in this environment from 1999 to 2012.

Gaining a clear understanding of what is meant by an innovation champion is an important aspect of this research. These "boundary-spanning individuals" or "stars" (Michael & Thomas 1981) are known for seeking out ideas of others and then spreading them across the organization. Howell and Higgins (1990b, p. 317) bring together elements from various sources (Achilladelis et al. 1971; Michael & Thomas 1981; Schon 1963) to provide a comprehensive definition of an innovation champion, noting that "this is an individual who informally [and formally] emerges in an organization and makes a decisive contribution to the innovation by actively and enthusiastically promoting its progress through critical organizational stages." The word "formally" has been inserted in the definition as companies such as CEMEX in Mexico and Lion in Australasia are examples of companies now giving formalized authority and acknowledgment to these roles. Formality has implications, and how such a nominated champion operates has interesting connotations that need additional research. Although much investigation has been conducted into innovation champions, there remains much to be learned about these individuals, including how much of an impact they have on innovation in a firm, sector, region, and nation. Additionally, we could ask, if organizations were given advice and skills in managing innovation champions, how would this intervention affect their innovation and growth?

Like much work in the management domain, most of the work on innovation champions has emanated from North America. What is exciting is therefore to investigate the global perspectives around such champions. Scott Shane (1995) has conducted several studies in which he has identified interesting similarities and differences of potential champions and their intentions across cultures. Nevertheless, to our knowledge no serious research has been conducted on champions in a country as culturally diverse as China. This chapter is a start toward addressing this imbalance.

There are two broad ways of researching cultures. Some like Hofstede et al. (1990), Trompenaars and Hampden-Turner (1997) and Shane (1995) take an etic or cross-cultural approach while others like Berry (1989), Fang (1999) and Kriz (2009) take an emic or within-culture perspective. Both have significant advantages, but ultimately a culture as complex and varied as China deserves culturally sensitive and focused investigation. The regional diversity and the way business is conducted on a provincial, local, and organizational level warrant such a review. Some have likened regional comparisons to comparing Germany to France or Britain, as there are stark variations and nuances among Chinese regions (Ambler et al. 2009).

The first objective of this chapter is to review the literature about innovation champions and accordingly, to set the scene. This includes identifying key concepts and synthesizing ideas. Such a synthesis is unlikely to do justice to the debate and complexity, but will provide a good overview and solid foundation. Next a cross-cultural analysis using current and previously published literature is undertaken to identify important implications and variants relating to China. Finally, the study focuses on China and its organizational diversity and discusses aspects like *guanxi* (connections and relationships) in order to identify key perspectives around China's likely innovation champions. The burgeoning growth of China makes this a particularly timely and important study.

## 2   Background

Innovation has been defined as creating a new idea, process, or product and then taking it forward to some form of implementation. Dodgson et al. (2008, p. 2) define it as follows: "Essentially, innovation is the successful commercial exploitation of new ideas. It includes the scientific, technological, organisational, financial, and business activities leading to the commercial introduction of a new (or improved) product or service." Innovation is a process linked to imagining, incubating, demonstrating, promoting, and sustaining (Jolly 1997). Notably, innovation is also now being defined more broadly to incorporate social enterprise and an increasingly liberalized approach to value that goes beyond commercial considerations. Research and development teams have grown in importance as modern society has advanced in science and technology, and a significant proportion of innovation is now generated by such groups. Nevertheless, individuals are equally responsible for innovation, whether they be the idea generators, tinkerers, or champions of others' innovations. Like *innovation, champion* is a common and often overused word. The idea of a product champion was mooted in 1963 by Schon. Since this time, researchers have afforded champions of innovation various titles: sponsor; technological gatekeeper; innovation director; promoter; chief innovation officer; broker; services champion; chief technology officer; and bridger. Various definitions have been utilized in the literature; however, all appear to have a consistent theme of identifying an individual within an organization who seriously pushes and supports a new idea.

Innovation champions are not necessarily the inventors or innovators and consequently are not the individuals normally coming up with the new ideas (Markham & Aiman-Smith 2001; Markham & Griffin 1998). However, they commonly take up worthwhile ideas of others. Most significant is their constant scanning of the environment, both inside and out. This allows for identification and development of ideas worthy of further attention (Anderson & Bateman 2000). Schon notes that "It is characteristic of champions of new developments that they identify with the idea as their own, and with its promotion as a cause, to a degree that goes far beyond the requirements of their job" (1963, p. 84). Champions use a variety of sources of information, such as colleagues, customers, conferences, papers, and journals (Anderson & Bateman 2000; Howell & Shea 2006). This scanning of information sources serves three purposes. First, external scanning enhances the champion's understanding of the marketplace, technological advances, customer needs, and competitor activities (Anderson & Bateman 2000; Howell & Shea 2001; Shane 1994a). Internal scanning provides champions with potential new ideas from colleagues. Finally, scanning the internal organization also keeps champions well versed on the organization's culture, processes, and standards (Frost & Egri 1991). This information is essential to the champion in communicating across domains and in effectively navigating organizational politics (Frost & Egri 1991).

## Why is the innovation champion so important?

Schon (1963, p. 84) was the first to note the importance of an innovation and its organizational champion, acknowledging that "the new idea either finds a champion or dies." As Tushman and Nadler (1986, p. 89) argue, without these champions, "organizations may have many ideas but few tangible innovations." Shane and Venkataraman (1996) highlight the common preference of most employees to avoid bending the rules or deviating from the norm. These individuals favor imitation or minor variations and inherently distrust ideas that disobey belief systems (Shane & Venkataraman 1996); they are keen on reducing risks rather than opening up ideas and instituting too much change. It is also natural for those in power to want to avoid innovations that have the potential to displace their position (Frost & Egri 1991). Modern businesses nonetheless have increasingly realized that innovation is critical to their survival. Failure to innovate is much less riskier than the alternative of doing nothing in the face of creative destruction. Accordingly, adding innovation into the firm is becoming both a top–down and bottom-up initiative. Finding individuals to foster innovation is therefore a key. Champions have an important role of scanning internally and externally for fresh ideas and, helping to push these ideas to some form of outcome, which is also important.

Overcoming organizational inertia and resistance to change is increasingly seen as a critical function of growing an effective organization. Howell et al. (2005) highlight impacts beyond successful innovation that a champion may have on the broader organization. Champions can cultivate cross-fertilization of ideas and improve cross-functional collaboration. Champions also bring in new information, which may lead to strategic opportunities and can alter resource allocation. It is argued that positive champion attitudes and behavior also stimulate and improve retention of other innovative employees. Markham (1998) has challenged some of the overtly positive assertions around champions and suggests that more empirical evidence is required. It could be argued, for example, that the innovation champion is potentially a "villain" if his or her objective is counterproductive to a firm's vision and there is lack of formal support. This position may have some traction, as what is good or bad for a firm is often subjective. It is why senior management is given responsibility for making leadership decisions. Radical innovations and new product initiatives have far more failures than successes, so abdicating these decisions to champions has some difficulties. The champion role is therefore likely to be scrutinized, as champions are naturally often involved in "riskier" innovative initiatives. Overall, the body of evidence is positive towards champions and the role they play, but there is still much to be learned. The nexus between those entrusted with the key business decisions and an innovation champion is worthy of a little more discussion.

## Bottom-up championing versus top–down management

Research into organizational determinants of innovation commonly identifies leadership as a determinant of organizational innovativeness (Crossan & Apaydin 2010; Mumford & Licuanan 2004). More specifically, transformational leadership is often found to be a key determinant of innovation, being more effective in encouraging and driving innovation than the transactional approach (Gumusluoglu & Ilsev 2009; Sarros et al. 2008). The distinction between innovation champions and those who facilitate innovation from the top is blurred and often is not clearly delineated in the literature. Some argue that CEOs and top management executives can also be innovation champions (Day 1994; Maidique 1980; Van de Ven et al. 1999), while others link the innovation champion with individuals working from middle to lower levels of the organization (Howell & Higgins 1990b). Day (1994) suggests that top management champions are required when the innovation is likely to reorient the strategic direction of the organization, is highly visible, or is costly. At this point, a distinction between the champion of innovation (at the top management level) and the innovation champion (who works at the middle to lower levels of the organization) can be proffered. Such a distinction is suggested in the literature but is not explored fully. The focus of this chapter is specifically on the innovation champions who operate at the middle management or lower levels. Much of the innovation champion literature appears to follow a similar vein, although without expressly identifying this or explaining why this is the case. Arguing against the importance of top managers in driving innovation within organizations is nonsensical. However, the use of "champions of innovation" makes the distinction a little clearer. Those promoting, supporting, and facilitating innovation and innovative ideas from upper levels of the organization have different responsibilities, challenges, and authority to those at lower levels.

Referring to those at upper levels of an organization as champions of innovation highlights the notion that these individuals, particularly the CEO, are responsible for cultivating a climate and culture that is conducive to innovation, developing and maintaining the processes around innovation, and providing access to resources. This signifies a broader level of responsibility. Where top management must decide on the allocation of resources, champions work to secure resources for their ideas (Howell & Boies 2004; Markham & Griffin 1998). As the necessity of innovation is increasingly recognized, the accountability in driving innovation is more likely to be formalized in top management teams. In contrast, while the role of the innovation champion may in some instances be formally appointed, champions most typically emerge informally (Markham & Griffin 1998). In fact, the informal nature of the role is identified as a key reason for champion success and why they

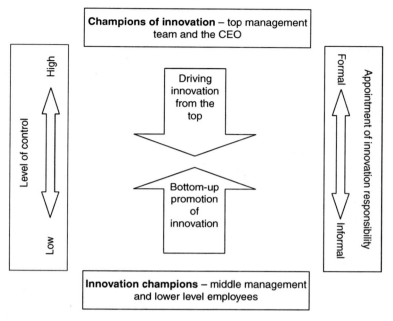

*Figure 12.1*   Innovation champions and champions of innovation
*Source*: Developed for this research.

choose to operate the way they do (Howell & Higgins 1990a; Howell & Shea 2006). Figure 12.1 provides a visual representation of these relationships. It is important to highlight that these are not necessarily either/or distinctions. Rather, the level of formalization and control is best described as a continuum. As Howell et al. (2005) argue, an individual should not be identified as a champion on the basis of displaying *all* of the prescribed behaviors; instead, championing should be seen as a continuous construct.

### Key characteristics of innovation champions

Champions have better skills than most in dealing with people (Frost & Egri 1991; Howell & Shea 2006). They have an ability to sell, persuade, negotiate, and enthuse (Sim et al. 2007), and are known for their capacity to communicate effectively with top management as well as with peers and subordinates (Anderson & Bateman 2000). They have capacity to handle organizational challenges – and as Morris et al. (2008) identify – to do this effectively means wearing many "hats." Successful champions have broad work experience across locations and functional areas (Chakrabarti 1974; Howell & Higgins 1990a; Kelley et al. 2011); their social networking skills (Taylor et al. 2011) provide for a broad array of friends and acquaintances within and beyond the organization; they often have extensive organizational knowledge and are politically astute (Chakrabarti 1974; Howell et al. 2005). Champions are also aware of who is important and who presents problems (Howell et al. 2005; Sim et al. 2007); are enthusiastic and passionate (Howell, et al. 2005; Kelley et al. 2011; Sim et al. 2007); and display high levels of energy, confidence, and persistence (Howell & Higgins 1990c; Howell et al. 2005). Champions often confront organizational roadblocks and people who are reluctant to change but are able to negotiate hurdles by their diversity in thinking and openness to new ideas (Kelley et al. 2011; Taylor et al. 2011). A champion's approach to driving innovation has therefore been identified as closely aligned with transformational leadership (Howell & Higgins 1990a).

Upon hearing a description of an innovation champion, it is possible that some would envisage an individual who works *outside* organizational processes. "Bootlegging" resources and hiding the innovation are some of the stereotypes associated with such individuals. However, champions often operate within the boundaries of organizational processes, having acquired or been given management approval, and this may be the more typical approach (Howell & Higgins 1990a). Some champions therefore have formal management support while others have to go it alone and back their own judgment. These two varying perspectives represent what are described as the rational and renegade approach (Howell & Higgins 1990a; Shane & Venkataraman 1996). Both approaches generally take the same steps, but do so in a different order, with formal approval being sought prior to any serious action towards implementation during the rational approach, while renegades operate for longer outside official channels. Champions using either approach build support outside their own functional area from subordinates, colleagues, superiors, and even from external parties such as clients and suppliers. Existing networks are therefore of substantial importance. Coalitions developed by champions are necessary to gather resources, gain support, and develop and implement ideas.

### Champions know how to package ideas

How the champion sells the idea is particularly important (Anderson & Bateman 2000). Successful champions are able to package their message and highlight the "right" points to the "right" people (Anderson & Bateman 2000). Whether a champion frames the idea as an opportunity or a threat is identified as a key decision (Anderson & Bateman 2000; Dutton & Jackson 1987). Howell and Shea (2001) identify that executives are more likely to restrict control and centralize decision making when dealing with a threat, whereas they are more likely to seek involvement of lower levels when faced with an opportunity. Howell and Shea (2001) contend that framing an idea as a threat is potentially counterproductive and diminishes the perception of a champion's ability, credibility, and influence. Conversely, the use of information that includes facts and figures is commonly identified as particularly important in the idea-selling process (Anderson & Bateman 2000). Building an appropriate argument and supporting this with evidence and the right personnel are therefore key success factors for potential innovation champions.

A key to a champions' success is their knowledge of the organization (Kanter 1988). A deep and broad knowledge of the organization's culture, political climate, processes, and strategic approach is important. Frost and Egri (1991) discuss the central role that organizational politics plays in the innovation process and highlight the necessity for champions to employ strategies that are politically appropriate. As Kanter (1988) argues, it is the effectiveness of a champion's political activity that determines whether an idea moves through the later stages of the innovation process. They identify two political layers that comprise an organization: surface politics and deep-structure politics. Surface politics could be described as the games played by individuals, teams, or organizations to shift or sustain power. The deep structure relates to the core institutions; it shapes the individual and the collective "interpretive frame" of organizational members. Surface politics interconnect with deep-structure politics. Champions must be aware of how the idea or innovation fits with the organization's direction and culture. An innovation in conflict with the deep structure determines the strategies that a champion must employ, the level of support they must garner and the fallout they may encounter (Frost & Egri 1991).

### How champions are motivated

The role of the champion is highly contextual and is built around the nature of the organization and type of innovation involved. Of significant importance in developing an understanding of an innovation champion is gaining insight into *why* champions choose to undertake championing behavior. The work of innovation champions may comply with or even threaten the organization's power and relationship structures (Frost & Egri 1991), potentially causing additional stress for the champion (Howell & Higgins 1990c) and hindering career advancement (Howell 2005). Yet innovation champions do choose to undertake these out-of-the-ordinary behaviors (Shane 1995). Understanding why this occurs may be useful in helping organizations know how to attract, reward, retain, and support innovation champions. There is surprisingly limited attention to motivation in the championing literature. Studies that do explore motivation indicate that champions are motivated externally by an idea, technology or business issue, and organizational strategy; or are internally motivated by being involved in the change process itself and the challenges involved (Bartlett & Dibben 2002; Howell & Higgins 1990a). Much still remains to be understood about innovation champions and their motivations. More than one champion can be required, depending on the innovation, its strategic importance, and the organizational size and culture. The right approach is likely to depend on a multitude of key factors. Such aspects can go well beyond the firm to include aspects like national culture.

## 3   Innovation champions across cultures

Identification of the need for innovation champions is becoming increasingly universal. Previous studies have focused on individual characteristics and organizational variables that are likely to support and/or hinder the championing process. While certain motives and characteristics of the innovation champion may transcend national borders, others are known to be culturally bound (Shane 1994c). Cultural values can influence national rates of innovativeness and inventiveness. On an organizational level, cultural values shape corporate culture, norms, and routines. If the innovation champion is to successfully promote an innovation, the champion must be strategic in selecting a championing style that will be readily accepted in the given environment (Shane 1994b). While champions commonly adopt roles and behaviors that deviate from standard organizational routines, rules, and norms (Shane 1994a), an individual's preferred championing approach will tend to be culturally specific. Using a champion preference scale, Shane (1994b) identified the most acceptable methods to promote innovation in different cultural contexts. These findings were also later supported by Shane, Venkataraman, and MacMillan's (1995) extensive study of championing behavior in 30 countries. This study found that collectivist, high power-distance, and uncertainty avoidance cultures prefer champions to build cross-functional alliances, gain support from authority, and conform to accepted processes.

The dimensions of the champion-culture relationship are particularly relevant for multinational corporations, which must contend with the challenge of promoting championing behavior in environments that may not necessarily encourage individuals to adopt accepted home-country roles. For instance, in cultures where innovation is not readily embraced, such as high uncertainty-avoiding cultures (Shane 1995), corporations must either encourage cultural change to support effective championing behavior or align to the preferred championing approach that is culturally acceptable (Shane 1994b). Regardless of which approach is taken, selecting a championing style that is culturally appropriate will enhance the likelihood of successful implementation of the innovation initiative. China was notably not included in Shane's discussion, but what it did show is that there are important differences, but also similarities, among a number of what may otherwise be seen as diverse cultures.

### An emic versus etic view

Shane's broad cross-cultural and etic view is beneficial for innovation management on a global level, but lacks the detail necessary when probing complex cultures within countries like China. According to Fang (1999, p. 9), "there are two basic approaches to the study of cultures: emic and etic." The dichotomy was suggested by Pike (1967) as a contribution from linguistics to a theory of human behavior, and it was later adopted in ethnography. *Emic* refers to the member of a culture – the insider who is familiar with participating in the system. *Etic* refers to a stranger to the culture – the detached observer's perception; it is the outsider's view. By unpacking the layers of Chinese culture and seeking to understand the influence of emics, it is possible to reach core issues driving particular behaviors (Berry 1989). In adopting such a view, it is then possible to move beyond the "ecological fallacy" of believing that broad studies accurately represent business individuals – in this case the innovation champion (Smith & Bond 1993). The key to understanding China is to reconcile etic with more detailed descriptions around emic (Berry 1989; Smith & Bond 1993).

Therefore, the emphasis should be on discovering native classifications and conceptualizations in an inductive mode when investigating innovation champions in emerging markets like China (Sturtevant 1964). The diversity of China is stark. A traditional North-South dichotomy does not do justice to the magnitude of differing management and organizational styles. Figure 12.2 illustrates the emic approach. The area marked (A) represents an approach that suggests including a cross-section of individuals/firms in Chinese regions. The aim is to understand independent perceptions within a region and then build a model of collective representations. If two countries (A) and (B) have emic studies applied to the same construct – in this case,

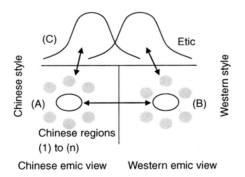

*Figure 12.2*   The cross-cultural emic approach

*Notes:* (A) represents an emic view with each region (shaded circles) being investigated individually. In this case it would be each Chinese region selected in the sample (1 to n) e.g. Shanghai, Zhejiang etc. These regions can the be compared for similarities and differences. A similar study could be conducted in the US for example with a number of regions selected e.g. Florida, Hawaii etc as in (B). Ultimately the two emic studies on a regional and then national level could be compared against each other. This bottom-up comparison is an advance on the simplistic approach of broadly comparing national characteristics on an etic level as depicted in (C).

*Source:* Developed for this research.

innovation champion attitudes and/or behavior – it is possible to compare and contrast. This means drawing on emic or within-culture insights to build an improved etic understanding (area marked C). This theory is in accordance with Berry's view (1989) that emic studies can be compared to other emic studies, rather than simply comparing cultures only on broad cross-cultural parameters. This overview on a cross-cultural and emic level is important as researchers begin reviewing a new domain like China as part of the global pursuit to better understand innovation management. It is important before embarking on more comprehensive studies to review what is already known within the context of this domain.

## 4   Innovation champions and the story of China

### Chinese rules of the game

China is known for its complexity and paradoxes (Fang 2006). China is an amalgamation of subcultures with sometimes contradictory influences, which differ according to where people live, their age, their work, their personality, and how Confucianism, communism, and the new market economy are influencing minds and behavior (Tung et al. 2007). Management researchers need to be thoroughly cognizant of the deeper essence of such a culture. Adding history and philosophy to the standard environmental analysis of China is an appropriate start. Confucian belief is at the essence of such reviews. Increasingly, the importance of Tao and complex elements like stratagem are also being recognized. The geographic and subcultural underpinnings, which can add other layers of complexity, have received less investigation (Kriz & Keating 2010). Business practices, for example, within a radius of a few hundred kilometers, between areas like Chengdu and Chongqing (both provinces in western China), can differ markedly. Extend such comparisons to regions like Beijing (North), Shanghai (Central East) and Guangzhou (South East), and then extend this to second- and third-tier cities, and the challenge becomes multidimensional.

Add local nuances around opportunistic entrepreneurial networks for specific subregions like Wenzhou (Zhejiang province), and the puzzle of understanding innovation management in China becomes even more complex. These nuances run deep, and the perceptions of the Chinese make for interesting reading. For example: "It has been said for every dollar, a Shanghainese would save 50 cents and invest 50 cents. A Wenzhou businessman would invest his dollar and borrow 50 cents from the Shanghainese to invest" (Tan 2011, p. 104). Guangdong people in the South offer another example. People from this region have established a considerable reputation for risk-taking. However, unlike their close geographic neighbors in Zhejiang, these people are known to be less calculating in how they invest.

The geographic variations defy the simple stereotype of a majority Han population. These nuances also play a major role in aspects like networking and negotiation style. Being further from the center of power (Beijing) is known to have an influence on regional politics and behavior. The transition from rural to urban is a very recent phenomenon in China and is having its own profound regional impacts.

The complexity and paradox becomes even more intriguing when one starts adding in aspects like organizational dynamics. China is a burgeoning economy with "one country – two systems" (Communist political structure with a relatively open economy). It is important to remember that China is a Communist state, irrespective of what the market dynamics convey. People like Rupert Murdoch and organizations like Microsoft, Facebook, and Google have all had lessons here. China has an array of organizational and business styles. State-owned enterprises (SOEs) dominated the Mao era and remain major players in modern China. However, privately owned enterprises (POEs) and town and village enterprises (TVEs) have now made major inroads. This has interesting implications for innovation management in China. Known for its stealth and a predominantly top–down paternal collective approach, SOEs have peculiarities that are known to inhibit innovation. This is also known to stem from a juxtaposition of Confucian values added to what is a large bureaucracy.

Confucian values extend from a long tradition that encourages ritual, obedience, and harmony. Such moralistic imperatives have extended to a bureaucratic tradition of meritocracy and scholastic traditions that are focused much more on rote than on critical thinking. Kriz and Keating (2010) highlight that the innovative spirit still exists, but environment and institutions play a large role in how these qualities are ultimately displayed. R&D is not new to SOEs, but having subordinates promoting ideas up the chain of command is contradictory to China's long-held traditions. Add the importance of face or *mianzi* to the hierarchical top–down approach, and it is easy to understand why Chinese are known to be extremely cautious about overtly saying "no" and challenging convention.

### The role of guanxi

One aspect of China in relation to innovation champions that is worthy of more focused discussion is the concept of *guanxi*. *Guanxi*, or "relationships," "connections," or "personal connections," have been widely regarded as a key success factor in Chinese management literature. *Guanxi* is an important element in how Chinese gather information and exchange ideas. Chinese traditionally rely on family and close friends for key relational bonds (Kriz & Keating 2010), but such networks extend to other relatives, friends, and ultimately to business acquaintances. Each individual's set of connections works both on a horizontal and vertical level (Kriz 2009). It then resembles a game of Chinese checkers. Traditionally it has been important for the business individual to firmly obey hierarchy. Face has already been discussed, and in this case innovation champions have to be careful how they negotiate through such hierarchies. Innovation champions will also have many other branches of relationship, relating to their lifelong personal *guanxi* links. Western innovation champions following the rational approach often use a sponsor, but in the Chinese situation it would likely require withdrawing or calling on a *guanxi* referral to enable access up the chain. Chinese know how to use such ties to leverage favors and form strategic alliances (Kriz 2009).

Understanding *guanxi* and the role it plays in a firm's development and innovation practices is a complex aspect of Chinese society; to do it justice would go well beyond the scope of this chapter. These links can be powerful and can extend into various areas. At the individual innovation champion level, it is therefore a case of utilizing a mix of vertical and horizontal ties. Finding the right person or *guanjianrenwu* to get something done is critical in China. The Chinese innovation champion will have many layers of connections that provide information and opportunity, both vertically and horizontally. Stronger ties and bonds open up considerable opportunities, and boundary-spanning without such ties is likely to be limited. Ultimately, *guanxi* ties can link innovation champions to POEs, SOEs, government personnel, and a range of informants. Utilizing such information sources is important in assessing opportunities within the organization.

### Changing institutional dynamics

China's top–down approach is practiced extensively and is difficult to change. "In China, the boss should have absolute power. No one can have any authority except you, because people will abuse any authority you delegate to them" (Tan 2011, p. 109). In the context of innovation management and innovation champions, developing alternatives to historical SOEs is important. Labeling all SOEs as conservative and subject to outdated management traditions is also too simplistic. There are a number of SOEs wanting to act globally; engaging management consultants like McKinsey and BCG to revamp HR practices and approaches to innovation (Fernandez & Underwood 2006) is now not unusual. The additional range of companies known as POEs is pertinent to this chapter. These are more aggressive, as they need to push the boundaries even further and need to seek more advanced management techniques and practices. Hiring of insiders and family versus professional managers has been the norm for many of these firms. Increasingly, a proportion of such firms are seeking more advanced management practices and techniques (Nie et al. 2009). A theme among the larger, more developed Chinese POEs is the need for training and continuous improvement. Innovation champions are likely to have more success in some of these more aggressive operations.

Another key aspect is the return of the educated overseas Chinese (sea turtles). This group is more likely to adopt some techniques from abroad and to be outspoken and even challenge convention (Tan 2011). Modern companies like C-Trip (an online trip organizer) are typical of a new breed of such innovative companies. They have fostered a strong internal climate for creativity through their open layout and modern headquarters in Shanghai. Haier's leader Zhang Ruimin has attempted successfully to spread innovation across the firm. Fetscherin (2010, p. 159) noted that Haier emphasizes that every employee is an independent and innovative strategic business unit with the collective goal of achieving primacy in the marketplace. iPartment is another such company where "employees are required to write about at least one thing they do innovatively during the term [annual review]. It can be a change to merely a small part of the administration work, a change to the technological approach, or even a change to the product design itself" (Tan 2011, p. 111).

Creativity and innovation did not prove to be obstacles to the Chinese in earlier periods (Kriz & Cunneen 2011; Kriz & Keating 2010). As Drucker (1994) and others like Kanter (1988) highlight, these skills can be taught and learned. Breaking traditions around hierarchy, observance, face, and harmony, however, is going to take considerable revamping of organizations and leadership practices. The Chinese have a propensity for innovation, so undoubtedly innovation champion behavior exists. Deng Xiaoping provides an interesting example. He confronted Mao himself and was prepared to tackle and champion serious issues he believed important to the welfare and prosperity of the Chinese people. He was eventually made leader after Mao's death and later proved to be quite a visionary. His earlier exploits were not in vain, as his rebuttal of previous policies helped build a level of credibility that enabled his rise and a new pathway for China.

## Harnessing the university and government innovation champions

An area where innovation champions are potentially prevalent in China is in the university sector. China has an extensive program to promote innovation, including allocation of 2.5 percent of GDP to R&D. The relationship between university and industry is among the known strengths of China. Lenovo is an example, having been established from seed funds (US $25,000) given from the Chinese Academy of Sciences to a group of researchers (Zhou et al. 2010). Universities are encouraged to work with industry to promote innovation and R&D. Prominent universities like Tsinghua, Fudan, and Zhejiang are known to have strong networks, with academics consulting for business and using these links to foster development of their institutions. However, considerable slippage seems to be occurring. Instead of playing the role of innovation champion or corporate entrepreneur for the university, many academics are becoming independent entrepreneurs in their own right. The list of companies that have emerged out of such behavior and R&D is extensive. The positive spillover effects should not be underestimated. Silicon Valley and Cambridge in the United Kingdom are examples of how clusters have emerged from similar links. Discouraging such independent start-ups to protect potential university intellectual property (IP) would likely be a mistake for China. Independent start-ups are one way of encouraging important spillover outcomes without being trapped in top–down hierarchies.

A unique aspect of innovation champions in China is related to the complex interaction between government and industry. Economic development of China in the last 30 years has been largely due to government initiatives to privatize land and promote foreign direct investment (FDI). Inevitably the roles of government and entrepreneurs have been "messy." Many ventures have stemmed from innovation champions in the public sector. Chery Automobile provides one of many examples. "Eight Guardians," or government-cum-entrepreneurial officials, were responsible for helping this very successful automobile initiative out of Wuhu (Lairson 2010). This intertwined relationship between entrepreneurship and government is complex. Huawei, for example, is a private company that was established by Ren Zhegfei in Shenzhen. Although this is a private company, many of their contracts are linked to telecommunication strategies of the national government and the military and its research institute. This crossover of government to entrepreneur is not unusual in China; it allows for a new breed of innovation champion. These champions-cum-brokers initiate activities that lead to new business opportunities. Thus far, the work on Chinese innovation champions is largely anecdotal; more emic empirical investigations of SOEs, POEs, university, and government are overdue.

## 5   Managing innovation champions in China

Identifying innovation–champion behavior is worthwhile and can have its own benefits, but the real benefits come from nurturing, harnessing, and managing such activities. This becomes particularly interesting when we investigate other cultures and environments. Management styles differ, as discussed above, in China, and fostering innovation is known to be hierarchical, with an emphasis on power-distance. This type of innovation is not unusual, and is associated with greater supervision and a more top–down, hands-on focus (Kelley & Lee 2010). A reward system that encourages innovation has the potential to work in China. However some paradigm shifting may be needed and it requires good leadership. "Innovation champion teams" are likely to be more effective as they support in-group behavior and equalize aspects like "face" (Fernandez & Underwood 2006). Performance-based measurement for innovation (Kelley et al. 2011) is a way of incentivizing innovation–champion behavior. A similar incentive norm has become common among Chinese firms, particularly among foreign-owned MNCs. This can be effective in China, if managed carefully with appropriate rewards and due diligence over aspects like risk.

As mentioned above, the concept of *guanxi* is endemic in China and trade secrets are easily exchanged and lost. Corporate-related activity often ends in independent entrepreneurship in China. One aspect of China that is not difficult to foster is cross-boundary exchange and boundary spanning (Marion & Uhl-Bien 2001; Van Velsor 2008). *Guanxi* encourages connections to workplaces. Encouraging similar links in the West has been suggested as a way of improving innovation champions' performance (Taylor et al. 2011). Moving people through roles in the organization to increase the exchange of ideas and to encourage champion activities has also been advocated in the West (Greene et al. 1999). Job rotation may be difficult in China, but has merit. If China is going to be able to achieve its ambitious targets for innovation, setting up better structures for innovation champions to flourish is a likely key.

Kanter (1988) used the analogy of "When a Thousand Flowers Bloom: Structural, Collective, and Social Conditions for Innovation in Organizations" as a way of describing the many permutations and combinations involved in the innovation process for a firm. The top–down approach to management in China has a long history and will be difficult to change. If it leads to champions of innovation, this is likely to be highly advantageous. Nevertheless, it is also important for managers in China to start empowering staff and to reward creativity and innovation (Kriz & Cunneen 2011). Chinese executives must support their words with tangible actions. Recognizing the important role played by innovation champions and fostering rewards to encourage such activities will have significant benefits. Fostering innovation champion teams would be an excellent start.

Risk-taking is an activity that most Western companies also find difficult (Hornsby et al. 1993; Howell & Higgins 1990b). The Chinese are not averse to risk in entrepreneurship, but encouraging this in an organization is a whole new concept. Obedience to the top–down hierarchy makes this a radical pursuit. However, paradox is common in China and it would not be out of the ordinary to see the odd company breaking with convention. Individuals with a predisposition for championing have been identified in the West by analyzing their personality characteristics and leadership behavior (Howell & Higgins 1990a). Champion "breeding" companies in the West are now known to be selecting and recruiting champions, despite the difficulty of managing such individuals. These firms focus on how they hire, picking people with the appropriate attitude. The characteristics a Western company looks for in a champion include flexibility, self-awareness, internal drive and responsibility, a wide-ranging knowledge base, and substantial networking skills. Some of these factors are not inherent in the cultural norms of China, but would be apparent in some individuals. Such propensities can be tested through appropriate diagnostic tools. Pushing the champions to take on challenging assignments and roles and mentoring them in handling politics and strategic roadblocks is part of the process (Howell 2005). These are all aspects China can investigate in the pursuit of encouraging champion behavior.

Innovation champions and, in the China case, innovation champion teams need to be encouraged to stay within organizational objectives if firms want to benefit from champion behavior. Managing champions is ultimately done around balancing the organizational focus but allowing adequate scope. Retaining innovation champions in China is likely to be closely linked with extrinsic and intrinsic rewards for champions, like being valued by their CEOs and their immediate bosses (Tian 2007). As already mentioned, the Chinese are experts at networking, and this should be encouraged (Howell & Boies 2004), but the networks need fostering around *guanxi* toward the organization rather than simply toward the individual. The problem in China does not seem to be a propensity for entrepreneurial behavior, but the lack of corporate control and management over such activities.

## 6   Future research directions

Research on innovation champions is disparate and lacks a consistent body of empirical support. Much of the current research relies on self-reported data and cross-sectional studies and lacks replication (Howell et al. 2005). The findings presented by key authors in the innovation champion space provide excellent starting points; it is important for future research to extend and replicate these studies. Specifically, the work of Shane (1994c, 1994, 1995) offers important insights regarding the influence of culture on championing behavior. As Shane (1995) highlights, studies that uncover *actual* champion behavior, rather than preferences, in the cross-cultural context are needed. Championing activities and success appear to be highly contextual. It is important to consider the mediating effects that organizational structure, culture, and politics have on championing activities. In addition, as Frost and Egri (1991) suggest, how the innovation itself fits into the organizational context is a highly relevant matter for review. Closer examination of championing activities that includes a comprehensive analysis of the internal and external setting is a key area for future research.

The identification of champions is a difficult process; their actions are not easily quantifiable and identifying champion behavior can be quite subjective. It is important nonetheless to work towards a model or models to identify champions working in organizations and those with the potential to be successful. This will have many theoretical and practical implications. It may also be pertinent for researchers to further evaluate the effect that formal recognition and promotion have on champion behavior and success (Howell & Shea 2006). Such a schema could be useful in recruitment, providing organizations with a basis for selecting individuals who may help stimulate innovative activity. Developing one model that identifies innovation champions may not be possible, and as discussed above, it is essential to include context. More research is required before conclusions can be drawn on this possibility. Testing the models developed in one organizational and/or cultural context against another context will be an important step forward. Markham and Aiman-Smith (2001) identify the lack of empirical support linking the involvement of a champion with successful outcomes as a significant gap in the literature. While it is consistently understood that champions are important, the exact nature of their contribution is yet to be defined.

A review of the China domain brings in many new elements about innovation champion research. In many ways, the waters are uncharted. This has inherent problems, but allows for a more inductive review and for building theory from the ground up. The notion of an emic-to-emic review, starting across regions as a first step, is critical, given the variations apparent. Theory testing in this domain requires further evidence in the organizational areas discussed. Expertise in the Chinese market and cooperation with appropriate research bodies are required before embarking on such an ambitious program. The area obviously makes for an exciting field to explore and has significant implications.

## 7  Conclusions

Innovation management is often investigated from the perspective of processes, products, and elements like coordination and structure. Ultimately, people are responsible for innovation management; too often this is forgotten in the pursuit of underlying enterprise and institutional elements. One of the key protagonists or agents of innovation management has been identified as the innovation champion. Although much research has been undertaken on these key change agents, there is still considerable work to be undertaken to identify the important elements that make these people do what they do. This chapter has reviewed and investigated a cross-section of the literature that is available within this area. Like entrepreneurs, innovation champions are known to deviate from more common organizational norms. This has the potential to ostracize them from the mainstream. However, as innovation becomes increasingly accepted as critical for organizational, regional, and national development, more companies will seek out the exception rather than the mainstream. Innovation champions have an inherent capacity for seeing what others do not and for making connections where others cannot.

Tapping into these individuals and their behaviors is not simple. These complex behaviors require considerable further investigation. Thus far, most of this research has occurred in North America. Increasingly, the value of innovation champions has taken researchers into new domains. The international market is a natural evolution for such a pursuit. Interesting findings are coming from this expansion of the literature. However, as the chapter has shown, too much of this investigation has been at the etic, cross-cultural level. This is an appropriate lens from which to begin; but to understand markets like China, it is time to probe further and go within (emic). This chapter has commenced such an investigation, uncovering some important institutional and cultural variants and idiosyncrasies around innovation champions in China.

The Chinese have a distinctive hierarchical emphasis in their organizational dynamics. Uncovering this is not new, but the implications for China are extensive. Innovation champions in the West do confront similar organizations, but rarely do the conventions relate to such deep philosophical undertones. China wants to be innovative and has identified the need to push for more radical innovation as it pursues more advanced economies like the United States. How Chinese organizations can combat this weakness has been discussed. The use of innovation champion teams has been recommended as an alternative where cultural norms are likely to be slow to change. Supporting the group has key benefits in terms of face and in-group harmony. The juxtaposed roles of government and business have important implications, and the innovation champion is as likely to be found in this domain as in other forms of enterprise. Organizational structures have also seen major recent changes in China, and private enterprise is more common. The role of innovation champion is likely to become more important as these companies develop. Spillover impacts and absorption of new ideas from "sea turtles" and foreign MNCs is going to speed up acceptance. Empirical research is still required in the innovation champion domain. Bringing China into view will broaden and deepen the results of such endeavors.

## References

Achilladelis, B., Jervis, P., & Robertson, A. (1971) *A Study of Success and Failure in Industrial Innovation*. Sussex, England: University of Sussex Press.

Ambler, T., Witzel, M., & Xi, C. (2009) *Doing Business in China* (3rd edn). New York: Routledge.

Anderson, L. M., & Bateman, T. S. (2000) Individual environmental initiative: Championing natural environmental issues in U.S. business organizations. *Academy of Management Journal*, 43(4), 548–570, doi: 10.2307/1556355

Bartlett, D., & Dibben, P. (2002) Public sector innovation and entrepreneurship: Case studies from local government. *Local Government Studies*, 28(4), 107–121, doi: 10.1080/714004159

Berry, J. W. (1989) Imposed etics-emics-derived etics: The operationalization of a compelling idea. *International Journal of Psychology*, 24, 721–735.

Chakrabarti, A. (1974) The role of champion in product innovation. *California Management Review*, 17(2), 58–62.

Crossan, M. M., & Apaydin, M. (2010) A multi-dimensional framework of organizational innovation: A systematic review of the literature. *Journal of Management Studies*, 47(6), 1154–1191, doi: 10.1111/j.1467–6486.2009.00880.x

Day, D. L. (1994) Raising radicals: Different processes for championing innovative corporate ventures. *Organization Science*, 5(2), 148–172.

Dodgson, M., Gann, D. M., & Salter, A. (2008) *Management of Technological Innovation: Strategy and Practice*: Oxford University Press.

Drucker, P. (1994) *Innovation and Entrepreneurship : Practice and Principles* (Rev edn). Oxford: Butterworth-Heinemann.

Dutton, J. E., & Jackson, S. E. (1987) Categorizing strategic issues: Links to organizational action. *Academy of Management Review*, 12(1), 76–90.

Fang, T. (1999) *Chinese Business Negotiating Style*. Thousand Oaks, California: Sage Publications.

Fang, T. (2006) From "onion" to "ocean". *International Studies of Management & Organization*, 35(4), 71–90.

Fernandez, J. A., & Underwood, L. (2006) *China CEO: Voices of Experience from 20 International Business Leaders*. Singapore: John Wiley & Sons (Asia) Pte Ltd.

Fetscherin, M. (2010) Haier. In W. Zhang & I. Alon (Eds.), *A Guide to the Top 100 Companies in China*. Singapore: World Scientific Publishing Co. Pte. Ltd.

Frost, P. J., & Egri, C. P. (1991) The political process of innovation. *Research in Organizational Behavior*, 13, 229–295.

Greene, P. O., Brush, C. G., & Hart, M. M. (1999) The corporate venture champion: A resource-based approach to role and process. *Entrepreneurship: Theory & Practice*, 23(3), 103–122.

Gumusluoglu, L., & Ilsev, A. (2009) Transformational leadership, creativity, and organizational innovation. *Journal of Business Research*, 62(4), 461–473, doi: 10.1016/j.jbusres.2007.07.032

Hofstede, G., Neuijen, B., Ohayv, D. D., & Sanders, G. (1990) Measuring organizational cultures: A qualitative and quantitative study across twenty cases. *Administrative Science Quarterly*, 35(2), 286.

Hornsby, J. S., Naffziger, D. W., Kuratko, D. F., & Montagno, R. V. (1993) An interactive model of the corporate entrepreneuring process. *Entrepreneurship Theory and Practice*, 17, 29–37.

Howell, J. M. (2005) The right stuff: Identifying and developing effective champions of innovation. *Academy of Management Executive*, 19(2), 108–119.

Howell, J. M., & Boies, K. (2004) Champions of technological innovation: The influence of contextual knowledge, role orientation, idea generation, and idea promotion on champion emergence. *Leadership Quarterly*, 15(1), 123–143.

Howell, J. M., & Higgins, C. A. (1990a) Champions of change: Identifying, understanding, and supporting champions of technological innovations. *Organizational Dynamics*, 19(1), 40–55.

Howell, J. M., & Higgins, C. A. (1990b) Champions of technological innovation. *Administrative Science Quarterly*, 35(2), 317.

Howell, J. M., & Higgins, C. A. (1990c) Leadership behaviors, influence tactics, and career experiences of champions of technological innovation. *The Leadership Quarterly*, 1(4), 249–264.

288 *Anton Kriz et al.*

Howell, J. M., & Shea, C. M. (2001) Individual differences, environmental scanning, innovation framing, and champion behavior: Key predictors of project performance. *Journal of Product Innovation Management,* 18(1), 15–27.

Howell, J. M., & Shea, C. M. (2006) Effects of champion behavior, team potency, and external communication activities on predicting team performance. *Group and Organization Management,* 31(2), 180–211.

Howell, J. M., Shea, C. M., & Higgins, C. A. (2005) Champions of product innovations: Defining, developing, and validating a measure of champion behavior. *Journal of Business Venturing,* 20(5), 641–661.

Jolly, V. K. (1997) *Commercializing New Technologies: Getting from Mind to Market.* Boston: Harvard Business School Press.

Kanter, R. M. (1988) When a thousand flowers bloom: Structural, collective, and social conditions for innovation in organizations. In B. M. Staw & L. L. Cummings (Eds), *Research in Organizational Behavior.* Greenwich: J.A.I. Press.

Kelley, D., & Lee, H. (2010) Managing innovation champions: The impact of project characteristics on the direct manager role. *Journal of Product Innovation Management,* 27(7), 1007–1019.

Kelley, D., O' Connor, G., Neck, H., & Peters, L. (2011) Building an organizational capability for radical innovation: The direct managerial role. *Journal of Engineering and Technology Management,* 28, 249–267.

Kriz, A. (2009) *Secrets to Building Personal Trust in China: An In-Depth Investigation of the Chinese Business Landscape.* Saarbrucken: Lambert Academic Publishing.

Kriz, A., & Cunneen, D. (2011) China's next big challenge: Mastering radical technology. *Journal of Science and Technology Policy in China,* 3(1), 6–25.

Kriz, A., & Keating, B. (2010) Lessons from China about deep trust. *Asia Pacific Business Review,* 16(3), 299–318.

Lairson, T. D. (2010) Chery automobile. In W. Zhang & I. Alon (Eds), *A Guide to the Top 100 Companies in China.* Singapore: World Scientific Publishing Co. Pte. Ltd.

Maidique, M., A. (1980) Entrepreneurs, champions, and technological Innovation. *Sloan Management Review (pre-1986),* 21(2), 59.

Marion, R., & Uhl-Bien, M. (2001) Leadership in complex organizations. *The Leadership Quarterly,* 12(4), 389–418. doi: 10.1016/s1048–9843(01)00092–3

Markham, S. K. (1998) A longitudinal examination of how champions influence others to support their projects. *Journal of Product Innovation Management,* 15(6), 490–504.

Markham, S. K., & Aiman-Smith, L. (2001) Product champions: Truths, myths and management. *Research Technology Management,* 44(3), 44–50.

Markham, S. K., & Griffin, A. (1998) The breakfast of champions: Associations between champions and product development environments, practices and performance. *Journal of Product Innovation Management,* 15(5), 436–454.

Michael, L. T., & Thomas, J. S. (1981) Characteristics and external orientations of boundary spanning individuals. *Academy of Management Journal,* 24(1), 83.

Morris, M., Kuratko, D., & Covin, J. (2008) *Corporate entrepreneurship and innovation : entrepreneurial development within organizations* (2nd edn). Mason: Thomson South-Western.

Mumford, M. D., & Licuanan, B. (2004) Leading for innovation: Conclusions, issues, and directions. *The Leadership Quarterly,* 15(1), 163–171, doi: 10.1016/j.leaqua.2003.12.010

Nie, W., Xin, K., & Zhang, L. (2009) *Made in China: Secrets of China's Dynamic Entrepreneurs.* Singapore: John Wiley & Sons (Asia) Pte. Ltd.

Pike, K. L. (1967) *Language in Relation to a Unified Theory of Structure of Human Behavior* (2nd edn). The Hague: Mouton.

Sarros, J. C., Cooper, B. K., & Santora, J. C. (2008) Building a climate for innovation through transformational leadership and organizational culture. *Journal of Leadership & Organizational Studies*, 15(2), 145–158, doi: 10.1177/1548051808324100

Schon, D. A. (1963) Champions for radical new inventions. *Harvard Business Review*, 41(2), 77–86.

Shane, S. A. (1994a) Are champions different from non-champions? *Journal of Business Venturing*, 9(5), 397–421.

Shane, S. A. (1994b) Championing innovation in the global corporation. *Research Technology Management*, 37(4), 29–35.

Shane, S. A., (1994c) Cultural values and the championing process. *Entrepreneurship: Theory & Practice*, 18(4), 25–41.

Shane, S. A., (1995) Uncertainty avoidance and the preference for innovation championing roles. *Journal of International Business Studies*, 26(1), 47.

Shane S. A., & Venkataraman, S. (1996) Renegade and rational championing strategies. *Organization Studies (Walter de Gruyter GmbH & Co. KG.)*, 17(5), 751.

Shane, S. A., Venkataraman, S., & MacMillan, I. C. (1995) Cultural differences in innovation championing strategies. *Journal of Management*, 21(5), 931–952.

Sim, E. W., Griffin, A., Price, R. L., & Vojak, B. A. (2007) Exploring differences between inventors, champions, implementers and innovators in creating and developing new products in large, mature firms. *Creativity and Innovation Management*, 16(4), 422–436.

Smith, P. B., & Bond, M. H. (1993) *Social Psychology across Cultures: Analysis and Perspectives*. Boston: Allyn & Bacon.

Sturtevant, W. C. (1964) Studies in ethnoscience. *Transcultural Studies in Cognition [special issue]*, 66, 99–131.

Styles, C., & Ambler, T. (2003) The co-existence of transaction and relationship marketing: Insights from the Chinese business context. *Journal of Industrial Marketing Management*, 32(8), 633–642.

Tan, Y. (2011) *Chinnovation: How Chinese Innovators Are Changing the World*. Singapore: John Wiley & Sons (Asia) Pte. Ltd.

Taylor, A., Cocklin, C., Brown, R., & Wilson-Evered, E. (2011) An investigation of champion-driven leadership processes. *The Leadership Quarterly*, 22(2), 412–433.

Tian, X. (2007) *Managing International Business in China*. Cambridge: Cambridge University Press.

Trompenaars, F., & Hampden-Turner, C. (1997) *Riding the Waves of Culture : Understanding Cultural Diversity in Business* (2nd edn). London: Nicholas Brealey Publishing.

Tung, R., Worm, V., & Fang, T. ( 2007) Sino-Western business negotiations revisited-30 years after China's open door policy. *Organizational Dynamics*, 37(1), 60–74.

Tushman, M. and Nadler, D. (1986) Organizing for innovation. *California Management Review*, 28(3), 74–92.

Van de Ven, A. H., Polley, D. E. P., Garud, R., & Venkataraman, S. (1999) *The Innovation Journey*. Oxford University Press.

Van Velsor, E. (2008) A complexity perspective on leadership development. In M. Uhl-Bien & R. Marion (Eds), *Complexity Leadership, Part 1: Conceptual Foundations* (pp. 333–346). Charlotte, North Carolina: Information Age Publishing Inc.

Zhou, S., Ren, B., & Sun, S. L. (2010) Lenovo group. In W. Zhang & I. Alon (Eds), *A Guide to the Top 100 Companies in China*. Singapore: World Scientific Publishing Co. Pte. Ltd.

# 13

# The "Frugal" in Frugal Innovation

*Preeta M. Banerjee*

## 1 Introduction

Frugal innovation is a new paradigm for appropriate design in developing markets like India. As India becomes a rising source of global innovation, the practice of frugality in innovation has become an important development in the field of innovation and technology management for academics, practitioners, and policy makers. Govindarajan and Ramamurti (2011, p. 191) write: "poor, emerging markets no longer just borrow innovations from developed countries; from time to time they also contribute innovations to the rest of the world, including developed countries." Frugal innovation captures the economic, physical, emotional, and experiential realities of domestic markets in emerging economies by creating affordable, basic innovations that every person can use (Kumar & Puranam 2011). Frugal innovation has been examined as the India Way (Cappelli et al. 2010) and India Inside (Kumar & Puranam 2011).

In the United States, the spread of frugal innovation began with a 2009 *Business Week* article by Reena Jana, which compared frugal innovation to a Hindi slang word, *jugaad* (pronounced "joo-gaardh"), which translates to an improvisational style of innovation that's driven by scarce resources and attention to a customer's immediate needs, not his or her lifestyle wants ... Moreover, because *jugaad* essentially means "inexpensive invention on the fly," it can imply cutting corners, disregarding safety, or providing shoddy service. "Jugaad means 'Somehow, get it done,' even if it involves corruption," cautions M. S. Krishnan, a Ross Business School professor. "Companies have to be careful. They have to pursue *jugaad* with regulations and ethics in mind."[1] Also known as Indovation (Indian innovations), frugal innovation is brought about by a complex environment, diversity, interconnectivity, velocity, ambiguity, and scarcity (Radjou et al. 2012). As frugal innovation addresses the need to do more with less, there have been two additional concepts in this space that seem related to *jugaad* and merit being discussed and distilled in order to understand the essence of frugal innovation. These two additional concepts include reverse innovation (Immelt et al. 2009) and Gandhian engineering (Prahalad & Mashelkar 2010). Reverse innovation is the sourcing of innovation from the global South to the global North. Gandhian engineering is the science of breaking up complex engineering processes and products into basic components and then rebuilding in the most economical way. Comparing *jugaad* to these two other similar concepts, propels us to ask "What is the 'frugal' in frugal innovation?" Is it just economical efficiency or are there social and environmental aspects as well?

By answering this question, this chapter makes the contribution of understanding innovation under conditions of scarcity. By comparing and contrasting three important concepts in frugal innovation (reverse innovation, Gandhian engineering, and *jugaad*), innovation and technology management can find an alternative to the more resource-rich innovation models of developed countries. At a time when innovation is a key issue in our world economic and political landscape, frugal innovation is a shining example of what is possible. While most developed-country innovators have the fortune to able to choose to work within tight constraints, those in developing countries have to grapple with the problem every day. In addition, by finding the essence of this new model of frugal innovation, innovation can move beyond economic efficiency towards the triple bottom line. The triple bottom line (Elkington 1994; Savitz 2006) involves capturing values and criteria for measuring success beyond economic ones and has changed the way businesses, nonprofits, and governments measure the performance of projects or policies. Beyond measuring sustainability on three fronts – people, planet, and profits – the flexibility of triple-bottom-line thinking allows organizations to apply the concept to processes as well as the outcomes of those process. This includes the innovation process, which can be improved by looking beyond economic efficiency. The frugal innovation process can embody adaptability and constant analysis with an eye towards sustainability: economic, social, and environmental.

In what follows, we review the three concepts related to frugal innovation: reverse innovation, Gandhian engineering, and *jugaad,* to summarize what we do know about frugality. Then, given that concepts of frugal innovation grew from the popular press and are not yet prevalent in academia, we follow Mazza and Alvarez (2000) to view the popular press as the arena where the legitimacy of management ideas are confirmed. The popular press has gone beyond the mere diffusion of ideas to the co-production and legitimation of management practices. Thus, we analyze popular press articles using LexisNexis[2] to understand the essence of frugal innovation. We conclude by summarizing the findings and providing avenues for future research as well as noting implications for academics, practitioners, and policy makers.

## 2   Background

The topic of innovation has typically been the realm of developed countries, like the United States, Japan, and the United Kingdom. However, developing countries are quickly catching up, not out of greed, but out of need. As Vossoughi (2012) writes: "When times are good, businesses tend to stick with what's working. "It's served us well so far," goes the reasoning, "so why mess with success?" We fall into ruts where we do not question whether the approach we are using is truly the best one – we just work to push it further. It's during these periods that needless layers are piled on: product lines expand, often without good reason; processes become more detailed and entrenched; project teams grow larger; and more people are involved in approving and completing every task. Companies, like households, accumulate stuff in times of abundance. When an economic downturn hits, the first response is to keep doing all the same things, but faster, with fewer people. To a certain point, this is good. It forces us to jettison extraneous expenses and activities and become more efficient. But in a long downturn like the one we are currently living through, we eventually run out of things to cut or streamline. That is when smart companies shift focus. Instead of trying desperately to get more efficient, they realize they have to get more effective. They seek out fundamental change. They question core assumptions: how and why they pursue every activity, who their customers are, even why they exist as a company in the first place. They treat planning efforts as creative opportunities for real improvement. They put lean, focused teams on each project, involving only the most necessary people. They urge these teams to make bold decisions and move forward. They behave, in other words, like innovators.

Developing countries are nations with a low living standard, undeveloped industrial base, and low Human Development Index (HDI) relative to other countries (Sullivan & Sheffrin 2003). For example, there are an estimated 4 billion people in the developing world who live on less than $2 per day, known as the bottom of the pyramid. Misconceptions about the bottom of the pyramid have kept this population largely underserved of products and services (Prahalad 2005). In such cases, scarcity is the mother of innovation, frugal innovation. Frugal innovation is a business model characterized by the use of limited resources to create low-cost products – from $35 tablet computers to $3,000 cars – that are sustainable for the environment and individual communities. Perhaps the most important characteristic of this business model is that it challenges the established innovation process. In developed countries, companies use highly structured and costly R&D to create new products (Table 13.1). To meet the price restrictions of developing markets, companies typically downsize those products with cheaper – and often flimsier – materials.

As an alternative to downsized innovation from the Western world, frugal technology has been characterized by eight core competencies (Table 13.2): ruggedization, affordability, simplification, adaptation, reliance on local

*Table 13.1* Alternative, complementary models of innovation

| Characteristics | Frugal innovation | Traditional innovation |
|---|---|---|
| Where | Developing, emerging markets | Developed markets |
| Starting line | What do they need? | What would be great to have? |
| Product | Functionality – rugged, lightweight, adaptable, simple | Desirability and design |
| Process | Bottom–up | Top–down |

*Table 13.2* Eight core competencies of frugal technologies

| Core competency | Definition | Example |
|---|---|---|
| Ruggedization | Designed with materials that can consistently operate in harsh physical environments (with heat, moisture, pests) | Redesigning a medical instrument that is used in US hospitals so that it can be carried on a backpack to a remote village that is without grid power. |
| Affordability | Designed to be purchased by very low-income communities, where economic markets are still developing | Distribution and marketing strategies based on high volume, low unit cost. |
| Simplification | Designed without the added features and functionality that are used to market products in the United States and Europe. | A water purifier that uses an agricultural waste product impregnated with nano-silver particles, thus requiring no electric power or running water |
| Adaptation | Designed by adapting from existing products | A bicycle-powered dynamo originally used to charge a headlamp is modified to charge mobile phone batteries. |
| Reliance on local materials, manufacturing | Designed and manufactured without importing equipment or materials | A prosthetic limb using locally available resources and the skills of local artisans. |
| Renewability | Designed to be powered by renewable resources | Solar-powered ATMs and mobile phones; wind turbines for communities without grid power. |
| User-centric design | Designed to be used by semi-literate people | Mobile health information technologies that use symbols and colors rather than text. |
| Portability | Designed to be carried by human beings through unreliable transportation systems | A disaster relief kit that can be carried in a suitcase |

*Source:* adapted from http://www.scu.edu/socialbenefit/innovation/frugal/corecomp.cfm

materials and manufacturing, renewability, user-centric design, and portability. Ruggedization refers to technology designed with materials that can consistently operate in harsh physical environments (with heat, moisture, pests) – for example, redesigning a medical instrument that is used in US hospitals so that it can be carried on a backpack to a remote village without grid power. Affordability refers to technology designed to be purchased by very low-income communities, where economic markets are still developing; this might mean, for example, having distribution and marketing strategies based on high volume and low unit cost. Simplification refers to technology designed without the added features and functionality that are used to market products in the United States and Europe – for example, a water purifier that uses an agricultural waste product impregnated with nano-silver particles, thus requiring no electric power or running water. Adaptation refers to technologies that can be adapted from existing products. For example, a bicycle-powered dynamo originally used to charge a headlamp may be modified to charge mobile phone batteries. Local sourcing refers to technology can be designed and manufactured without importing equipment or materials – for example a prosthetic limb made using locally available resources and the skills of local artisans. Renewability refers to technologies that can be powered by renewable resources, for example solar-powered automated teller machines (ATMs) and mobile phones or wind turbines for communities without grid power. User-centric refers to technology that can be used by semi-literate people – for example, mobile health information technologies that use symbols and colors rather than text. Portability refers to technology that can be carried by human beings through unreliable transportation systems – for example, a disaster relief kit that can be carried in a suitcase. In summary, frugal technology results in no-frills, good quality, functional products that are also affordable to the customer with modest means.

However, these eight core competencies only focus on outcomes of technologies and resulting products, not on processes and the resulting frugal services. While frugal innovation has been linked with *jugaad* (Radjou et al. 2012) and the creativity that results in economical efficiency, we are left to wonder about the social and environmental aspects. Is it just the economical efficiency of *jugaad* that defines frugal innovation? We are still left wondering what is the essence of frugal innovation. Thus, we need to review the practice of frugal innovation related to three separate but connected phenomena in innovation: (1) reverse innovation ; (2) Gandhian engineering and (3) *jugaad*.

Reverse innovation is the opposite of "glocalization," which postulates that great products are created and produced in developed countries and then distributed worldwide with some adaptations to local conditions, including making cheaper (low price, low performance) models for developing countries. Instead, the directionality of international capital, knowledge flows, and human talent shifts and is from the global South to the global North (Sarkar 2011). Innovations occuring in emerging economies tend not to involve technological breakthroughs of the kind that drive innovation in developed countries (Govindarajan & Ramamurti 2011). Instead, they involve novel and innovative combinations of existing knowledge and technologies to solve pressing local problems. General Electric (GE) is one of the main benefactors of reverse innovation; in 2009, GE announced $3 billion to create at least 100 health-care innovations that would substantially lower costs, increase access, and improve quality (Immelt et al. 2009). Two of the GE products that have already been developed are a $1,000 handheld electrocardiogram (ECG) device for rural India and a $15,000 ultrasound machine for rural China. GE needs such innovations not only to expand beyond serving high-end segments in India and China but also to preempt local companies in those countries from creating similar products and using them to disrupt GE in developed countries.

Innovation can also elevate all of humanity, including the poor, to be seen as both creative entrepreneurs and as value-demanding consumers. The phrase "bottom of the pyramid" is used in particular by researchers developing new models of doing business that deliberately target the "poor" demographic class, often using new product innovation (Prahalad 2005). As aforementioned, the bottom of the pyramid is the largest and poorest socio-economic group, which in global terms constitutes the four billion people who live on less than $2 per day, typically in developing countries. In the case of the bottom of the pyramid, the outcome of product innovation can bring quality of life improvements if innovative firms work with civil society organizations and local governments to create new local business models. However, bringing materialism to the masses gives fuel to the fire of unsustainable consumption. Bottom-of-the-pyramid product innovation needs to develop sustainable products and services so that human needs can be met not only in the present, but also for future generations.

Prahalad and Mashelkar (2010) suggest that a new way of making new products and services is through getting more from less for more people, or "Gandhian innovations." In other words, each innovation should be for the benefit of the common person (inclusive), and not only for people that can afford it at a high price (exclusive). Innovation must be MLM (more from less for many).The challenge with Gandhian innovation is trying to overcome the concept of "low price, low performance" by having "low price, high performance." Success stories such as Tata's Nano (the cheapest car in the world today, price a little more than US $2,000) or the Jaipur foot (a high-performance prosthetic leg at a cost of approximately US $30) demonstrate the validity of the Gandhian innovation concept (Anderson & Markides 2007).

The bottom-of-the-pyramid approach to targeting the poorest people in developing countries appears counterintuitive to the fundamental business philosophy of capturing and sustaining value. In comparing the three concepts (reverse innovation, Gandhian engineering, and *jugaad*) to understand frugal innovation, we can look at the model examples and summarize their main characteristics as illustrated in Table 13.3. GE's ECG machine is a model of reverse innovation. The ECG is the most widely performed cardiac test in the developed world, and GE Healthcare is the market leader. A premium ECG machine costs about $5,000 and a scan costs about $20. This complex equipment is heavy and bulky and requires a skilled technician to operate it, as well as elaborate service support. GE's premium ECG machines were nonstarters in rural India, because patients did not have the money to pay for the test and small clinics and physicians could not afford the machine or the support costs. These constraints defined the sandbox for GE Healthcare to develop an $800 ECG machine for rural India that is portable, battery-operated, easy to use, and easy to repair. GE found many ways to cut costs. The high-end machine was custom-designed, so GE built a machine using commodity components, realizing huge cost advantages. For a cost-effective printer to print out the ECG results, GE used the kind of ticket printer found on public buses and in movie theaters. Since these printers are produced in the millions, GE could enjoy significantly lower costs, due to economies of scale. The small printer reduced the weight of the machine – it weighs less than a can of Coca-Cola – and helped make it portable. By eliminating the monitor, GE reduced the need for huge power consumption. This, in turn, contributed to longer life for the rechargeable battery. GE also found ways to build a machine that can employ the same signals as traffic signals: A green button indicates START and a red button means STOP. GE did away with the need for extensive service support, designing the machine with a few standard modules that can be easily replaced. If the device fails, users can swap modules.

Tata's Nano is an example of Gandhian engineering. Tata Motors focused on the two main ways Gandhian innovators solve problems – acquiring or developing technologies and altering business models or capabilities – when it created the world's cheapest car. In developing the Nano, Tata broke down the complex automobile into fundamental parts and rebuilt it in the most efficient manner possible. Ratan Tata, the head of Tata Group, had a dream of getting poor Indians off the motorbikes to which whole families precariously cling and into shiny new cars they could buy for 1 lakh (100,000 rupees), or about $2,200. Although the car is assembled at plants, it is much easier to assemble and disassemble than other complex automobiles. Thus Tata envisioned distributing flat packs of parts to rural mechanics who would become successful entrepreneurs by assembling the kits into complete cars in the heartlands, where 750 million poor Indians live.

For jugaad innovation, an exemplar is YES Bank's credit appraisal toolkit (CAT). This innovation was one of the many targeted at the 600 million

*Table 13.3* Literature review of frugal innovation concepts

| Key concept | Author(s) | Year | Example | Innovation outcome |
|---|---|---|---|---|
| Reverse innovation | Immelt, Govindarajan, and Trimble | 2009 | GE's electrocardiogram (ECG) | Ability to add bells and whistles to developing country innovation that competes in developed countries. |
| Gandhian Engineering | Prahalad and Mashelkar | 2010 | Tata's Nano | Ability to break down a complex process/product and rebuild in the most economical way. |
| Jugaad | Radjou, Prabhu and Ahuja | 2012 | YES Bank's credit approval toolkit (CAT) | Ability to innovate with an eye toward economical efficiency but also in response to social and environmental scarcities |

Indians who had no access to a bank. The CAT toolkit was a solution to the problem of credit appraisal of micro-entrepreneurs, who neither maintain formal business records nor file business details with the authorities (Radjou et al. 2012). CAT is an Excel-based data analysis tool that compares details orally provided by the applicant to information from their peers. But beyond economic efficiency and the resulting profitablity is the social sustainability aspect – this innovation is not CSR (corporate social responsibility) but a core part of the business model.

## 3   The essence of frugal innovation

### Issues, controversies, problems

In comparing the three concepts found in management literature, we can see that there is something beyond economic efficiency that characterizes frugal innovation. As aforementioned, the phrase "frugal innovation" is often used interchangeably with the Hindi word "*jugaad*," which originally referred to North Indian cars that were pieced together out of spare parts. For this reason, commentators often disparage *jugaad* as a "quick fix." Beyond economic efficiency, what about the social and environmental aspects of frugal innovation? Social and environmental aspects are particularly important to investigate in addition to economic aspects in light of the triple-bottom-line performance requirements, i.e. measurement and improvement of economic, social and environmental sustainability require-ments (Hubbard 2009).

*Process and product innovation*

Innovation has typically been dichotomized into product or process inno-vation (Francis & Bessant 2005). Product innovation is exemplified by novel physical outputs of an organization, or changes in things (products/ services) that an organization offers. These are what we have focused on so far in comparing and contrasting the three key concepts of reverse innova-tion, Gandhian engineering, and *jugaad* to understand frugal innovation. Process innovation is exemplified by a change in the way an organization conducts its business, such as the way it manufactures its products. A novel way to do something that previously was unchanged for years, process inno-vation focuses on changing how something is done, and not on what is done. For example, Groupon has joined the consumer's desire for a good deal (i.e., traditional coupons) and the vendor's need to get new customers (i.e., group/mass marketing) into an online way of getting good deals to mass markets. Frugal innovation is both product and process innovation. However, by focusing on the product only, we miss understanding the true essence of frugality. Focusing on the frugal innovation process, as well as product, allows us to move away from the traditional, resource-rich models of innovation and focus on the customers, the people who buy and/or use the innovations. Customer base is truly the starting point of the differences between frugal innovation and other models of innovation.

# 4   Methodology

In order to understand the essence of frugal innovation, we turn to the popular press as an autonomous channel of production and diffusion of knowledge (Mazza & Alvarez 2000). More than a populizer of ideas gener-ated and provided to it by academia, the popular press reflects sources of legitimacy of theories and practices different from those adopted by standard academic outlets. Usually of low technical detail and high ideological stand-ards (Mazza & Alvarez 2000), popular press articles are essential to under-standing the essence of frugal innovation in practice.

Using LexisNexis search of All News, English, which includes Major World Publications[3] on the terms *jugaad* and "frugal innovation" resulted in 347 articles returned from sources (as described in Table 13.4), from January 1995 to January 2012. With removal of multiple mentions of the same article (reprinted or shared across outlets) as well as those that were about a movie named Jugaad, not the innovation practice, the total count of articles was 318. The first articles are on the simple, low-cost car that gave name to the practice: "a chassis and four wheels, an engine and a gear-box; it has no cabin, and the driver sits on a wooden plank behind the steering wheel;

in its cheapest form, it has a wooden floor on which are fixed two benches to seat passengers; *jugaad* does not boast any electronics; price of *jugaad* varies greatly; if owner insists on all new components and a 10-horsepower engine, the price would be around $1,000."[4]

## 5  Findings

In 2002, the origin of *jugaad* as the practice it is known as today came to attention: "One has to put it down to the creative individuality of the Indian mind, which is unfortunately not very organised in a community sense," Mr. Robin Grewal, chairman of New Delhi-based Strategic Technologies & Implementation Group (Sting), told *Sunday Review*. New Delhi-based information technology entrepreneur Karan Vir Singh, managing director of Voxtron Dezign Lab, called it the *jugaad* factor – the improvised quick fix. "It's like putting two spoons of turmeric powder into your radiator if you spring a small leak," he said. "It works, it will seal the leak. In Punjab, I have seen villagers buying an agricultural water pump at government subsidised rates, cannibalising some other parts from here and there, and turning it into a vehicle. This talent for lateral thinking is useful in developing software, which requires multiple solutions within the parameters of mathematical logic."[5]

Today, frugal innovation has found acceptance by innovators in developed countries. Indian *jugaad* or innovation for local use has found another fan. The €2.6 billion outdoor power tools maker, Andreas Stihl & Co. KG, has found that paddy farmers in south India have special needs, which the local Stihl dealer is supporting by supplying components. "We always look for such innovative ideas and if our engineers find them suitable and the volumes justify it, we even productionise them," said Dr. Nikolas Stihl, chairman-designate of the family-owned business.[6]

### The social aspect of frugal

What becomes evident as the essence of frugal innovation is the practice of bricolage (Baker & Nelson 2005; Lévi-Strauss 1966). *Bricolage* refers to doing what has to be done with the resources available at a given moment, and is not intentionally the overall "optimal" utilization solution. As Cappelli et al. (2010) write: "In a complex, often volatile environment with few resources and much red tape, business leaders have learned to rely on their wits to circumvent the innumerable hurdles they recurrently confront... Anyone who has seen outdated equipment nursed along a generation past its expected lifetime with retrofitted spare parts and jerry-rigged solutions has witnessed *jugaad* in action."

Most important to frugal innovation is the creative use of people and their knowledge, skills, and capabilities in novel ways, or human capital bricolage (Banerjee & Campbell 2009). Human capital bricolage is essential not just

*Table 13.4*  English-language newspaper articles on *jugaad* and frugal innovation (Jan. 1995–Jan. 2012)

| Over 20 mentions | 1 mention |
| --- | --- |
| Hindustan Times (38) | AFP – RELAXNEWS (English International Version) (1) |
| The Times of India (TOI) (41) | BusinessWorld (1) |
| The Economic Times (35) | The Christian Science Monitor (1) |
| Indian Express (24) | City A.M. (1) |
| **6–20 mentions** | The Daily Mirror (Sri Lanka) (1) |
| DNA (19) | Daily News Egypt (1) |
| MINT (18) | The Daily Telegraph (London) (1) |
| The Financial Express (17) | Daily the Pak Banker (1) |
| The Pioneer (India) (13) | Deseret Morning News (Salt Lake City) (1) |
| The Telegraph (India) (7) | DLA AM (1) |
| International Herald Tribune (6) | The DQ Week (1) |
| **2–5 mentions** | The Edge Malaysia (1) |
| Mirror Publications (5) | The Express (1) |
| New India Express (5) | The Gazette (Montreal) (1) |
| Mail Today (4) | The Globe and Mail (Canada) (1) |
| McClatchy-Tribune Business News (4) | The Gravesend Messenger (1) |
| The Northlines (4) | Gravesend Reporter (1) |
| The Straits Times (Singapore) (4) | The Guelph Mercury (1) |
| The Age (Melbourne, Australia) (3) | The Hamilton Spectator (1) |
| The Business Times Singapore (3) | The Independent (London) (1) |
| IPR (3) | Kamloops Daily News (British Columbia) (1) |
| Right Vision News (3) | Kashmir Monitor (India) (1) |
| Toronto Star – Metroland Newspapers (3) | Korea Times (1) |
| Contra Costa Times (2) | The National (1) |
| Crest (2) | National Business Review (New Zealand) (1) |
| Economic Times (E-Paper Edition) (2) | New Straits Times (Malaysia) (1) |
| Information Bank Abstracts (2) | Political & Business Daily (India) (1) |
| The Irish Times (2) | The San Francisco Chronicle (1) |
| Money Life (2) | San Jose Mercury News (California) (1) |
| The Nation (Thailand) (2) | Scottish Express (1) |
| The New York Times (2) | South China Morning Post (1) |
| Newcastle Chronicle & Journal Ltd. publications (2) | St. Paul Pioneer Press (Minnesota) (1) |
| The Observer (2) | Star Tribune (Minneapolis MN) (1) |
| The Toronto Star (2) | Sunday Observer (Sri Lanka) (1) |
|  | The Times (London) (1) |
|  | The Washington Post (1) |
|  | Waterloo Region Record (1) |

for economic efficiency, but for the positive social impact of giving people new opportunities to elevate themselves and their communities. For example, Banerjee of Mu Sigma commented, "When we started, the industry was hiring analytics experts to do this business. 'We said we do not necessarily require analytics expertise,'" Banerjee says of the Mu Sigma bet on processes over individuals. Mu Sigma University, as the firm's training facility is now called, which grooms talent from the ground up, is the company's silver bullet and something the rest of the industry missed.[7] Banerjee is giving people who would otherwise not be employed in such white-collar jobs a chance to be educated, trained, and to improve their quality of life. As another article describes: "India's smart cities might consist of a patchwork of technology modules that use people as infrastructure and employ analytics to create the information flows that empower citizens and improve city life."[8] Additionally, a recent article interviewed Earl Daum, Professor at the Tuck School of Business, who commented: "out of the world's seven billion people, four billion do not have a house. So I issued a challenge: why can't we create a $300 house for the poor? We created a global design contest; every entry there can design and build a house for $290. We have six design winners; we will come up with a workable scale of their designs and build a prototype in Haiti. Then, we will build a model village using that prototype. My hope is, once private sector companies see this is possible, they will jump in. Also, this is not just a house; the house is a metaphor for delivering health, education, jobs, water, electricity, etc."[9] In fact, 115 of 318 articles surveyed (36 percent) mention the creative use of people in achieving frugal innovation.

### Environmental aspects of frugal innovation

Yet from the experimentalism inherent in frugal innovation, there is an overall concern about the lack of robustness in the resulting products and services of frugal innovation, where robustness refers to stability in the face of variations in the operating environment. To achieve this robustness, Krishnan (2010) argues that *jugaad* needs to be systematized. Kumar and Puranam (2011) write that frugal innovation (frugal engineering in particular) should be more than mere *jugaad*, and should be a systematic approach to making constraints irrelevant or at least less important; mere *jugaad* may in fact impede the ability to fundamentally transform a situation.

Not all perspectives of frugal innovation see *jugaad* as "useful, reusable, durable, beautiful things."[10] In fact, 43 of the 318 articles (14 percent) on frugal innovation or *jugaad* were related to problems like safety and security. For example:

India is to impose safety standards on its jugaads, the home-made transport vehicles built from used car, motorcycle, tractor and bullock cart

302 Preeta M. Banerjee

parts. Jugaads are dangerous and safety rules are routinely flouted leading to numerous accidents, officials said. Indian businessmen and academics claimed that the move would hamper the spirit of innovation which helped the country develop space rockets and the world's cheapest car.[11]

Another article notes:

> Jugaad, whatever be the de rigueur definition for it, is only a short-term solution. ...Jugaad is not, I am afraid, the "antidote to the complexity of India" any more than wearing a long kurta is the antidote to the complexity of wearing a pair of pants with a broken zipper that won't close. What we need are solutions that are scalable, repeatable and can be institutionalized.[12]

In these articles, *jugaad* (more than frugal innovation) is seen as a sometimes illegal, makeshift, temporary, crude patchwork, "tossed together, salad-like from scrap and wood."[13]

The negative view of *jugaad* is most eloquently summarized by comparing it with the notion of just-in-time (JIT):

> India has the patent on the original JIT – *jugaad* in time... In the *jugaad* system of thinking, forecasting, planning, strategy, project management, compliance, and due diligence are wasteful... Building sustainable success requires systems and methodologies that can be institutionalized. Our divisive social structures are perhaps the reason why we are not able to build institutions or organisations.[14]

Those who write about bricolage also acknowledge the just-in-time nature of making do with resources at hand. Bricolage is a necessary, but at times not sufficient strategy for survival (Banerjee et al. 2011). However, bricolage can lead to better solutions than resource-rich alternatives, and as Garud and Karnøe (2003) highlighted in their writing about Danish managers with a practical interest for wind turbines and limited financial resources, who decided to search for possible materials from scrap dealers. The solution that resulted from the materials recovered from scrap proved to be superior to the resources of the American firms.

Does frugal innovation need to be systemitized? What is the solution to the lack of robustness found in *jugaad* or bricolaged products and services?

## 6 Solutions and recommendations

We have answered our question regarding the "frugal" in frugal innovation by finding the existence of social and environmental aspects beyond economic efficiency. However, we are still left with a question about the lack of robustness of frugal innovation. In answering this question, sustainable design (McDonough & Braungart 2002) provides an answer to the the question of robustness raised by frugal innovation or *jugaad* and its essential practice of bricolage. Sustainable design merges the efficiency of doing more with less with the practice of long-term thinking about the environment from multiple stakeholder perspectives. Sustainable design was initially developed for environmental sustainability through green product innovation or life cycle assessment (LCA), a method that allows systematic analysis of a manufactured item into its components and their subsidiary industrial processes, and allows measurement of impacts on nature from the beginning of the product to the final disposal (Goleman 2009). The result of LCA is creating products that are nontoxic, energy and water-efficient, and relatively harmless to the environment.

In designing for sustainability, the entire life cycle of the product is analyzed from birth of the innovation to rebirth, in an infinite cycle. This approach highlights all the impacts that a product will have from a sustainability standpoint, and this insight helps people to make more informed decisions very early in the product's life cycle, at the design phase. The outcome (in this case, sustainability) of a product cannot be determined on the basis of an individual firm but rather in the context of the ecosystem that the product is a part of in terms of its social, economic, and environmental impacts. This ecosystem has distinct components, which include the processes in place, the people and the products, a system of organization, technology, activities, information, and resources involved in moving a product or service from raw materials, to manufacturer, to supplier, to customer. In keeping with LCA, frugal innovation can integrate issues and flows that extend beyond the traditional innovator view to include multiple stakeholder perspectives, for example SPICE (society, partners, investors, customers, employees) (Sisodia et al. 2007). By including multiple stakeholder perspectives, frugal innovation can provide more for less for many for longer (MLML), thereby making MLML through managing the by-products produced during product use, the product's life extension, the product's end-of-life, and the recovery process at end-of-life (Figure 13.1). As an article in *India Express* highlights: "Indian innovations in the unorganised sector are often termed as *jugaad*, but we need to give them importance. For example, the Indian brick industry can contribute immensely in reducing carbon emission by producing perforated bricks."[15]

For example, Bihar-based Husk Power Systems (HPS) has developed novel uses of rice husks – one of India's most common waste products and perhaps the only bio-waste available to rural populations. Husk Power Systems has adapted and converted an existing method of biomass gasification using diesel technology into a single-fuel rice husk gasifier for rural electrification. The result is that households stop using dim kerosene lamps when they get Husk Power Systems electricity, saving on kerosene (with associated

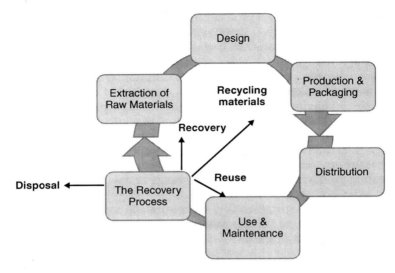

*Figure 13.1*   Life cycle assessment (LCA)

reductions in $CO_2$ emissions) and facilitating evening studying and other productive activities. The company has significantly reduced the use of kerosene power in rural housing. On an average, each power plant serves about 400 households and replaces about 42,000 litres of kerosene and 18,000 liters of diesel fuel per year. As of August 2010, HPS has already sequestered 50,000 tons of $CO_2$. HPS lit its first village from its first 100 percent biomass-based power plant that uses discarded rice husks to generate electricity. Today it has installed 60 mini-power plants that power about 25,000 households in more than 250 villages and hamlets. HPS's technology impacts the lives of approximately 150,000 people in rural India. Most importantly, HPS employs more than 300 local people in rural India to run and manage its power plants, and has saved $1.25 million for the households it serves.

Similarly, Tata has created the $24 Swach water filter, targeted at rural households with no electricity or running water, using ash from rice milling to filter out bacteria. Tata's Swach Bulb is the main purifying unit of the product. The Swach Bulb can purify about 3000 liters of water, depending on the quality of water, after which the bulb must be replaced. The bulb has a "fuse" indicating when a cartridge change is required. Swach was designed as a low-cost purifier for Indian low-income groups, and it has given access to safe drinking water to millions. It is a replaceable filter-based product, which is entirely portable and is based on low-cost natural ingredients, which safeguards safe drinking water at a new benchmark of Rs 30 (50 cents) per month for a family of five.

Moreover, from local connections with NGOs to public–private partnerships, engaging stakeholders in conversation, sense making and decision making to taking action together can be transformational (Cooperrider & Fry 2010). The potential impact of high-quality connections from stakeholder engagement is the ability to create and cultivate positive emotions and trust, which contribute to higher coping, greater resilience in the face of setbacks, more creativity, greater attention, and a broadening of the thought-action repertoire (Dutton & Heaphy 2003; Fredrickson 1998). In other words, engaging stakeholders meaningfully creates more robust solutions when employing frugal innovation or *jugaad*.

## 7 Limitations of this study

It is important to note that this chapter is an exploratory, illustrative analysis of the relation between reverse innovation, Gandhian engineering, and *jugaad* in understanding the essence of frugality in frugal innovation. As the academic literature on frugal innovation is just starting to appear, it was essential to examine the popular press, as opposed to academic journals. In addition, frugal innovation has been identified as an Indian practice, *jugaad*, the India Way. By necessity, Indians must be resourceful. A little more than 407 million people – nearly one-third of India's population – live below the poverty line. Many of India's public buildings are dilapidated; its power grid is unreliable; its highways are congested, and its roads are riddled with potholes. However, frugal innovation also exists in a different cultural contexts and we would benefit our understanding of the practice to examine it across all these cultural contexts. It would be wrong to think that only innovators in India can develop frugal innovations.

## 8 Future research directions

The future of frugal innovation is in merging the essential practices of economic efficiency, human capital bricolage, and sustainable design from the perspective of multiple stakeholders, i.e., stakeholder analysis. Stakeholder analysis is the process of identifying the individuals or groups that are likely to affect or be affected by a proposed action and sorting them according to their impact on the action and the impact the action will have on them. This information is used to assess how the interests of those stakeholders should be addressed. In order to develop multiple stakeholder perspectives, the power of social media can be harnessed to provide previously unavailable personal experiences and accounts. Additionally, innovators can facilitate connectivity through global alliances and supplier development linkages for global, sustainable value chains so that resource utilization can be green. Lastly, policy makers can look at ways to harness the rule of law, contract enforcement, and other methods to enforce the positive aspects of *jugaad* or frugal innovation. All of these facets provide vast areas for improvements in practice and theory.

There are challenges to putting the triple bottom line into practice. These challenges include measuring each of the three categories (economic, social, and environmental), finding applicable data, and calculating a project's or policy's contribution to sustainability. These challenges aside, the triple-bottom-line framework allows organizations to evaluate the ramifications of their decisions from a truly long-run perspective. Future research can focus on the best way for educators to provide education on frugal innovation and measuring the triple-bottom-line in both theoretical and practical ways, including innovation training and other technical and vocational education and training.

Most importantly, the concept of bricolage (Baker & Nelson 2005; Lévi-Strauss 1966) describes the efforts of entrepreneurs who often face substantial resource constraints to make use of the resources they do control in novel ways. It describes the creative process by which bricoleurs devise new uses for the resources available within their organization, making it possible for them to overcome their limitations. Firms can maximize the value from their existing limited resources by innovatively changing their current configuration. Whether the resources were never satisfactorily configured for the business challenges in the environment, or whether conditions changed, which caused the resource configuration to become suboptimal, bricolage represents an approach where organizations survey their resources at hand and reassess how to bundle and allocate the resources to generate more value for the organization (Banerjee et al. 2011). This may involve the use of materials or procedures considered by others to be strategically worthless, a selective refusal to acknowledge socially constructed obstacles, and the combination of existing resources to produce new benefits.

Thus, bricolage, in the sense of this chapter, is truly about making do with the resources at hand through recombination and redeployment. Bricoleurs, i.e., those who implement bricolage, are forced to test the limitations of existing social constructions (Baker & Nelson 2005). The uses to which resources can be applied are limited under normal circumstances by law, initial conditions, social constructions about their utility, and the dominant mindset in the organization (Prahalad & Bettis 1986), which filters out information that might suggest alternate uses for resources that do not fit within the current paradigm. For organizations in favorable resource positions, this is not a problem; working within the limitations will achieve acceptable results. Resource-constrained organizations, such as social entrepreneurs, however, must adopt other approaches. In other words, further research can more explicitly connect bricolage, a concept in entrepreneurship, to innovation.

Lastly, teaching *jugaad*, human capital bricolage, and sustainable design ensures that students, academics, entrepreneurs, and policy makers alike understand the difference between economic efficiency alone and social and environmental impact in the form of meeting the triple bottom line (Hubbard 2009). More work needs to be done on finding ways to expose learners to new experiences, new resources, and new contexts so that frugal innovation is not seen as a tremendously difficult, or even foreign, task. Further research on what contexts around innovation can help an organization to switch directions, akin to Tyre and von Hippel's (1997) work, but pertaining to frugality would be helpful. Additionally, as frugal innovation becomes more prevalent, larger data sets can be gathered and analyzed systematically to provide trends about those strategies that might provide better results in frugal innovation endeavors.

## 9  Conclusions

We do indeed find that frugality is not just economical efficiency but social and environmental sustainability as well. Socially, frugality involves human capital bricolage (or utilizing people and their knowledge, skills, and capabilities in novel ways). Environmentally, the problem of lack of robustness of *jugaad* is articulated by many popular press articles as well as by academics. Authors like Krishnan (2010) call for systematization of *jugaad* in response to this lack of robustness. However, systematization is not required. Instead, by incorporating sustainable design, frugal innovation can be supported and made more robust. We conclude that ideally, frugal innovation should be a combination of *jugaad*, human capital bricolage, and sustainable design.

Frugal innovation when just focusing on economic efficiency or *jugaad* can be seen as just a creative ad hoc solution to a vexing issue, making existing things work and/or creating new things with scarce resources. However, such practices can be seen in a negative light as makeshift, or as cheap fixes that are short-lived. With a move to view management practices with a triple bottom line, we can integrate concepts from entrepreneurship, stakeholder analysis, sustainability, and innovation together. Fundamental to frugal innovation is human capital bricolage, or making do with the resources at hand, which allows people to take control of bettering their quality of life while stretching the boundaries of their knowledge, skills, and capabilities. Also essential is the incorporation of sustainable design from multiple stakeholder perspectives. With the integration of *jugaad*, human capital bricolage, and sustainable design, frugal innovation can become the art of holistic lateral thinking that results in long-term resilient creativity under severe constraints.

Frugal innovation is the means to developing innovative products under extreme constraints, but with good value to the consumer. Developing and emerging markets, especially bottom-of-the-pyramid populations, hold great potential for companies seeking new growth and economic efficiency. To reach these markets, company leaders must develop a firm understanding of people in the developing world and also adjust pre-existing business metrics. They must implement the strategies of frugal innovation with new approaches to serve the needs of the bottom-of-the-pyramid population. In doing so, companies not only improve the bottom-of-the-pyramid living conditions, but also create a new source of growth, cost-saving opportunities, and access to innovation. While a departure from traditional ways of innovating and doing business, there is much promise in this alternative, complementary model (see Table 13.1). Traditional ways of innovating have not been sustainable in developing and emerging markets, as stripped-down versions of developed-country innovations have not held up under the tougher conditions of use.

The practice of frugality is at the heart of India's innovation culture. In the West, innovation is the purview of scientists and inventors working in labs filled with high-priced equipment and computers. In India, innovation is done by farmers, transporters, traders, and housewives; it happens in homes, in gardens, and even in the slums. In India, innovation does not require costly R&D or a degree from a fancy school. It simply needs ingenuity and vision. At a time when innovation is a key issue in our world economic and political landscape, frugal innovation is a shining example of what is possible. While most developed-country innovators have the fortune to be able to choose to work within tight constraints, those in developing countries have to grapple with the problem every day. The essence of the new model of innovation is adaptability and constant analysis with an eye to sustainability: economic, social and environmental.

## Notes

The author would like to thank the Fulbright-Nehru Senior Scholar Research Fellowship for supporting this research.

1. India's Next Global Export: Innovation, December 2, 2009, 3:51 PM EST as found at http://www.businessweek.com/innovate/content/dec2009/id2009121_864965. htm (accessed January 20, 2012).
2. LexisNexis is a trusted source of popular press articles in management, for example Pollock and Rindova (2003). LexisNexis is accessible to nearly seven million students and faculty members at more than 1,700 colleges and universities across the United States and provides access to information from the most trusted and authoritative business, legal, and news sources worldwide (http://academic.lexis-nexis.com/) (accessed on January 29, 2012).
3. The Major World Publications group file, MWP, contains full-text news sources from around the world which are held in high esteem for their content reliability. This includes the world's major newspapers, magazines and trade publications which are relied upon for the accuracy and integrity of their reporting.

4. Barun S. Mitra, For rural India, simple low-cost car is the way to go, *Asian Wall Street Journal*, January 30, 1995, Monday, section A; p. 12, column 1.
5. What's culture got to do with IT? *The Straits Times* (Singapore), September 29, 2002 Sunday, Analysis.
6. Gouri Agtey Athale India's 'jugaad' finds acceptance from German tool maker The Economic Times January 10, 2012.
7. Goutam Das, Master of data universe, *Business Today*, February 5, 2012, BPO.
8. Ayesha Khanna, Is your city smart enough? *Indian Express*, January 2, 2012 (Monday).
9. Alokesh Bhattacharyya and J. Anand, Business Buddhas, *Business Today*, Management. January 22, 2012.
10. Akash Raman (author) On a trash course; "Kabaad Se Jugaad" – an unusual workshop – saw Bangaloreans discovering ways to transform household kabaad (trash) into jugaad (useful, reusable, durable, beautiful things),, *DNA*, June 14, 2011 (Tuesday).
11. The Daily Telegraph, Safety drive on spare-part trikes; World Bulletin, *The Daily Telegraph* (London), January 19, 2012 (Thursday), News; p. 22.
12. Mint (Author), Cubiclenama die Jugaad, die December 10, 2011 Saturday. Mint publication.
13. Jyothi Prabhakar, Govinda's next stuck in 1.3 cr spat. *The Times of India* (TOI), March 9, 2011 (Wednesday), News & Interviews.
14. Yatish Rajawat, India's own JIT – Jugaad-in-time; Being Indian, *DNA*, September 18, 2010 (Saturday).
15. Indian Express, Go Green: Experts ask firms to invest in eco-friendly ways, *Indian Express*, June 9, 2010 (Wednesday), Express News Service.

## References

Anderson, J. & Markides, C. (2007) Strategic innovation at the base of the pyramid. *MIT Sloan Management Review*, 49(1), 83–88.
Baker, T. & Nelson, R. (2005) Creating something from nothing: Resource construction through entrepreneurial bricolage. *Administrative Science Quarterly*, 50(3), 329–366.
Banerjee, P. M., & Campbell, B. (2009) Inventor bricolage and firm technology and development. *R&D Management*, 39(5), 473–487.
Banerjee, P. M., Campbell, B. & Saxton, B. (June 2011) The Role of Bricolage in Sustained Competitive Advantage. Paper presented at the meeting of Babson College Entrepreneurship Research Conference (BCERC), Syracuse, NY.
Cappelli, P., Singh, H., Singh, J., & Useem, M. (2010) *India Way: How India's Top Business Leaders Are Revolutionizing Management*. Cambridge, MA: Harvard Business Press Books.
Cooperrider, D., & Fry, R. (2010) Can stakeholder engagement be generative? *Journal of Corporate Citizenship*, 38, 3–6.
Dutton, J. E., and Heaphy, D.H. (2003) The power of high quality connections. In K. S. Cameron, J. E. Dutton and R. E. Quinn (Eds), *Positive Organizational Scholarship*. San Francisco: Berrett-Koehler.
Elkington, J. (1994) Towards the sustainable corporation: Win-win-win business strategies for sustainable development. *California Management Review*, 36(2), 90–100.
Francis, D., and Bessant, J. (2005) Targeting innovation and implications for capability development. *Technovation*, 25(3), 171–183.

Fredrickson, B. L. (1998) What good are positive emotions? *Review of General Psychology*, 2(3), 300–19.

Garud, R., and Karnøe, P. (2003) Bricolage versus breakthrough: Distributed and embedded agency in technology entrepreneurship. *Research Policy*, 32(2), 277–300. doi:10.1016/S0048-7333(02)00100-2.

Goleman, D. (2009) *Ecological Intelligence*. New York, NY: Penguin Books.

Govindarajan, V. & Ramamurti, R. (2011) Reverse innovation, emerging markets, and global strategy. *Global Strategy Journal*, 1(3–4), 191–205.

Hubbard, G. (2009) Measuring organizational performance: Beyond the triple bottom line. *Business Strategy and the Environment*, 19, 177–191.

Immelt, J., Govindarajan, V. & Trimble, C. (2009) How GE is disrupting itself. *Harvard Business Review*, 87(10), 56–65.

Krishnan, R. T. (2010) *From Jugaad to Systematic Innovation: The Challenge for India*. Bangalore: The Utpreraka Foundation.

Kumar, N. & Puranam, P. (2011) *India Inside: The Emerging Innovation Challenge to the West*. Cambridge, MA: Harvard Business Press Books.

Lévi-Strauss, C. (1966) *The Savage Mind*. Chicago: University of Chicago Press.

Mazza, C. & Alvarez, J. L. (2000) Haute couture and Prêt-à-Porter: The popular press and the diffusion of management practices. *Organization Studies*, 21(3), 567–588.

McDonough, W., & Braungart, M. (2002) *Cradle to Cradle: Remaking the Way We Make Things*. New York: North Point Press.

Pollock, T. G., & Rindova, V. P. (2003) Media legitimation effects in the market for initial public offerings. *Academy of Management Journal*, 46(5), 631–642.

Prahalad, C. K. (2005) *The Fortune at the Base of the Pyramid. Eradicating Poverty through Profits*. Philadelphia, PA: Wharton School Publishing.

Prahalad, C. K. and Bettis, R. (1986) The dominant logic: Retrospective and extension. *Strategic Management Journal*, 16, 5–14.

Prahalad, C. K. and Mashelkar, R. A. (2010) Innovation's holy grail. *Harvard Business Review*, July–August, 132–141.

Radjou, N., Prabhu, J., & A. Simone (2012) *Jugaad Innovation: Think Frugal, Be Flexible, Generate Breakthrough Growth*. New York, NY: Jossey-Bass.

Sarkar, M. B. (2011) Moving forward by going in reverse: Emerging trends in global innovation and knowledge strategies. *Global Strategy Journal*, 1(3–4), 237–242.

Savitz, A. (2006) *The Triple Bottom Line*. San Francisco: Jossey-Bass.

Sisodia, R. Wolfe, D., & J. Sheth (2007) *Firms of Endearment: How World-Class Companies Profit from Passion and Purpose*. New York, NY: Pearson Prentice Hall.

Sullivan, A., & S. Sheffrin (2003) *Economics: Principles in Action*. Upper Saddle River, New Jersey: Pearson Prentice Hall.

Tyre, M. J., & Von Hippel, E. (1997) The situated nature of adaptive learning in organizations. *Organization Science*, 8(1), 71–83.

Vossoughi, S. (2012) The silver lining to scarcity: It drives innovation. HBR Blog Network as accessed on May 25, 2012 at http://blogs.hbr.org/cs/2012/03/the_silver_lining_to_scarcity.html

# 14
## Flexible Working, Mobility and IT Innovation, and ICT in 2012: The Case of Flexible Working

*Ramon Costa-i-Pujol*

## 1 Background and starting point

### The information and knowledge society

To talk about and understand the needs of the information and knowledge worker, it is necessary to understand and define what we mean by the "information and knowledge society."

One of the main authors in this matter is Manuel Castells, who began using the term "information society" to define modern social realities, on the premise that the attribute *information* is a specific "form of 'special' social organization, in which information generation, processing, transmission, and conversion are fundamental sources of productivity and power" (Castells 1999).

From another point of view, the Institute for the Development of the Information Society (Russia) proposed this other definition of the Information Society:

> The information society is a step in the development of modern civilization characterized by an increasing function of information and knowledge in the life of society, with the support and implementation of tools and communication technologies, a growing proportion of information products and services in GDP, and the creation of a global information infrastructure that ensures the effective interaction of people, their access to information, and meeting their social needs. (Yershova et al. 2008)

The Canadian theorist Don Tapscott (1998, pp. 60–81) identified twelve features that characterize what is known as the Information Society:

- Orientation to knowledge; therefore, the new society is an actually a knowledge society.

- Digitalization, which means also that we are in a digital society.
- Virtualization of physical objects, which can become "virtual objects," but also virtualization of offices and workplaces, companies, and jobs
- Molecularization, so the concept of mass is replaced by the concept of "molecularized."
- Work on the Internet, where the company operates through networks.
- Disintermediation between consumers and producers.
- Convergence between computing, communications, and content industry.
- Origin of the innovation through creative processes.
- Combination of information producers and consumers into one.
- Immediacy becoming a key factor in business and its success.
- Globalization of an economy, which works on a global scale.

## Trends in the new world of work

The information society and intensive use of tools and information and communications technologies (ICT) have led to the emergence of new models in the workplace, but also new challenges that organizations must face.

In the research work of Tom Austin (Gartner) (Austin 2010), they analyzed continuous changes in the nature of work and proposed a list of aspects to consider for the 2010–2020 period:

- The nonroutine nature of work.
- "Work swarms" (working in a swarm). "Swarming" means work characterized by a flow of collective activity, carried out by many individuals, where everyone is available and is able to add value.
- Weak bonds between members of team; this is tied to the concept of swarming.
- Collective work. There will be informal groups of people (stakeholders), beyond the control of the company, who may impact the success or failure of the organization, becoming more and more important, using and applying social networks, about the companies and professionals.
- "Work sketch-ups." Most nonroutine processes will be highly informal, without following a standard pattern.
- Spontaneous labor. Spontaneity implies reactive activity, but also includes proactive work (e.g., the search for new opportunities and the creation of new designs and models).
- Simulation and experimentation. Users will move through different dimensions (social media, video, documents, and transactional data) to analyze and work with different data types.
- "Pattern sensitivity." The business world will become more volatile.
- Hyperconnectivity. The world of work is (and will increasingly be) hyperconnected inside and outside companies through networks of networks, including many beyond the control of companies.

– Workplace. The workplace becomes more and more virtual, with inter-actions (meetings, communications, collaborations) among people in different time zones, geographical areas, and companies.

Among these trends, we want to highlight a basic one to understand this new way of working and its implications: workplace virtualization.

As the study points out, planet earth will become the workplace of professionals. Many workers will not have a physical office space or a specific physical workplace and their work will be scheduled 24 hours a day, 7 days a week. We are heading towards a reconciliation of working life and family, adjusting schedules, to the extent possible, for the different professionals.

This means that a part of our professional work will be developed out of the office (at home, in a third place) and that schedules will allow us to avoid being stuck in rush hour, make meetings virtually, save travel time

Some of the concepts that will shortly appear, or will be consolidated, deal with issues such as telepresence, work in the virtual world, co-working or working in a third place:

- **Virtual presence.** Increasingly, we'll find professionals who will work remotely (from home, usually), but they will be present, thanks to technology, virtually in the office. Telepresence systems that work in real time (for example, videoconferencing), will allow these professionals to interact with their teammates as if they were present in the office. The use of high-resolution screens provide the difference between real and virtual presence of these workers almost imperceptible to their fellows.
- **Working in a virtual world.** While phenomena such as Second Life have been redirected to a 'normal situation' in their application, the benefits that this type of virtual world will have in the near future for the world of work are not negligible. Many of the relationships between companies and professionals will be conducted in such virtual environments, where virtual reality will merge with actual reality. Many companies have already established business and professional activities carried out through these applications, interacting with customers, suppliers, and current and future employees.
- **Hyperconnection.** Use of social networks and virtual communication tools allow professionals to interact with their peers, but also with customers, suppliers, and potential users with the company's own tools, as well as through third-party applications (e.g., Facebook and LinkedIn). This hyperconnection ability (through Internet hyperlinks) brings significant challenges in the way business will be done and in how we'll know and will interact with our customers and potential users, as we do in marketing in positioning our products, or how we manage corporate online reputation online, for example.

- **Cloud computing.** Cloud computing (computing distributed on the Internet), will change the way we manage and work with computer software companies. Instead of having our applications and their associated infrastructure (computers, servers, etc.) in our office or in a third place, these will be available on the Internet. Through a few suppliers who are able to offer us data storage and communications infrastructure, dynamically and according to our needs, we will have access remotely to our applications, data, etc. And therefore we will not have to worry about the maintenance and development of infrastructure, or the purchase of hardware, software, and updates, but simply the use of cloud computing, wherever we are.
- **Co-working.** The ability of working anywhere brings a new dimension to the work shift. We can work in the office, at home or in third place, wherever we are. Gradually, we will see workspaces or co-working spaces, where different professionals from different companies will find themselves physically, to telework (or do displaced work), sharing resources. The concept is not new. In fact, it has been a well-established trend in the United States for several years and is now starting to happen in different countries, like Spain. We could define it as a variant of telecommuting, but it consists in sharing space with other professionals working in different industries, computer people, designers and architects, journalists and writers, for example. The sharing of these areas means workers will have a set of benefits and services at a very low cost, which is shared by all users of space. But it also means working with other professionals with whom people share concerns, needs, and, in many cases, professional synergies.

The last important concepts that have emerged strongly in recent years, the third workplace, flexible working, and the flexible office, are the basis of this analysis.

### Information and knowledge workers and their needs

These new models of work and changes in the way to communicate, collaborate, and interact, also bring new needs for information workers and businesses. First, it is worth noting that the information and knowledge worker and knowledge is not a job description, responsibility, or profession within companies but describes a role or set of features that are increasingly refer to most business professionals – management and middle management, and administrative staff professional support in departments such as sales, marketing, R&D (research and design) and communications. Knowledge worker also refers to the majority of workers in the areas of operations.

Graphically summarized, all professionals not obligated to be in a production chain are considered in some extent to be information and knowledge workers.

Earlier I listed the main considerations for this type of worker (Costa-i-Pujol 2010):

- Most of the worker's time is spent with professional documentation, creating documents, working with spreadsheets, presentations – all known as "productivity tools"
- Electronic mail (e-mail) has become a common tool for a number of important communications. The schedule, calendar, and other tasks are managed with e-mail.
- Communication channels have been expanded, and beyond e-mail and telephone, at our disposal there are mobile telephony, instant messaging tools, and audio- and video-conferencing tools.
- Communication is an important point for workers as well as for collaboration and teamwork. Participation in working groups (more than one) at the departmental level of internal projects and also externally with other companies is a must.
- Other activities of these workers are projects. First of all, the project manager or project leader is an example of an information worker, as this person has the responsibility to coordinate and manage people and also to manage all information and project documentation. People who participate in a project have to share information and documentation and collaborate with each other.
- A lot of time is spent finding information (on our computer, on the company's servers, on the intranet, extranet, and on the Internet in general). A lot of time is devoted to this, but it is reasonable to search for information that we ourselves and other professionals have saved about the company.
- We all have relationships with people who are working out of the office – customers and suppliers, companies to whom we have outsourced part of our operations, and even professionals in our company who are displaced. These people have to share documents and information, and need to work together and communicate.
- And finally most of us have to work out of the office, sometimes for business trips such as visits to customers. During the time we are out of office, we have to be connected with our information systems, collaborate with our colleagues, and communicate with our team.

### Mobile workers and flexible working

The mobile worker, as we stated in the last section, is increasingly common in our workplace. Although mobility has been traditionally linked to the business profiles of directors or consultants, who require trips to their business clients domestically or abroad, mobile working is an option that may be available to a large group of other roles and professional occupations.

According to a research study published by Forrester (Pelino 2010) of over 5,500 workers in 2010, 34 percent of professionals answered they had worked in a remote way, at home, during trips, or from other places (bars or hotels) once a week.

Beyond the concept of mobile working is the term "flexible working." It means the option of working at different times and places and may include, for example, working at home, telecommuting, or working in "roaming" and doing it independently of the standard divisions of working hours or free time.

Hence the term, "flexible working" includes more situations than "mobile working," and that's why it has been chosen for this study.

Studies on the effects of flexible work have demonstrated its positive impact for workers, organizations, and even for the health of the environment.

Thus the concept of flexible and comprehensive work recognizes that there are people who telecommute and consider their main workplace to be at home, the second one to be at the company's office, and the third place to be public spaces that the city offers where one may telecommute.

### Flexible work and flexible offices

Flexible working is considered to be any change made in the time or place where we work, and so it could include, for example, working at home, telecommuting, mobile working, and work done on different or free timetables.

Flexible working means, therefore, to work at a combination of different locations: at home, remotely, or on the move (in airports, cars, at the client's house) and at the headquarters of the company.

According to a study done in the United Kingdom in 2005 (Clake 2005), the flexible worker distributed his or her time working at home (35 percent), in motion (45 percent), and the rest of the time (20 percent) working at the office, compared to a distribution of 25 percent / 20 percent / 55 percent in 2002.

This new model is a reality. It may mean changing the way professionals work, in many cases and in others, new ways of behaving should be adopted. These changes, however, have a number of effects, positive and negative, with some associated risks, which must be known:

– *Improved productivity;* 84 percent of flexible workers feel they have improved their productivity.

  • *Risk of loss of sense of belonging;* 40 percent think they are "socially disconnected" from their peers.
  • A new type of leadership of flexible workers by their directors / managers is needed. Managers must be educated and encouraged to have some meetings in a virtual way and have some in a face-to-face way at the office.

- Operating costs are less because of reduced space needed per employee at the office, at ratios of 7 square meters per employee (75 square feet).

Companies, then, provide technology, services, and working conditions more suitable to the employee wherever he or she is, giving attention to tending things 24/7/365 (24 hours, 7 days a week, 365 days a year), which represents a set of challenges for organizations:

- Providing technology so people are able to work from anywhere at anytime: personal digital assistants (PDAs), smartphones, 3G connections.
- Providing access to other corporate areas, with the concept of satellite offices.
- Developing web-based support for employees that includes support areas of finance, human resources, and technology.
- Providing grants for installation of each job: asymmetric digital subscriber line (ADSL), workspace, etc.
- Having a new conception of space at the company's headquarters and breaking the concept of fixed working hours.
- Strengthening internal communications to avoid people's feeling isolated or socially disconnected.

It is also true, however, that mobile working reduces the size of offices needed around 50 percent and reduces operating costs, in many cases, around 30 percent.

Finally, in view of these changes, we are also faced with a new workspace, as mentioned above (Oulasvirta & Sumari 2007), which leads to the flexible office.

Offices will not be the places to work in. Other places will become the main ones to work.

As I noted earlier (Costa-i-Pujol 2010), we will all increasingly work in the third place, in the office, at home, or wherever we are, at the airport, the office of a client or vendor, at a library, at a bar, at a business center, ... Wherever we are, the office will be there.

Traditional offices will become workspaces to meet, where project teams will meet to share their work, collaborate, and work together. In these offices, spaces will be more and more open, without being assigned, and there will be multipurpose rooms and offices that will be used when needed.

The main purpose of the flexible design of the office is to foster communication and collaboration, giving much more importance to face-to-face communication.

It is therefore necessary to consider a series of actions and change management before, during, and after implementing this new model of the office:

- Develop a technology that enables mobility.
- Define the group.

- Define staff ratio and adjust through monitoring.
- Review current design of the office space and its services.
- Defining criteria for the use of space: how space is used and if each space will stay or not.
- Consider a variety of types of spaces: rooms for group work, individual workrooms, jobs in open spaces, meeting rooms, etc.

It seems, therefore, that it's important for the space to always be appropriate to the changing functional needs of the worker.

For example, anyone who needs a confidential work space or individual office at some time should have it, temporarily, regardless of his or her hierarchical position in the company.

Antti Oulasvirta (2007) has already pointed out the need to modify the classical office spaces, and has suggested new types of work spaces:

- *Drop-in points:* tables for short sessions
- *Hot desks and work desks:* tables for longer sessions that have to be booked in advance
- *Silent rooms:* places to work in silence
- *Community rooms:* semi-open spaces in which to conduct meetings
- *Virtual meeting rooms:* (rooms equipped for virtual meetings

### Working at third place

Finally, we introduce a last concept, the third place. This concept of the third place was introduced by Ray Oldenburgh in 1989 in his book *The Great Good Place*, By "the third place," he referred to those locations where people work and socialize other than the home or the office. The home was considered the first place and the office was considered the second one.

Ray Oldenburg's view was not too far from what has really happened. As other studies and authors point out (Austin 2010; Matson & Prusak 2010), mobile working is increasingly evident in our society, with benefits for both the company and the worker – and let's not forget the environment.

Mobile workers could be defined as professional who have no fixed place of work; they can perform their work from various locations, including: offices of customers, the corporate office, from home, at airports, in bars, etc.

## 2  ICT, a key for innovation in the mobile environment

In this new model, employees of companies that change their office model are called "information workers;" they focus on unstructured processes and, therefore, they require the use of other tools and technologies that help increase their skills and let them be more efficient in some scenarios of work (meetings, teamwork, search and information management, decision making, effective communication, etc.).

## i-Productivity technologies

In these scenarios, i-productivity technologies, new technologies tools, play an important role, complementary to the enterprise management systems (ERP, CRM [customer relationship management], etc.). These technologies are all those tools, platforms, and solutions that are aimed to improve the productivity of professionals, mainly in the fields of communication, teamwork, collaboration, search, storage, and information management.

These tools, such as collaboration spaces, instant messaging, mobility solutions and tools to search for information, have already demonstrated their impact on the income of companies and have shown their ability to innovate and to be more competitive. That's why it is essential that companies adopt them in an efficient and effective way.

However, unlike information technology and information and communications technology (ICT), whose functions are to automate structured processes, these i-productivity tools need to be integrated into organizations by someone with a strategic vision and used in a shared corporate culture, in order to maximize their application. These tools play a fundamental role in mobile working scenarios.

Technological advances in recent years (late 1990s and early 2000s) have completely changed the way of work.

The Internet, along with the increased capacity and portability of computers and cell phones, has allowed administrative people, professionals, and companies to work productively from anywhere at anytime. These capabilities have led, for example to the explosion of outsourcing (moving offshore) of many jobs to countries like India and China. And different online productivity software, such as blogs and wikis, are changing the way in which teams work together, communicate, interact, and collaborate all over the world.

## Use and application of i-productivity technologies

The incorporation of information technology (IT), particularly in the personal productivity technologies of people in organizations, will lead to an improvement in the productivity of professionals if the IT is accompanied by the necessary investments in training and there is corresponding change management and proper adoption of IT by all members of the organization. These technologies are tools, and it is necessary that professionals apply a set of best practices in their use. These practices must be homogeneous throughout the organization and must come from management, with a strategic vision.

To implement a tool for our businesses does not mean or guarantee a success or productivity improvement. Even in the case of productivity tools for individuals, or i-productivity technologies, they can become the opposite; tools misused can become a thief of time and a source of nonproductivity.

The application of these tools in different ways by different people leads to malpractice involving nonproductivity and inefficiencies in their work. Finally, the lack of understanding of the features and benefits of these tools leads people to not use their full potential or not to know how to apply them properly in their daily lives.

Here is where role play and training scenarios can be useful. For example, a spreadsheet management tool could have huge potential power, but we have to know what features it offers for the tasks that we develop at the organization and how to use the spreadsheet. Some of the clearest examples of the need for practice are the use of e-mail and instant messaging.

Businesses and organizations must innovate in the way of incorporating new technology, so they will be able to use and apply computer tools to improve their competitiveness, efficiency, and productivity.

### Working outside the office

As mentioned, work has been done outside the office for many years, especially those jobs related to sales and representation of products or services. Such professional tasks were performed without the existence of technologies that permitted the definition of workplace to change drastically.

Several research studies on workers in the information and knowledge field and mobile workers have analyzed different aspects of everyday life of these professionals, such as the type of traveling and remote work that they do.

- Oulasvirta (2007) distinguished between mobile work (including traveling – short or long distances) and traveling to a city, going to offices of a company, or visiting customers, suppliers, and mobile work at the same office, areas, departments and buildings. In all cases, it appeared that there was the need for remote collaboration, telecommuting and working in roaming.
- In a study published in *McKinsey Quarterly* (Matson 2010), the authors analyze the barriers that hinder the interaction of knowledge workers, such as physical and technological barriers resulting from mobile work, and in different places the authors look at the use of technology tools for secure communication and collaboration among professionals.

Before the widespread existence of mobile phones, laptops, and the Internet, mobile workers developed their work using cars and telephones as their primary tools.

The emergence, spread, and improvement of technological tools (smartphones, PDAs, laptops) and data networks radically altered the landscape of the mobile worker. Not only that, but the shape of work systems were modified, new professional profiles appeared, and companies are even reassessing the need for costly equipment and offices, as we have ntoed earlier.

Mobile workers must have the tools to execute their work from any physical location and to function without the rigidity of a fixed schedule. The job is no longer based only on "office" skills, but requires new skills: querying sources of information, access to company resources at any location and time, availability of various communication tools, etc.

Mobility needs of companies have increased gradually due to several factors, the most important of which are globalization and enlargement of the scope of places where companies can offer their products or services, as well as others factors like:

- The greater geographic dispersion of customers and greater scope of markets precisely because of technology
- Increased displacement of commercial forces
- The deployment of sales subsidiaries
- Products and services subcontracted or sold to third parties
- The relocation of production
- The increase in trips of managers and executives

In some cases, a new organizational culture with a clear focus on goals, has allowed certain people a new flexibility in the way they work, combining the need for a more efficient, fair, and productive professional life and time for the family lives of workers.

Clearly, the benefits of working outside the office will depend directly on the activity of the company and the type of work its members do.

All in all, we emphasize the following:

- Better execution of the tasks to be performed in situ
- Elimination of barriers related to the availability of workers
- Reducing costs of chattel or related office equipment (see below examples of the Flexible Office)
- Improvement in execution of professional duties, as open access to information and productive habits related to information management are gained
- Trend to aim for work-based goals rather than filling working hours.
- Employee satisfaction in order to have more information available and the feeling of "connection" with the company or team when away from the office or moving around.

## Best practices, threats and risks

Any change in the way of organizing and working in a company must be identified, analyzed, and must deal with a range of threats and risks that may inhibit the adoption of this new way of working; and we must apply a set of best practices to reinforce the change. Among the threats, risks, and barriers, and inhibitors, we highlight the following:

- *Tools (hardware and software).* The lack of appropriate tools or tools that are not well integrated can produce the experience that labor mobility is not positive, making the worker uncomfortable with this model.
- *Personal habits.* Job mobility requires a high level of personal discipline, organizational skills of good work habits and teamwork, as it reduces personal contact. Adapting to this new work model requires continuous practice and support by the company to facilitate the transition.
- *Cultural barriers.* Traditional models of work schedules, based on the workers' physical presence in the office, rather than models based on performance objectives and collaborative work; the feeling of lack of control by the company or, conversely, the feeling of control over by the worker (suspicion of presence and of synchronous communication tools) may hinder the launch of these practices.

Among good practices, we highlight the following:

- Availability of technology that facilitates work outside the office:
- Communication devices: PDA, smartphone.
- Devices for work: laptop.
- Connectivity: ADSL at home, 4G mobile, WiFi networks, public or private, free or paid.
- Devices for mobile phone communication: wireless microphone and headset, hand's-free setup for car.
- Availability of technologies that enable access to resources and tools that involve corporate security (corporate mail system, unified communications, intranet, document management, ERP, CRM)
- Availability of adequate working space and conditions: minimum space and comfort, adequate lighting that works with your laptop, cell phone coverage, electrical outlets, adequate level of silence, other services (cafeteria, toilets, etc.).
- Knowledge and flexible working practices.

### Tools for mobility

Finally, there is the identification and analysis of the various ICT tools and solutions that will allow these workers (mobile, roaming, or telecommuters) to effectively pursue their responsibilities and assigned tasks.

Several research studies on workers in the information and knowledge fields have analyzed different aspects of everyday life of these professionals as described below. The use of fixed and mobile devices (laptops, notebooks, mobile phones, and smartphones) were studied as follows:

- Tokuda et al. (2009) studied and showed that information workers manage different devices to balance their work, mobility. and efficiency during the day.
- Oulasvirta (2007) analyzed and characterized the different digital devices that a set of mobile information workers using IT in a company employed, and the benefit that its application brought.
- A study performed by the Fraunhofer Institute for Industrial Engineering (Haner et al. 2009) analyzed the need to define new work environments and equipment for information workers, as well as the availability of different resources and devices.

The worker, now, needs new tools to continue working on the go or, regardless of the location, to provide access to the same information and documents that would have been accessible from the office.

The simple and secure access to e-mails, documents, and shared calendars, instant messaging and intranets, combined with remote access to applications make sure that the move out of the office is productive.

The new work set-up requires tools that provide connectivity so any device (laptop, PDA or mobile phone) can use interfaces and functionality common to all, to be integrated with the agenda, both the user's own and those of third parties; can perform management tasks and priorities, and collaborate online and make calls from mobile devices with VoIP (voiceover IP), directly from a contact list of corporate personnel. The components of mobile solutions can be divided into three groups:

- Hardware devices and their connectivity
- Management software
- Corporate information systems

Mobile hardware devices should allow the user to properly profile and access information, so the worker can easily and securely access his information sources through a data connection:

- Mobile telephony: Available since the late 1990s, it allows certain employees, who were frequently traveling or making visits, to continue to be accessible to make and receive voice calls or get SMS messages.
- PDAs (personal digital assistants) were the first devices of small size of management that allowed the execution of certain tasks without lugging around a laptop, which was at that time a device with certain limitations of battery weight and volume and thus not too portable.

- The evolution of advanced mobile phones created the category of smart-phones, which are terminals with capabilities for data connection and access to secure networks. The smartphone can also install other programs to extend into new uses and functions.
- The evolution of laptops has allowed their capabilities and capacity to take on to those of desktop computers, and in many cases making them the first choice in choosing a personal computer. Apart from power, other aspects that have considerably improved the laptop have been the reduction of its weight and the improving the life of the batteries. These improvements have meant that any company can actually have computers whose work can be carried from one place to another without losing or reducing their capacities. Almost all notebooks have an integrated camera and microphone or speaker inputs and audio outputs, so that voice and video communications are now available without adding more hardware components.
- New devices such as netbooks, the UMPC (ultra mobile personal computer), and the tablet market appear as intermediate solutions between a smart-phone and a laptop, and have uses that tend to a greater or lesser intensity in production capacity or the use of entertainment or media consumption to the use of new uses and functions.

Having the device and the connection only gives one the possibility for mobile work, but we cannot really talk about labor mobility until there is some integration between software tools in which the worker manages the tasks and corporate systems containing information and documents required.

There are a number of i-productivity technologies that can improve the capabilities of workers and the results and agility of the company. Most of these technologies and tools have the ability to be used in mobile scenarios; either it allows the worker to be fully integrated with the corporate systems or there are some limitations.

The main ingredients that should compose a mobility solution are as follows:

- E-mail/calendar/tasks, as this package is known, forms the core of the management of personal productivity, and good use can produce dramatic improvements in personal productivity and company
- Communications: Instant messaging, voice, video, webconferencing, etc. The availability of communication tools through computer help when we communicate, make decisions, or work with clients or a team, represent a leap forward in terms of integration of communication channels for different uses.
- Publication of documents offline and online. The tools for editing and sharing documents are in most cases the applications used to perform

tasks related to information. Able to work in sync with our local news papers and corporate systems is a big advantage.

- Tools and Services 2.0, securely accessible via the web browser. Increasingly witnessing the conversion between the model systems have the business or use online systems that can be rented for use on secure servers hosted and play a high degree of safety or free applications or services which are used in general in exchange for advertising.

Apart from personal productivity systems and collaborative work, companies have their central systems located where the knowledge and digital information is generated.

Such knowledge is usually within corporate systems that allow the company to dispose of its assets safely and efficiently. Access to these systems securely from the outside has been one of the biggest challenges for companies.

In many cases firms are concerned about the costs of infrastructure and systems to ensure safe and controlled access by users to corporate services. Sometimes not taken into account is a way that benefits would quantified benefits and what the loss of business opportunity is because the firm does not have these features. In the middle of this debate there are the users and the corporate culture. In many cases, these problems or lack of planning for these costs means that the systems become obsolete and lose business opportunities.

The most common systems at corporations are mail servers and ways of access the web, Intranet, access to web security, and document management systems or files.

Depending on the type of company, it may have numerous complementary systems, from order management systems and accounting systems, to logistics, warehouse and distribution systems, and customer relationship management.

One of the latest trends in the way information technology is consumed and used, so-called cloud computing, should be the catalyst, in terms of cost and ease of management and administration, for all companies, including small and medium-sized ones, to adopt such systems as security, accessibility, and availability.

## References

Austin, T. (2010) Watchlist: Continuing changes in the nature of work, 2010–2020. Gartner. March 20.

Castells, M. (1999) *The Information Age: Economy, Society and Culture.* Wiley-Blackwell, p. 42.

Clake, R. (February 2005) Survey report. Flexible working: Impact and implementation. An employer survey. CIPD (www.cipd.co.uk).

Costa-i-Pujol, R. (2010) iProductivitat. Les TIC per incrementar el rendiment de les persones. ACC1O (www.acc10.cat).

Haner U., Kelter, J., Bauer, W., & Rief, S. (2009) Increasing information worker productivity through information work infrastructure. Ergonomics and Health Aspects, HCII 2009, LNCS 5624, pp. 39–48.

Matson, E., & Prusak, L. (September 2010). Booting the productivity of knowledge workers. McKinsey Quarterly.

Oldenburgh, R. (1989, 1997, 1999) The great good place: Cafes, coffee shops, bookstores, bars, hair salons, and other hangouts at the heart of a community. Marlowe & Company.

Oulasvirta, A., & Sumari, L. (2007) Mobile kits and laptop trays: Managing multiple devices in mobile information work. CHI 2007, April 28–May 3. San Jose California.

Pelino, M. (2010) The rise of wannable and maverick mobile workers. Forrester.com, February 16.

Tapscott, D. (1998) *Growing Up Digital. The Rise of the Net Generation.* New York: McGraw-Hill.

Tokuda, H., Beigl, M., Friday, A., Brush, A. J. B., & Tobe Y. (2009) Working overtime: Patterns of smartphone and PC usage in the day of an information worker. Pervasive 2009, LNCS, pp. 398–405.

Yershova, T. V., Khokhlov, Yu. Ye., and Shapotchnik, S. B. (2008) Information society for everybody today and tomorrow: Cooperation of stakeholders aimed at the implementation of the strategy for the development of information society. *Informatsionnoye Obshtchestvo,* nos 5–6, p. 18.

# 15
## Creating an Environment for Successful Innovation: A Management Consultant's Perspective

*Koen Klokgieters and Robin Chu*

## 1  Introduction

Innovation is everywhere, it always has been, and it is becoming continuously more apparent to the everyday consumer. Moreover, 43 percent of respondents in Capgemini Consulting's Global Innovation survey[1] in 2012 indicated they have someone appointed as chief innovation executive or innovation manager, compared to 33 percent who made this response in 2010. Also, 76 percent of executives indicated that innovation is among the top three priorities of their organization and the main lever for growth. It is apparent that companies are increasingly putting more emphasis on innovation; however, what does innovation really mean?

In 2007, the release of the Apple iPhone was groundbreaking and the first time a computer and software manufacturer stepped out of its core business boundaries and leaped into the mobile phone industry with a smartphone. Or, take a look at the firms that lease cars to companies and consumers alike; they innovated a business model that went from ownership to use.

Not all innovations are as apparent to the consumer as product innovations (e.g., Apple's iPhone) or business model innovations (e.g., leasing cars). Other innovations are more subtle and occur under the surface of companies in form of process innovations, which intend to increase cost efficiency and attain a more sustainable footprint. Figure 15.1 diagrams the three types of innovation.

To illustrate product, process, and business model innovation, let us take Ford Motor Company as an example. Henry Ford designed and invented the Ford automobile (product innovation) and it was a success among wealthier people, since it was quite expensive and thus not affordable for the average consumer. After several years, the invention of the assembly line (process innovation) helped to reduce costs dramatically, which made the cars available for a much larger market. Finally, if we look at Ford today, like many

327

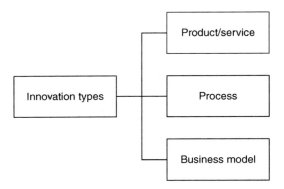

*Figure 15.1*  Types of innovations
Source: Abernathy & Utterback (1978), Capgemini Consulting.

of its competitors, it offers financial services and leasing options for both consumers and businesses (business model innovation). To put it in simpler terms, Henry Ford invented a product that met a demand for personal transportation. When he produced it, he found that a certain market segment could not afford it and he went on to find a way to reduce costs in manufacturing it in order to reduce prices. This enabled Ford to access a new market. Finally, as many customers' preference moved from ownership to use, automotive manufacturers redesigned their business model to accommodate.

The example of Ford is a relatively straightforward one; however, nowadays there are a set of factors that influence companies with regard to innovation (see Figure 15.2). First of all, the speed at which innovations are introduced – whether they be product, process, or business model innovations – seems to have increased over the past years. Every year, new models of cars with new gadgets and innovative solutions are introduced, whereas this would be unimaginable a couple of decades ago. Second, the intensity of competition has increased in some industries due to unexpected competitors (e.g., Spotify, an online platform that allows users to listen to unlimited songs for a monthly fixed fee instead of buying audio CDs) and start-ups (e.g., fast following Asian pharmaceutical firms, which quickly formulate and produce generics or off-patent drugs). Finally, there is an increase in availability and accessibility of information. Not only can companies have the means to better leverage existing knowledge and technology internally and externally (e.g., an electronic database of patents, intrafirm ideation platforms, strategic alliances), they can also quickly gather market intelligence, customer requirements, and engage directly with end-consumers through online open innovation platforms.

The basic principles of Charles Darwin's evolution theory affect business as well: If the current environment does not allow a company to survive

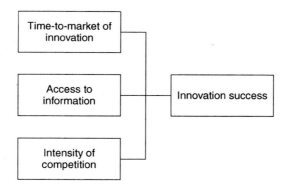

*Figure 15.2*   Variables of turbulence that affect innovation success
*Source*: Capgemini Consulting.

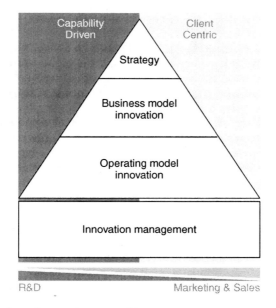

*Figure 15.3*   Business innovation pyramid
*Source*: Capgemini Consulting.

in its present shape, that company will need to adapt to the environment. But what if that company does not have the right internal environment for innovation to succeed? What environment should one create?

This chapter is structured around Capgemini Consulting's business innovation framework (see Figure 15.3). In Section 2 ("Innovation strategy"), we will elaborate on the importance and difference of innovation strategy compared to organizational strategy. Thereafter, Section 3 will cover the "Innovation operating model and innovation management;" Section 4 will cover "Innovation leadership;" Section 5 will cover "Innovation culture" and the implications for successful innovation. Section 6 conveys the growing need for collaborative capabilities and a shift from capability-driven and client-centric innovation. Finally, Section 7 expands on business model innovation and its relevance to innovation success.

## 2   Innovation strategy

Before they delve deeper into what environment they should create in order to be successful in innovation, executives need to establish what it is that they want to achieve. This is the top triangle in Figure 15.3 business innovation model – namely, strategy. The premise of developing an innovation strategy seems logical: without a clear destination of innovation within the organization, how does one expect to measure results, let alone achieve results?

In Capgemini Consulting and IESE Business School's joint Innovation Leadership Study,[2] it becomes apparent that the highest-ranking constraint for innovation success is the absence of a well-articulated and/or well-communicated innovation strategy (24 percent of respondents ranked it as the top constraint for innovation).

What is more striking is that only 42 percent of respondents indicated that their companies have an explicit strategy for innovation, whereas the remaining 58 percent do not. These figures by themselves mean relatively little, because it could well be that a company that has no formal strategy for innovation is much more successful than a company that does have one. However, from the study, it becomes apparent that 65 percent of innovation leaders have an innovation strategy as opposed to only 29 percent of innovation laggards. (Innovation leaders have indicated that over 75 percent of their innovation efforts are successful. Innovation laggards have indicated that under 25 percent of their innovation efforts are successful.) A formal articulation of innovation strategy is essential to success.

### Strategy development

Having established that strategy is of utmost relevance in innovation success, how does one develop one? Looking at the results from the Innovation Leadership Study, it becomes clear that the majority of an organization's innovation strategy is developed top–down and is heavily intertwined with corporate strategy. Only 11 percent of respondents indicated that innovation strategy was developed with involvement of employees. Note, however, that respondents state that the innovation strategy at top level is defined as more abstract and is a broad range of targets. Flowing from that, people in the organization have more freedom in how to organize themselves in order to achieve results.

In our experience an innovation strategy that is developed top–down is half of the equation to successful innovation. Due to increasingly turbulent environments, as depicted in the introduction of this chapter, developing a strategy centrally (with help of customer data, market data, and financial projections) will not incorporate all the intricate changes that employees see and does not permit inclusion of intangible experiences of people in the field. As Albert Einstein once professed: "Not everything that can be counted counts, and not everything that counts can be counted." In innovation, and especially in radical innovation, this holds all the more true. Firms need to involve the people who work at the outer rims of the firm in order to develop a well-informed and, consequently, effective innovation strategy.

## 3 Innovation operating model and innovation management

Given that innovation ranks among the top three priorities in the Innovation Leadership Study, one would expect that there is certain degree of governance to manage innovation. On the contrary, however, only 30 percent of respondents in the study convey that their organizations have an effective innovation governance (or operating model, as depicted in Figure 15.3) in place for innovation. As shown in Figure 15.4, we have divided innovation governance into high-level/long-term and operating-level/short-term purposes.

### Organizational structure

According to one of the respondents, it is difficult for large companies to instate an organizational structure for innovation, because of their existing structure of business units and other layers in the organization. This premise becomes clearer with the Greiner curve. as depicted in Figure 15.5.

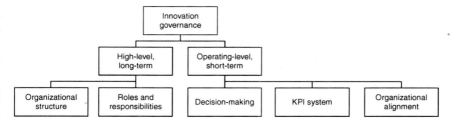

*Figure 15.4*  Elements of innovation governance

*Notes*: KPI = Key performance indicator.

*Source*: Innovation Leadership Study (2012), Capgemini Consulting.

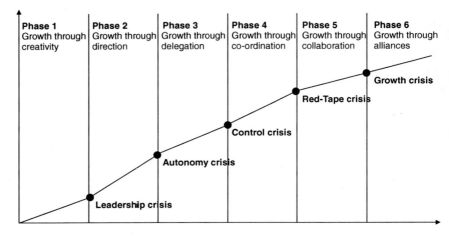

*Figure 15.5*   The Greiner curve
*Source*: Greiner (1972).

Many of the older large firms today have to change according to an ever-faster changing business environment. Looking at the priority placed on innovation, it seems that the direction is relatively clear. Like a large cargo ship, it is relatively easy to set a new course. However, if the ship is facing the exact opposite direction to where it needs to go, it will take time and effort to turn it into the right direction. The Greiner curve (see Figure 15.5) conveys why today it seems that smaller start-ups are introducing many more of the innovative products/services and business models than the larger incumbent firms are. The majority of large firms that exist today, once started in Phase 1 (the phase that start-ups begin in). Take, for instance, our example of Henry Ford. He invented and design a product (Phase 1), grew the company, and today we see that it has reached phase 4, 5, or 6.

In our experience, many companies struggle with what is popularly called "red tape." Red tape is what occurs when bureaucracy and regulations increasingly become more important than creativity (see Phase 1). To relate back to the large cargo ship analogy, red tape means obstacles that make it more difficult to change direction.

With two forces pulling at opposite ends of the rope (high priority on innovation at one end and organizational rigidity on the other), we see that companies organize innovation in a wide variety of places across the organization. depending on the type of innovation. For instance, an incremental product improvement may be assigned to the business unit that oversees the existing product, whereas more radical innovations may be organized centrally. This may seem logical; however, the issue that most commonly arises from this is that business units and centrally organized R&D units become increasingly disconnected. The issue, and possible solutions, of organizational alignment in innovation governance will be discussed more elaborately later in this chapter.

## Roles and responsibilities

Perhaps even more striking than the absence of an organizational structure for innovation is that 40 percent of respondents in the Innovation Leadership Study indicate that there are no clear roles and responsibilities for innovation in their firm. It seems that successfully developing and marketing new products/services is more attributable to luck in these organizations, considering the limited organizational design. An organization fit for innovation has its people accountable and responsible for it.

## Operating-level, short-term innovation governance

Imagine a hypothetical firm that has been able to identify the high priority of innovation. Moreover, it has been able to develop a strategy with involvement of its employees. Finally, it has been able to organize itself in terms of organizational structure and roles and responsibilities. Innovation projects within the company are sprouting and, as an executive, you just have to wait and reap what you have sowed.

Even though this is hypothetical, in our experience, some companies' executives actually believe this is true. However, what is overlooked is the greater context of the organization and the pitfalls it brings with regard to innovation.

### Decision-making

In the Innovation Leadership Study, 39 percent of respondents indicated that their organization does not have an effective process for decision-making in place for innovation. The underlying problem is mainly the decision-making process or, better said, the lack thereof, with regard to prioritizing and allocating time, resources, and funding to innovation projects (49 percent).

In our experience, the decision-making process is of utmost importance in innovation and the context of the company. What many firms decide to do in economic downturns is to cut spending on long-term projects such as innovation projects, because of its indirect nature: "If I spend 1 dollar more on R&D, I will only see the returns in 5–10 years. But if I spend 1 dollar more on marketing, maybe I can sell more of what I already have, today." This, however, is devastating for the success rate of innovation and for future growth, because of the intermittent and fluctuating nature of assigned budgets for innovation. Moreover, it can have negative implications for the firm's culture and motivation, since the commitment to innovation seems to wither as soon as there is some turbulence.

More often than not, we see that the daily operations within a firm absorb all the attention of executives, leaving very little or even no attention for innovation projects. This causes innovation projects to be delayed, due to lack of material and/or immaterial commitment. One of the respondents in the Innovation Leadership Study proposes the following:

> There are certain things that have to be sequestered, because if you hold these up to the conventional tools for prioritization, you'd kill these projects before they get started. These are sort of in a protected environment. And what we do here is, we have what we call a new business board of half a dozen of the top-top people, and they make calls on this, separate from their actual job.

The above is one of the ways companies can structure and manage decision-making for innovation projects. Without such decision-making processes being made formal, it is hard for team members of innovation projects to feel that there is a commitment. Moreover, in terms of portfolio management, clear decision-making processes provide insights into innovation projects which, in turn, can then be compared. Based on the status of projects, executives can make better informed decisions with regard to which projects seem most promising and should get additional funding, and which are less promising and should perhaps be stopped entirely.

### KPI system

A stunning 54 percent of respondents in the Innovation Leadership Study indicated that they do not have a formal key performance indicator (KPI) system for promoting innovation. The main challenge that was found with regard to KPIs is that respondents find it difficult to measure innovation effectively. Examples of KPIs are: number of patents granted, number of projects, and sales in terms of innovation products versus existing products.

Even though it may be difficult to set targets, in our experience, having any target for innovation is better than having none at all. Targets can go beyond financial results, as mentioned above; they could go into the number of partnerships one has started, how quickly one has been able to develop a concept, how many value chain entities are involved, et cetera. The reason for having softer targets in, for instance, partnerships, is that external involvement, e.g. from partnerships, has proven to accelerate speed and success of innovation.

Furthermore, it is especially important to instate some type of KPI for new products/services for the sales division in a company. What often happens is that salespeople have a set target with associated bonus schemes. As soon as a new product enters the company's portfolio through a successful innovation project, there is no real incentive for salespeople to start selling it with great effort and results. Now this may seem counterintuitive at first, however, consider the effort a salesperson would have to put into learning what the new product does, how to sell it, and who to sell it to. The effort required to sell the new product does not compute back into total incentives. In other words, without innovation KPIs, a salesperson would be better off selling more of the product he or she is already an expert in, which ultimately makes it seem that new products are less successful than their potential.

## Organizational alignment

The example described previously with regard to KPIs for salespeople implies that there is a misalignment between innovation and the existing business. From the Innovation Leadership Study, it becomes apparent that only 24 percent of respondents believe that their organizations have an effective organizational alignment for innovation efforts.

Organizational alignment can be described as how a company's different functions collaborate to make innovation projects successful. Innovation is not something that can work just from an R&D department; involving cross-functional teams (e.g., teams consisting of members from R&D, manufacturing, marketing, sales, etc.) in the process is considered greatly advantageous. This is closely related to the fact that the more diverse and cross-functional a team is, the higher the variety of information that is available to make decisions with. In turn, more information reduces uncertainty and thus increases efficiency and effectiveness. Moreover, interfunctional collaboration or cross-functional teams can be helpful in identifying issues such as manufacturing difficulties or the potential disconnect of a new product with customer requirements.

Some of the respondents indicated that they intend to diffuse innovation better throughout the organization by centralizing decision-making and promoting innovation throughout the entire company. Others have created a list of the top 50 innovation projects across the organization, which are continuously monitored, supported, and boosted.

Operational-level or short-term innovation governance seems to be the most underdeveloped area, compared to innovation strategy and high-level or long-term innovation governance. In our experience, companies try to organize innovation in many different ways, which leads to a higher degree of complexity. Add to this the lack of clear roles and responsibilities, KPIs, and decision-making processes and it is a self-fulfilling prophecy that some companies will have a hard time in bringing innovative products to market. Innovation leaders have already tried out and developed best practices for their innovation governance, and have seen success rates increase over the years. The question remains whether innovation laggards can quickly adapt and catch up.

---

*Exhibit 1*: Life sciences and the pharmaceutical industry

The industry for life sciences and pharmaceuticals is one that is characterized by heavy legal and safety procedures before new products can be introduced to the market. Organizing innovation effectively is even more important than in other fields, considering the high initial investment and relatively low success rate due to these legal and safety procedures (e.g., clinical trials). How can one sustain high level of innovation when such high risks are associated with it?

One pharmaceutical company employs a team that provides updates on ongoing activities in the external technical and scientific environment and on the competition faced by the company. The team is used for in-licensing opportunities to judge which external drugs would complement the internal R&D programs and contribute to the company's drug pipeline. One or two professionals associated with this team are aligned to a specific therapeutic area and are supported by information specialists.

By consolidating specialists in a team that play a key role in external market information and internal knowledge for innovation, the pharmaceutical company was able to quickly identify, create, and diffuse synergies. Furthermore, the team tracks changes that occur in the external environment during the entire drug development phase of the product. It has been able to reduce the number of potential innovations projects that go through the entire new product development process, thereby greatly reducing the amount of time and investment spent on evaluating each project.

---

## 4 Innovation leadership

Back to the hypothetical firm; it has now resolved its issues regarding operating-level, short-term innovation governance in terms of decision-making and KPIs and has properly aligned the organization for innovation. The mechanisms of innovation strategy and innovation governance are formal. On the other end of the spectrum, a company needs to consider the informal mechanisms that are important for managing innovation successfully. One of these mechanisms is innovation leadership (Operating Model Innovation and Innovation Management in Figure 15.3).

One of the findings of the Innovation Leadership Study, is that smaller firms (those with less than 500 million euros in annual revenues) were over-represented in the group that found that 75 percent of their innovation efforts were successful. In other words, smaller firms are more successful in innovation than large firms. In our experience, smaller firms require less effort to bridge the communication gap between executives and employees. What is inherent in the majority of large firms is that the larger the company, the more distance exists between executives and employees. This distance is not only a barrier for communication, it also creates a disconnect between executives and employees.

Also, what we found is that the motivation for executives is very different from the motivation among employees regarding innovation. Executives feel motivated by extrinsic drivers such as feeling accountable for realizing growth. Second, they feel responsible for advancing innovation in the organization. Now, even though a sense of accountability and responsibility are virtues one would like the executive leadership to have, they is not exactly inspirational. Employees, on the other hand, feel motivated by intrinsic drivers such as their perception that innovation is exciting, that they like to be part of a team or taskforce for something new, or they just have an intrinsic motivation to improve things. This disconnect between motivations leads to several complications. First of all, firms that have larger distances between executives and employees start to lose oversight and control. To cope with that control, KPIs are formulated. However, the rigidity and short-term nature of KPIs for regular business are incomparable to the softer, longer-term innovation KPIs. Also, usually, the newly introduced innovation KPIs come on top of existing KPIs, which implies extra tasks and responsibilities on top of the ones an employee might already have. Furthermore, existing KPIs may get more attention than new innovation KPIs, because of distrust, internal competition for resources, and the desire for stability.

## 5  Innovation culture

Besides innovation leadership and the connection between executives and employees, innovation culture is the second informal mechanism. Something as elusive and fluid as culture within a company is the hardest thing to change. In practice, we see many large organizations struggle with the existing culture within a firm. These organization cultures may have originated from a long and established history, country-of-origin effect, or may have simply emerged over time because of KPIs. To add to the complexity of organizational culture, we found that there are several elements that are specific to innovation culture.

How is it that innovation culture is sustained throughout an entire organization? The elements of and the source of innovation culture are elaborated upon in this section.

### Elements of innovation culture

*Openness*

It may seem relatively obvious – however, openness is seen as the most important driver for innovation culture according to respondents in the Innovation Leadership Study. Openness entails sharing information, being open to new ideas, and being open to change. The main barrier to establishing openness is the tendency to stability. If trust is not optimal in a firm, its employees will not be able to be open in sharing information because of the possible consequences for themselves, their team, or their business unit. An example: business unit A does not want to share specific information with business unit B, since this might benefit the business unit B at the expense of business unit A (e.g., extra budget). Internal competition in innovation is devastating when collaboration is not incentivized in some way.

Second, collaboration with externals (e.g., customers, suppliers, research institutes) is nearly impossible without the element of openness. If the innovation culture within a firm is closed, it is harder to move outside and collect knowledge and insights from the market or establish partnerships. Furthermore, a lack of trust within a firm will lead to a lack of trust towards externals. Because of the growing importance of external collaboration, a lack of openness can either delay new product launches due to absence of partners and market insights or can decrease the total number of ideas and innovation projects because of constraints in capabilities.

### Flexibility

The element of flexibility is closely related to decision-making. In practice, we see that some firms use existing lines of communication and decision-making for innovation. This means that innovation projects can only move forward when a decision in higher management is made with regard to budgets and resources.

By nature, innovation projects have a higher degree of uncertainty than existing business. The reasons for uncertainty can be lack of market insights, no comparable product in the market, and/or no clear customer (yet). Some of these uncertainties can be accounted for by collaborating with a partner or conducting market research. However, the degree of uncertainty will still remain higher than it is for regular business, despite thorough preparation. Furthermore, as increasingly more information is collected by the project team, decisions may have to be made that change the course of the original plan. The rigidity of the lines of command and communication within an existing company can slow down innovation projects significantly, because they are unable to provide the flexibility that innovation requires.

### Trust

A sense of trust is closely related to inspiration, commitment to that inspiration, and follow-through with action. If there is a lack of trust within a firm, innovation cannot occur. Executives need to understand that there will be innovation projects that will not succeed. Without employees' trust that failing will not lead to immediate demotion or automatic firing, it will be hard to find people within the organization who will want to take risk, whereas risk-taking is inherent to innovation (be it in a calculated way).

## Source of innovation culture

As it is closely related the elements of an innovation culture that we have identified, we were curious about where innovation culture originates within the firm. The Innovation Leadership Study's results showed that 69 percent of respondents feel that innovation culture is diffused by the CEO. Moreover, 59 percent also indicated that peers and people you work with are a source for innovation culture. Finally, mid-level management was indicated as a source of innovation by 51 percent of the respondents.

---

*Exhibit 2*: FMCG industry

A large multinational beverage producer, a leader in production and sales of soft drinks in the industry, struggled with the upcoming trend of interest in health and sustainability. Smaller, more agile firms came up with more sustainable bottling and healthier drinks. Why couldn't the larger firm, a leader in the industry, keep up with its smaller peers?

Inside the multinational company, one entrepreneurial-minded manager had picked up on the trend two decades ago. He started to lobby inside the firm to gain commitment in terms of budget and resources in order to develop an idea that he had, which concerned a "green" bottle. It took twenty years before his idea materialized and finally reached mass production, promotion, and sales.

The ground-breaking idea was not a standalone project that could be implemented besides the regular day-to-day production. It had implications for the entire supply chain, production, sales, and marketing function within the firm. These functions showed resistance to change and lacked openness to find a way to integrate the innovation into the rest of the existing organization.

One of the causes of the resistance to change was the implication that there would be extra work for individuals (both managers and employees) in order to make the changes. Furthermore, the existing KPIs created rigidity within the organization, which further slowed the progress of the idea.

The "greener" bottle has achieved major cost savings for the firm and has added to brand equity due to its sustainable nature. During the slow progress of the innovation project, executives of the firm decided that this delay could not occur again. Instead of opting for changing the current innovation strategy, governance, and informal mechanisms, executives decided to create a separate entity for radical innovations, which would be financially supported by their firm. Today, many more promising innovation projects are grown and nurtured in this newly created environment.

---

With regard to the CEO and mid-level management, respondents argued that even if an innovative culture existed among the employees of a firm, higher management could enable and encourage it. Without a sense of commitment from the leaders within the organization, innovation culture cannot thrive.

Formal and informal mechanisms behind innovation success tell us that the environment within the firm needs to be adequate in order for the firm to become a leader in innovation. It is likely that firms who cannot implement the changes required to enable adequate formal and informal mechanisms will choose other ways to establish a nurturing environment for innovation, such as setting up an entirely separate entity for innovation (see Exhibit 2). In essence, what these firms are doing is creating an entity that can exist in phase 1 in the Greiner curve (Figure 15.5), the phase in which creativity drives growth. One can then create an entirely new culture, which seems to be easier, compared to changing an incumbent culture. Additionally, more effective innovation KPIs can be formulated, separate from the existing entity. As soon as new products have been discovered and successfully developed, the innovation can be transferred from the start-up environment and integrated into to the incumbent company.

## 6   The growing importance of collaboration

Until now, we have discussed the formal (strategy, strategy development, and governance) and informal mechanisms (leadership and culture) behind innovation success in firms. They provide firms with the levers to create an environment that allows innovation to occur and succeed. If we look at Figure 15.3, we see that we have covered the top triangle of strategy development, as well as operating model innovation, and innovation management. This section will elaborate on the reasons why the abovementioned elements of successful innovation are relevant for the growing importance of collaboration and their relation to capability-driven and client-centric innovation.

### Collaboration to bridge the capabilities gap

A firm's capabilities and resources can lead to a competitive advantage. Li et al. (2006) describe a research framework of competitive advantage in which the following five advantages have been defined: cost, quality, delivery dependability, product innovation, and, last but not least, time to market (see Figure 15.6). In our experience, it is hard if not impossible for a firm to excel in all five elements. Either one's products have superb quality or have the lowest price in the industry; it rarely occurs that a product is both the best and the cheapest. The same holds, more or less, for the other elements in this framework.

Some companies choose to lead in product innovation and quality. However, in practice, it can mean that, since they are trying to introduce radical products of the highest quality, they are slower than their peers, who set their targets on more incremental innovations. How can the more innovative company increase the speed at which innovation is brought to market?

In our experience, the core capabilities of such a firm lie in product innovation and quality. To be able to accelerate the speed at which they bring

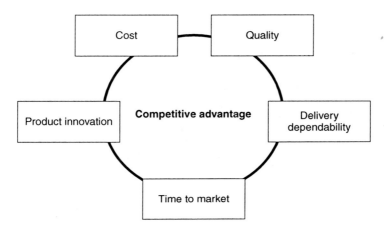

*Figure 15.6*   Framework of competitive advantage
*Source*: Li et al. (2006), Capgemini Consulting.

products to market, the firm can choose to grow that capability internally (make), acquire a firm that already has that capability (buy), or collaborate with firms that have that capability (ally). With regard to "make," the main disadvantage is time; growing a capability from the ground up takes time. "Buying" a capability through acquiring a company brings its own issues with it – namely, high lump-sum investment, post merger integration issues, and decrease in agility. The latter one, decrease in agility, is the main reason for the growing need for collaboration (ally). When a firm decides to acquire another company, in essence it invests heavily in acquisition and integration and, in times of economic crisis, for instance, it cannot just discard the newly acquired company if needed. This leaves the firm incurring costs and less flexible. To relate back to the cargo ship analogy, acquiring and integrating a firm is basically expanding the size of your current ship. It makes it harder to steer due to its weight, and discarding it is equally as difficult as integrating it. If we take the same analogy for collaboration, it is like having your own ship and being able to work together with another ship with a set of capabilities that are useful to you. Now, when economic times get rough, these ships may part and decide to go their own way, leaving them more agile while still benefiting from each other's capabilities when they needed to.

Considering the speed with which markets are changing nowadays, it is beneficial for a firm to be capable of collaborating for innovation in terms of flexibility. It can then create synergies without decreasing agility or increasing internal rigidity. The formal and informal mechanisms for innovation, as described in previous section, are fundamental to the capabilities of collaboration.

*Exhibit 3*: The airline industry

The airline industry has been struggling with margins on tickets for the past few decades. The reason for this lies in the rise of low-cost carriers and the shift of demand from customers (for business and pleasure purposes alike) to preferring cheaper flights over luxurious flights. To increase revenues, airliners have introduced many ancillary services, such as the possibility of buying products in flight (e.g., perfume, cigarettes, electronics).

The issue with ancillary services is that they are relatively easy to imitate by the airliner's competitors; today we see that all airliners carry an assortment of in-flight services. Second, ancillary services only provide existing customers with new products/services (the bottom right corner of the Ansoff matrix, Figure 15.7).

One airliner in particular decided to do things differently. Instead of expanding ancillary services, it decided to provide a new product to both existing and new customers. The new product would be a credit card with which consumers could earn points every time they used it to make a purchase. When a consumer had earned enough points, he or she could get discounts on tickets from the airliner.

Observe that the new product lies in the top right corner of the Ansoff matrix. The airliner could have opted to start its own financial services firm or acquire one. However, instead of doing so, it partnered with one of the major credit card companies, which enabled the airline to have a fast time to market of a new product with a relatively smaller investment.

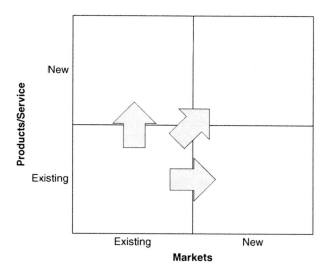

*Figure 15.7*   The Ansoff matrix: product and market growth strategies
*Source*: Ansoff (1957, 1989), Capgemini Consulting.

## Collaboration to shorten the learning curve

Other than capabilities, firms may seek knowledge and expertise in certain fields or markets that they currently does not possess internally. The more radical or farther out the innovation project lies outside the borders of the existing firm, the more likely it is that the firm will need external knowledge or technology to be able to successfully innovate. To illustrate, see the Ansoff matrix in Figure 15.7. The left bottom corner defines the firm's current products/services and the current markets it targets. Innovation in this quadrant may be typified as incremental innovation (e.g., adding new features or making a new model of an existing product). Innovation projects that intend to address new markets and/or new products are likely to need external knowledge or technology. Addressing new markets may require market intelligence, new distribution channels and infrastructure, specialized marketing and sales, et cetera. Again, not unlike what we saw with capabilities, a firm can decide to build or buy all the aforementioned knowledge and technological requirements; however, the disadvantages of "make" and "buy" are evident here. By collaborating with partners who have existing knowledge, a firm with new products/services for new markets can be faster in going up the learning curve.

## Customer collaboration

As mentioned, we have covered most of the conceptual model that is used by Capgemini Consulting (Figure 15.3) with exception of explicitly addressing capability-driven and client-centric innovation. In the past few years, we have observed a shift from capability-driven innovation to client-centric innovation. The availability and accessibility of information have been the underlying drivers for this shift; due to the rise of digital media (e.g., social media, increased access to Internet, and increased engagement with customers), it has become easier for firms to collect consumers' requirements for new products or new features for existing products.

Instead of partnering with other businesses to generate new products and services, companies are now using digital media not only as a collection mechanism for knowledge and market intelligence; instead, in open innovation, companies engage interactively with consumers, leading to innovation.

## 7   Business model innovation: a boundary condition for success

We have discussed the internal environment conditions, both formal and informal, required for innovation to thrive and the reasons why innovation culture elements are the key to collaborating and creating synergies with externals to boost innovation even further. Value creation through innovation is enabled once the aforementioned conditions are established, but how does one deliver that value and, more importantly, how can one capture the value of an innovative product? (see Figure 15.8)

Increasingly, we see game-changing business models arise that threaten existing business models: The airline industry is threatened by low-cost carriers; traditional postal services are seeing drops in demand for regular mail; traditional newspapers have had to change their business model to cope with changes in online behavior. And more seems to be coming: it is not unimaginable that our traditional gas stations will have to innovate their business model when electricity becomes the main source of fuel for cars.

To increase the success of innovation projects, companies cannot to pursue a "one-size-fits-all" business model; they must carefully specify business models to match market requirements and capabilities. For instance, one product might lend itself much better to licensing, whereas others can be sold directly to the end consumer. Prior to coming to a preferred business model, various alternatives need to be considered and compared with each other. The alternatives as they are generated vary in terms of value chain involvement, value creation strengths, and value capturing opportunities.

Since companies most often have an elaborate innovation portfolio of different product- market combinations, the same should apply to business models. However, most companies struggle with this. While a firm is heavily relying on the current skill set together with processes of bringing innovations to market, the opportunity to create new and game-changing business models is often overlooked. Companies with strong brands are able, more than others, to maintain and manage a portfolio of business models that sends a clear

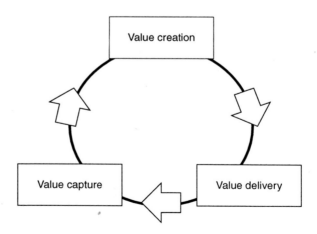

*Figure 15.8*   Value creation, delivery, and capture

*Source*: Osterwalder et al. (2009), Capgemini Consulting.

message to various segments of customers. The portfolio of business models drives decision making along with insights into attractiveness of business models, the firm's ability to execute, and of course, the business impact.

The obstacles to business model innovation are not very different from obstacles to implementing innovation governance. How to overcome the obstacles lies in how receptive the organization and its employees are to changes in their current way of working. Nevertheless, the impact and the effect of business model innovation are often underestimated.

## 8  Conclusions

All in all, innovation is increasingly becoming the focus for companies and its executives. The most successful companies already have implemented an adequate environment for innovation to flourish, and others at least have identified that there is a need to do so.

Also, as innovation rises to the top of executives' agendas, we expect to increasingly see the boundaries between organizational strategy and innovation strategy to fade.

An increasingly important key success factor for innovation and new product development is co-creation with externals of the firm. The environment that a firm creates to enable innovation to be successful, which has been the focal point of this chapter, establishes the capabilities for collaboration and co-creation.

Developments in digital tooling and the increased usage of online digital platforms by end consumers enable firms to co-create with customers and partners alike, further accelerating the pace of innovation.

We have discussed the implications of innovation environments for decision-making processes, KPIs, and measurement for success. Conventional decision-making processes and KPIs will need to be adapted for an innovation environment.

Finally, we foresee that business model innovation will be of essence to bridge the gap between innovation strategy and innovation operations in order to ensure and enhance innovation success.

It is not today's biggest or strongest firms who will shape the future business landscape; it is the one who can leverage and sustain its collaborative capabilities to remain agile and adaptable to external turbulence. New leaders in innovation will arise from unexpected corners, and the incumbent firms will need to adapt quickly in order to stay ahead of the curve.

## 9  Capgemini consulting: founded on a rich consulting heritage

The Capgemini Consulting business today is founded on a rich consulting history and heritage. The creation of Gemini Consulting in the 1990s, from an industry-defining merger combining the strategic expertise of the Mac Group and the operational improvement capability of United Research, signaled a new chapter in the consulting industry. Gemini Consulting's business transformation philosophy and capability delivered a new level of client value from strategy to implementation consulting, and it anticipated the industry's move to multidisciplinary consulting. Building on its early success, Gemini Consulting continued its growth and market leadership with the integration of Gruber Titze and Partners in Germany, Bossard in France, and the consulting business of Ernst & Young.

Capgemini Consulting continues to build on this strong strategy and management-consulting legacy and today, with its digital transformation focus, it is again driving the new business transformation agenda of its global clients.

## Notes

1. Methodology of the Global Innovation Survey. As the knowledge partner for the World Innovation Forum, Capgemini Consulting conducted an online survey using HSM's network of conference participants and attendees, plus a selection of Capgemini Consulting's client base. The online survey, in the field from April 07 to May 07, 2010, generated responses from 375 executives around the world, representing the full range of industries, regions, functional specialties, and seniority. In addition, 13 follow-up interviews were conducted to get an even better understanding of the context of the survey findings and to add depth to the survey result interpretation. The methodology differentiates between innovation leaders and laggards based on a self-assessment by survey respondents of their innovation success rate. The innovation success rate is determined by the percentage of innovation efforts that had a positive material impact on the company's business results. We distinguish among 4 categories of innovation success, based on this rate, namely: less than 25%, 25–49%, 50–74%, and over 75% of innovation efforts having a positive material impact on the company's business results. The "less than 25%" category represents the innovation laggard group and the over 75% category is the innovation leader group of analysis.
2. Methodology of the Innovation Leadership Study. The innovation leadership study carried out jointly by IESE Business School and Capgemini Consulting is Capgemini's third report in the innovation leader versus laggard series. It aims to understand how those leading and managing innovation in their organizations think about the innovation function and offers an insider perspective into both the formal and informal mechanisms for managing innovation. In order to secure this insider perspective, we specifically targeted innovation leaders and managers for participation in this study.

   Twenty-five in-depth interviews with innovation leaders have been conducted, most between July and September 2011, to explore their perspectives on the innovation function and mechanisms for managing innovation. The in-depth interview results informed our survey questions, provided a better understanding of the context of the survey findings, and added depth to the survey in the field from 12 September to 12 October 2011, and generated responses from 260 innovation executives around the world representing the full range of industries, regions, functional specialties, and seniority.

# References

Abernathy, W. J., & Utterback, J. M. (1978) Patterns of industrial innovation. *Technology Review*, 80(7), 40–47.

Ansoff, I. H. (1957) Strategies for diversification. *Harvard Business Review*, 35(2), 113–124.

Ansoff, I. H. (1989) *Corporate Strategy* (rev. edn). Penguin: Harmondsworth.

Duppen, F., & Inniss, D. (2010) Global Innovation Survey. Capgemini Consulting.

Greiner, L. E. (1972) Evolution and revolution as organizations grow. *Harvard Business Review*, 50(4), 1–11.

Li, S., Ragu-Nathan, B., Ragu-Nathan, T. S., & Subba Rao, S. (2006) The impact of supply chain management practices on competitive advantage and organizational performance. *Omega*, 3, 107–124.

Miller, P., Klokgieters, K., Brankovic, A., & Duppen, F. (2012) Innovation Leadership Study. Capgemini Consulting

Osterwalder, A., Pigneur, Y., Smith, A., & 470 practitioners from 45 countries (2009) Business Model Generation. Self-published.

# Conclusion

*Alexander Brem and Éric Viardot*

Innovation is sometimes compared to a journey – this principle applies to this book as well. The different contributions in these chapters help us to pinpoint new ways of innovating. Companies have to keep on moving into new territory, in terms of structure, management, and geography. Regarding structure, they have to promote collaborative innovation with customers and with a large variety of partners, whose list is expanding and changing according to time and to priority.

In terms of management, they have to consider that innovation management (IM) is a specific field of management with its own set of principles, methods, and tools. Finally, companies have to take into account the international evolution of the innovation challenge, as emergent countries are becoming serious contenders; companies have embraced the road to innovation lately, and there is no turning back.

In this chapter, we will comment about those avenues for the evolution of IM. But before that, it seems that another important message has come out of the various contributions to this book: at the beginning of the second decade of this new century, companies have to be able to simultaneously realize exploration and exploitation in the innovation process. Those two processes were considered fundamentally incompatible by March and others in the 1990s; on the contrary, they are complementary as illustrated in Figure C1: First, they have to be considered jointly when the company is willing to enlarge its network of partners, including other countries. Second, another challenge for IM is to identify, to balance, and to sequence the various steps of the innovation process in order to make sure that exploration will generate useful ideas, which are going to be transformed and exploited to become valuable products on the markets. Finally, it seems that emergent countries have already bridged the achievement gap with developed countries as far as exploitation is concerned, and they are quickly improving on or even leading the way in the exploration process.

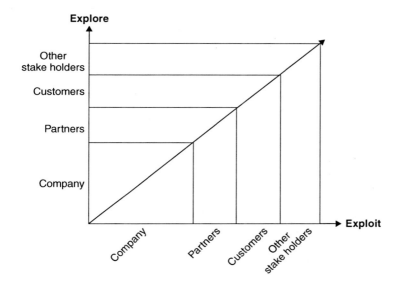

*Figure C1*  The success line of innovation management

## Increasing external collaboration

Today it is well known that collaboration with consumers and with a variety of strategic partners is mandatory in order to have a successful manage-ment process in the hypercompetitive world system of customers, suppliers, distributors, markets, governments, and institutions. Especially with meth-odologies like lead-user workshops, user toolkits, and crowdsourcing plat-forms, these kinds of integration are now at a new level, which could not be expected just ten years ago.

The early emphasis of the 1990s was on exploitation-side concepts such as lead-users, one-to-one marketing, and customer-centric marketing. The second wave in the first decade of the twenty-first century was more about exploration, with approaches favoring co-creation, open innovation, open source, and service-dominant logic. The challenge of collaboration or networked innovation is now to reconcile exploration and exploitation with the help of all the strategic partners, in order to deliver innovative solutions to the market and have a significant competitive advantage.

Of course, it would be naive to believe that the collaborative process is the silver bullet solution for companies that are not yet very efficient in their innovative process. They are some dark sides to the collaborative model, as illustrated in some chapters of this book. But overall, this extended and outstretched model of innovation management is required to survive in the current environment.

How firms can more effectively use collaboration to develop and manage innovation is one of the challenges of innovation management. The various contributions to this book provide some strong elements of answers. They emphasize the role of communication, but also the importance of an innovative culture that is open and transparent, as well as the need for governance in the management of innovation, in terms of structure and leadership. These factors will be briefly discussed in the following sections.

## Communication

One can say that if innovation always requires communication, open innovation requires even more communication. In the last 30 years, the concept of communication has evolved dramatically. For years, researchers had defined innovation communications as one element of the marketing mix of innovative products whose role was to accelerate the diffusion of innovation. Then the focus moved more internally, with the analysis of innovation communications as a competence easing the transmission of information from idea to launch, and also being a part of corporate communications. We now have a completely integrated view of innovation communications at a strategic level that embraces both the internal and the external perspective and that considers both exploration and exploitation for innovation. Moreover, the focus is now enlarged to both internal and external recipients: on the one hand, employees must be convinced that innovation is necessary and reasonable; on the other hand, all external stakeholders (like suppliers, but even public organizations) must be addressed as well. Here, again, new communications technologies like social media are changing the rules of the game.

Actually, it has also been noticed that successful innovation champions are master at communicating and highlighting the "right" points to the "right" people, inside and outside of a company. It is also quite apparent that if smaller firms are more successful at innovation than large firms, one reason is that they require less effort to make communications flow, notably between executives and employees, but also between the company and its various partners or stakeholders.

But communication is not enough. It has to be part of an organizational culture that stimulates – and does not inhibit – the appetite for innovation.

## Innovation culture

There are some common traits among the different innovative cultures. One is openness, because collaboration with external partners is nearly impossible without the element of openness. Another one is the employee's trust that failing will not lead to immediate demotion or automatic firing; without trust, none within the organization will want to take the risk inherent in innovation. A third feature is the passion to innovate: the collaborators must have the will to create and transmit knowledge, as opposite to fearing to take risks. In some cases, such an innovative culture is engrained in the mission and the values of a firm; but in the majority of the cases, it requires the deliberate willingness of top management to make the organizational culture more prone to innovation. Hence, people cannot be forced to follow a specific cultural approach, but they can be motivated.

## Leadership

This brings us to the leadership issue and its fundamental role in the management of innovation, even though there are still many companies that do not have clear roles and responsibilities for innovation. As innovation is becoming a critical factor for a firm to survive, many companies have identified, or even have sometimes appointed, "innovation champions." They are not necessarily the inventors, but their role is to foster innovation by spanning the firm's structure internally and externally for fresh ideas and to help push these ideas to some form of outcome. Those innovation champions,, who work at the middle to lower levels of the organization, are different from the "champion of innovation" at the top management level, who strategically allocates and manages the resources allocated to innovation.

One of the challenges for the future is, hence, to harmonize the relationships between those two categories of champions, who are very different. Top executives are motivated by extrinsic drivers, such as being accountable for realizing growth and advancing innovation in the organization. Employees or staff are motivated by more inspirational drivers, such as enjoying improving things, working on an exciting innovation, or being part of a team that is creating something radically new. The gap between motivations leads to several complications, which could be corrected with the right set of key performance indicators (KPIs), but those KPIs are usually not in place.

## KPIs and innovation model

There is still a long way to go to come up with a clear and practical conceptualization of innovation KPIs, both quantitatively and qualitatively. "IM [Innovation Management] cannot be measured" is a typical answer both from academia and practice. However, what you cannot measure, you cannot manage adequately. Hence, traditional management measures and methods are obsolete when considering the cooperative innovation strategy. New measures are needed that allow managers to monitor and control the innovation process and the associated risks and opportunities. As mentioned in one chapter, technology roadmapping and a balance scorecard are two interesting multidimensional pointers, but the reality is that few companies are using them so far.

The absence of specific innovation KPIs is also preventing many firms from learning about their past experiences in innovation. Various chapters of this book are dedicated to some specific tools of IM, such as the use of creativity, scenarios, mobile social networks, and early customer relationships. Those tools have to be integrated into a more general model such as the "Want, Find, Get, Manage" model commonly used by large companies or "Scan, Focus, Resource, Implement, Learn" coming from academia. The former model is all about action, while the latter incorporates the knowledge acquired in the different steps of the innovation process. Companies will undoubtedly gain a lot from building on their experience, but they are lacking the metrics to do it.

## Structure for innovation

In general management, the topic of structure is correlated with the issue of leadership. From the different contributions to this book, it does not seem obvious that there is an ideal structure for effective innovation, especially given the current necessity to open the firm to the outside. The model seems to be a structure that is loosely organized and where communities of practice (CoPs) and social networks have a significant role, both at the exploration and exploitation level. Interestingly, the influence of leadership is not absent in those structures, meaning that the top managers may have an influence, especially when the time comes to not only to create value but to capture it for the company. Finally, it depends on the industry, on the structure of the company, on the shareholders, etc.

## The international dimension of innovation

After the first internationalization wave dominated by Europe and the United States in the 1970s, and the second one led, by Japan and the other members of the "four Asian Tigers" in the 1980s, IM was also affected by the third wave, led by the BRICs (Brazil, Russia, India, and China). In all those countries, there is a strong consensus among all the actors, firms, government, customers, and universities that innovation is a necessity to achieve economic success both at micro and macro levels. Furthermore, there is a shift in the profile of the companies, as the number of state-owned enterprises has dropped significantly while the number of large privately owned companies and multinational companies has expanded dramatically. Those later categories of companies are less conservative and are more aggressive when it comes to innovating, as illustrated in some chapters of this book, e.g., when it comes to frugal innovation.

Furthermore in China, Russia, and India, creativity and inventiveness are not an obstacle to innovation, as we can see when we consider the history of those countries. The main issues are much more about leadership and organization. In China, what impedes innovation is related to the weight of tradition around hierarchy and observance, face and harmony. In Russia, it is the lack of entrepreneurial spirit and the weaknesses of business processes that are preventing good ideas from becoming successful products. In India, the main hurdle is the absence of infrastructure. But there is no doubt that those barriers will be overcome in the future.

Additionally, the innovation process in emerging economies is sometimes different that the one in developed countries. By necessity, it is based less on technological breakthroughs than on novel and innovative combinations of existing knowledge and technologies to solve local problems. This frugal innovation uses limited resources to create low-cost products that are environmentally sustainable and that work for individual communities. In some cases those frugal innovations are even finding their way into the developed market.

Significantly, the innovations coming from those parts of the world are taking place in locations where it is most urgent to solve economic, social, and ecological issues. They are paving the way for what is the clear road to innovation management for this twenty-first century: to deliver products and services that are not only valuable but also just in time, global, and sustainable.

# Index

CPSIA information can be obtained at www.ICGtesting.com
Printed in the USA
LVOW10*1923161113

361591LV00025B/571/P